P9-ARC-159

Intermediate Organic Chemistry

Intermediate Organic Chemistry

John C. Stowell

Department of Chemistry
University of New Orleans
New Orleans, Louisiana

A WILEY-INTERSCIENCE PUBLICATION

JOHN WILEY & SONS

New York • Chichester • Brisbane • Toronto • Singapore

Library of Congress Cataloging-in-Publication Data:

Stowell, John Charles, 1938–
Intermediate organic chemistry.

"A Wiley-Interscience publication."
Includes bibliographies and index.
1. Chemistry, Organic. I. Title.

QD251.2.S75 1987 547 87-21554
ISBN 0-471-09899-X

Printed in the United States of America

10 9 8 7 6 5 4 3 2 1

Preface

Consider the typical student who has finished the two-semester introductory course in organic chemistry and then picks up an issue of the *Journal of Organic Chemistry*. He or she finds the real world of the practicing chemist to be mostly out of reach, on a different level of understanding. This text is intended to bridge that gap and equip a student to delve into new material.

There are two things to learn in studying organic chemistry. One is the actual chemistry, that is, the behavior of compounds of carbon in various circumstances. The other is the edifice of theory, vocabulary, and symbolism that has been erected to organize the facts. In first-year texts there is an emphasis on the latter with little connection to the actual observables. Thus students are able to answer subtle questions about reactions without knowing quite how the information is obtained. Chemistry is anchored in observations of specific cases, which can be obscured by the abstractions. For this reason this text includes specific cases with more details and literature references to illustrate the general principles. Understanding these cases is also an exercise to ensure the understanding of those principles in a concrete way.

This book is by necessity a selection. Subjects that are generally well covered in introductory texts are omitted or briefly reviewed here. Advanced topics are treated to a functional level but not exhaustively. Specifically, the subjects are those necessary for understanding and searching the literature and some topics that are the elements of many current journal articles. The outcome is a selected sampling on a scale with latitude for creative lecturers to amplify with their own selections. Advanced texts give much attention to classical work that led to modern understanding. This text, with all due respect to the originators, does not cover that familiar ground but again covers current examples grounded in those classical ideas with modern interpretation.

There is a modicum of arbitrariness in the selections, and, considering this text as a new experiment, the author would welcome suggestions for substitutions and improvements.

JOHN C. STOWELL

New Orleans, Louisiana
November 1987

Acknowledgments

The material on pages 23 and 24 from the *Ring System* Handbook, and the data retrieved in the computer file search in Figure I, Chapter 2 is copyrighted by the American Chemical Society and is reprinted by permission. No further copying is allowed.

The computer search in Figure I, Chapter 2 is reprinted by permission of DIALOG® Information Services, Inc.

The material from *Beilstein's Handbuch der Organischen Chemie* on pages 30 and 31 is copyrighted by Springer-Verlag and reprinted by permission.

Permission for the publication herein of Sadtler Standard Spectra® has been granted, and all rights are reserved, by Sadtler Research Laboratories, Division of Bio-Rad Laboratories, Inc.

Figure VIII, Chapter 7 (an adaptation) is reprinted with permission from Nome, F.; Erbs, W.; Correia, V. R. *J. Org. Chem.* **1981,** *46,* 3802. Copyright 1981 American Chemical Society. Figure X, Chapter 7 (an adaptation) is reprinted with permission from Baumstark, A. L.; McClosky, C. J.; Williams, T. E.; Chrisope, D. R. *J. Org. Chem.* **1980,** *45,* 3593. Copyright 1980 American Chemical Society.

Figures XVII–XIX, Chapter 10 are adapted from Johnson, L. F. and Jankowski, W. C. *Carbon-13 NMR Spectra,* copyright© 1972 by John Wiley & Sons Inc.; reprinted by permission of John Wiley & Sons, Inc.

The graphs in Figures II and III of Chapter 8 were calculated, plotted, and graciously provided by S. L. Whittenburg, University of New Orleans.

The simulated spectra in Figures V–VIII as well as Figure X, Chapter 10 were graciously provided by R. F. Evilia, University of New Orleans.

An extraordinary note of thanks is due to David Lankin of Varian Associates, Park Ridge, Illinois for preparing essentially all the artwork for Chapter 8 as well as providing the NMR spectra for Figures III, IV, XII, and XX in Chapter 10.

Finally, I am grateful to Frances Caldwell, who typed this manuscript with patience and expertise.

J.C.S.

Contents

List of Abbreviations

Ac	acetyl
Aq	aqueous
Bu	butyl
DME	1,2-dimethoxyethane
DMF	dimethylformamide
DMSO	dimethyl sulfoxide
Et	ethyl
HMPA	hexamethylphosphoric triamide
LDA	lithium diisopropylamide
Me	methyl
Ph	phenyl
Pr	propyl
rt	room temperature
THF	tetrahydrofuran
Ts	*p*-toluenesulfonyl

Intermediate Organic Chemistry

1

Nomenclature

Organic chemistry is understood in terms of molecular structures as represented pictorially. Cataloguing, writing, and speaking about these structures requires a nomenclature system, the basics of which you have already gained in your introductory course. To go further with the subject you must begin reading journals, and this requires understanding of the nomenclature of complex molecules. This chapter includes a selection of examples to illustrate the translation of names to structural representations, including some IUPAC and *Chemical Abstracts* conventions.[1–4] They are known compounds published in recent issues. The more difficult task of naming complex structures is not covered here because each person's needs will be specialized and can be found in nomenclature guides.[1–4] Most of the naming rules are used to narrow the understandable alternatives to a unique (or nearly so) name for a particular structure and need not be known in order to simply read the names.

1.1 ACYCLIC POLYFUNCTIONAL MOLECULES

2,2-Dimethyl-3-hydroxypropyl *(E,E)-deca-2,7,9-trienoate*

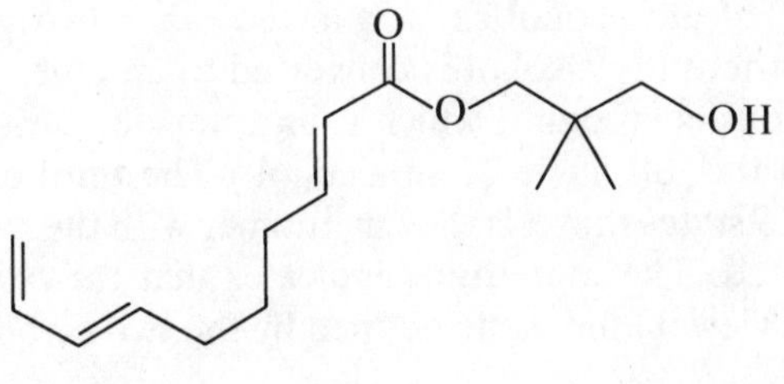

O

O

OH

This structural representation is arrived at by the following detailed analysis. The name consists of two separate words and ends with -oate, which indicates ester functionality. The alcohol part in the first word consists of propyl with two methyl groups and another alcohol function located as indicated by the numbers. The acid portion contains 10 carbons in a continuous chain with three double bonds located at the numbered positions with the carbonyl carbon numbered 1. Two of the double bonds are designated as trans isomers by *E,E*. Recall that the highest atomic number atoms attached at each carbon of the double bond are determined, and if they lie across the double bond from each other, it is the *E* (entgegen) diastereoisomer, or if they lie on the same side of the double bond, it is the *Z* (zusammen) diastereoisomer. A separate letter must be used for each one.

Considering that free rotation exists about all the single bonds, many alternative shapes may be drawn for the same compound, for example,

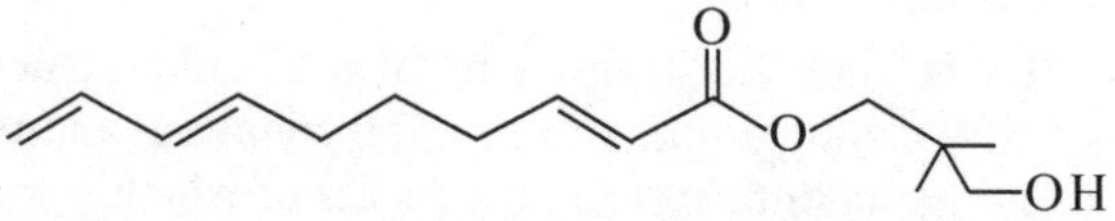

O
O
OH

In all of these structures each vertex or end of a line (with no letter) represents a carbon atom with an appropriate number of hydrogens to complete tetravalency. A line that leads directly to a letter representing an atom represents a bond to that atom, and there is no carbon at that end of that line; that is, there are five carbons in OH.

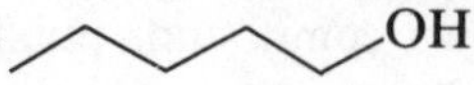

3-(*S*)-*trans*-1-Iodo-1-octen-3-ol methoxyisopropyl ether

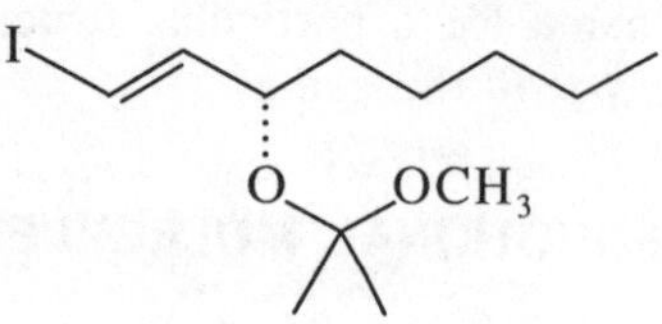

I
O
OCH_3

This is an example of a derivative name; that is, the first word is the complete name of an alcohol as is, and the other two words describe a derivatization where the alcohol is converted to an ether (ketal). (In contrast, the previous compound, which is an ester derivative of an alcohol, did not contain the full name of an alcohol.) The third carbon is a chiral center, and the *S* states that it is the enantiomer with the three-dimensional shape represented. The dotted line indicates that the oxygen sits behind (away from the viewer) the plane defined by the two carbon–carbon bonds shown as lines to the chiral center. It remains understood that the hydrogen on the chiral center is in front of (toward the viewer) the plane of the carbon–carbon bonds. The *R* enantiomer would be represented as

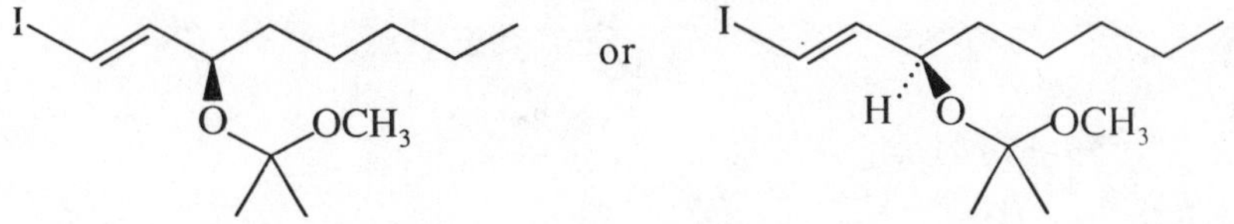

I O OCH_3 or I H O OCH_3

1.2 CARBOCYCLIC COMPOUNDS (MONOCYCLIC)

3,3-Dimethyl-6-oxo-1-cyclohexene-1-carboxaldehyde

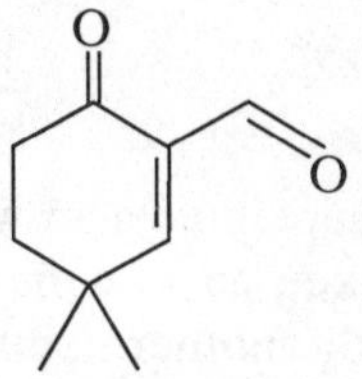

O O

The numbering of carbons in the ring begins at the double bond and continues, crossing the double bond. It starts with the carbon that also bears a substituent in order to give the lowest number to the first locant. An oxygen that substitutes for two hydrogens (i.e., a ketone) is called *oxo*. (Careful! The prefix "oxa-" has another meaning; see p. 12.) The group CHO is treated as a substituent but listed as as suffix instead of a prefix. In IUPAC rules it is called *carbaldehyde*. Similar suffixes are used for acids (. . . carboxylic acid) and nitriles (. . . carbonitrile) where they are one-carbon substituents. As a prefix, the aldehyde group would be called formyl. . . .

With rings comes the necessity of designating cis–trans relationships as well as chirality.

t-5-Chloro-*r*-1,*c*-3-cyclohexanedicarboxylic acid

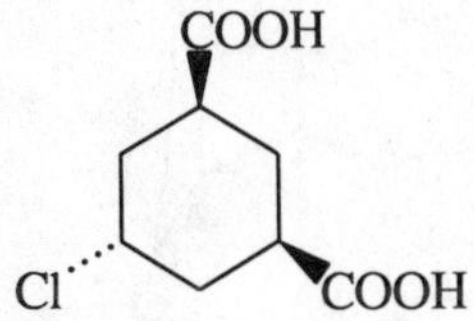

COOH

Cl COOH

Here one of the suffix substituents (lowest locant) is labeled *r* for reference, to which the others relate as trans (*t*) and cis (*c*).

Ethyl 7-[(1*R,2*S**)-2-(dimethoxymethyl)-5-oxocyclopentyl]heptanoate**

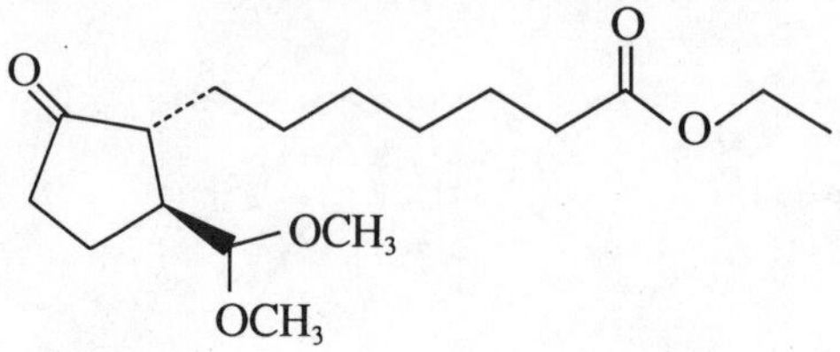

or

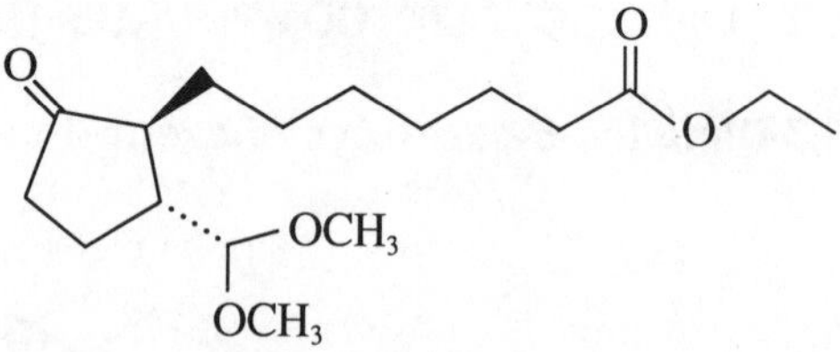

This is an ethyl ester of heptanoic acid with a complex substituent on the seventh carbon. The brackets enclose the description of that substituent that is a cyclopentyl ring with another numbering system. The number 1 carbon of the ring is the point of attachment as a substituent on the heptanoate. The two chiral centers, carbons 1 and 2, are specified as having the configurations *R* and *S* as drawn in the left-hand structure, but the stars indicate that only the relative configuration is specified; that is, the centers could also be (1*S*,2*R*) as shown on the right but not (1*R*,2*R*) or (1*S*,2*S*). This gives the same amount of information as simply calling it the trans isomer. The researchers who prepared this compound used the name to represent racemic material, (1*S*,2*R*) and (1*R*,2*S*), together.

1.3 BRIDGED POLYCYCLIC STRUCTURES

exo-2,2,4-Trimethylbicyclo[3.2.1]oct-6-en-3-one

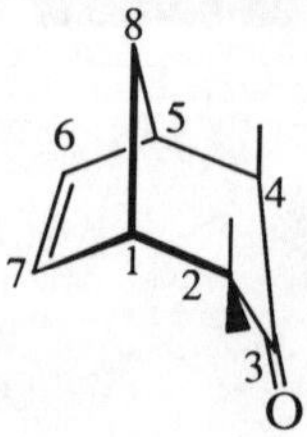

This molecule contains a ring with a bridge extending across it. It would require two bond breakings to open all rings; thus it is termed *bicyclo*. Rather than consider the rings (which can be viewed as five-, six-, or seven-membered), two carbons are considered as bridgeheads (atoms 1 and 5 in this case) from which three paths branch and recombine. These paths contain three, two, and one carbons each, as indicated by the bracketed numbers separated by periods. All bicyclo compounds require three numbers in the brackets, tricyclo require four, and so on. The numbering for locating substituents, heteroatoms, and unsaturation begins at one bridgehead and proceeds over the largest bridge to the other bridgehead and then continues with the next largest, and so on. The total atoms in the bridges and bridgeheads (excluding substituents) is given after the brackets, in this case as "oct." Finally, carbon 4 carries a methyl group that could be close to either the two-carbon bridge or the one-carbon bridge. If it is located stereochemically toward the smaller of the two choices, it is called *exo*. The opposite is the *endo* isomer.

In tricyclic compounds, the relative stereochemistry at the additional bridgeheads often allows two choices that are specified with Greek letters α and β in the *Chemical Abstracts* system.

(1α,2β,5β,6α)-Tricyclo[4.2.1.0^{2,5}]non-7-ene-3,4-dione

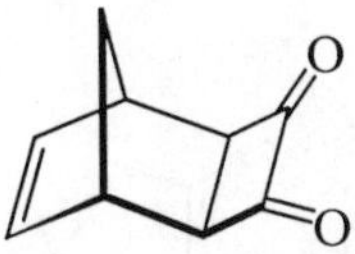

Starting with a pair of bridgeheads, the four-, two-, and one-carbon bridges are drawn. The zero bridge then connects carbons 2 and 5 according to the superscripts, thus making them bridgeheads also. At bridgeheads 1 and 6 the smallest bridge is considered a substituent and given the α designation. At bridgeheads 2 and 5 it is treated as a ring fusion and the hydrogen locations are α if cis to the C-9 bridge or β if trans as in this example.

(1α,2β,4β,5β)-5-Hydroxytricyclo[3.3.2.0^{2,4}]deca-7,9-dien-6-one

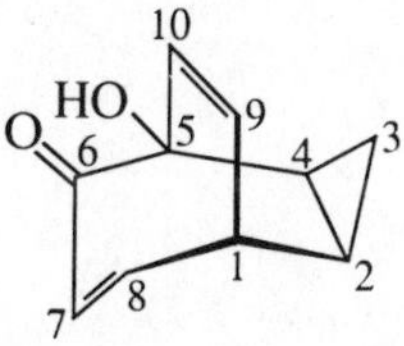

At bridgehead 1 the last numbered bridge is a substituent and designated α. The hydrogens at carbons 2 and 4 are trans to it and marked β. The OH group on carbon 5 is a higher priority substituent than the C-9–C-10 bridge, and it is trans to the bridge; thus it is labeled β.

1.4 FUSED POLYCYCLIC COMPOUNDS

Fused ring compounds have a pair or pairs of adjacent carbon atoms common to two rings. Over 35 examples have trivial names, some of which need to be memorized as building blocks for names of large or more complex examples. The names end with "-ene," indicating a maximum number of alternating double bonds. A selection is (one resonance form drawn):

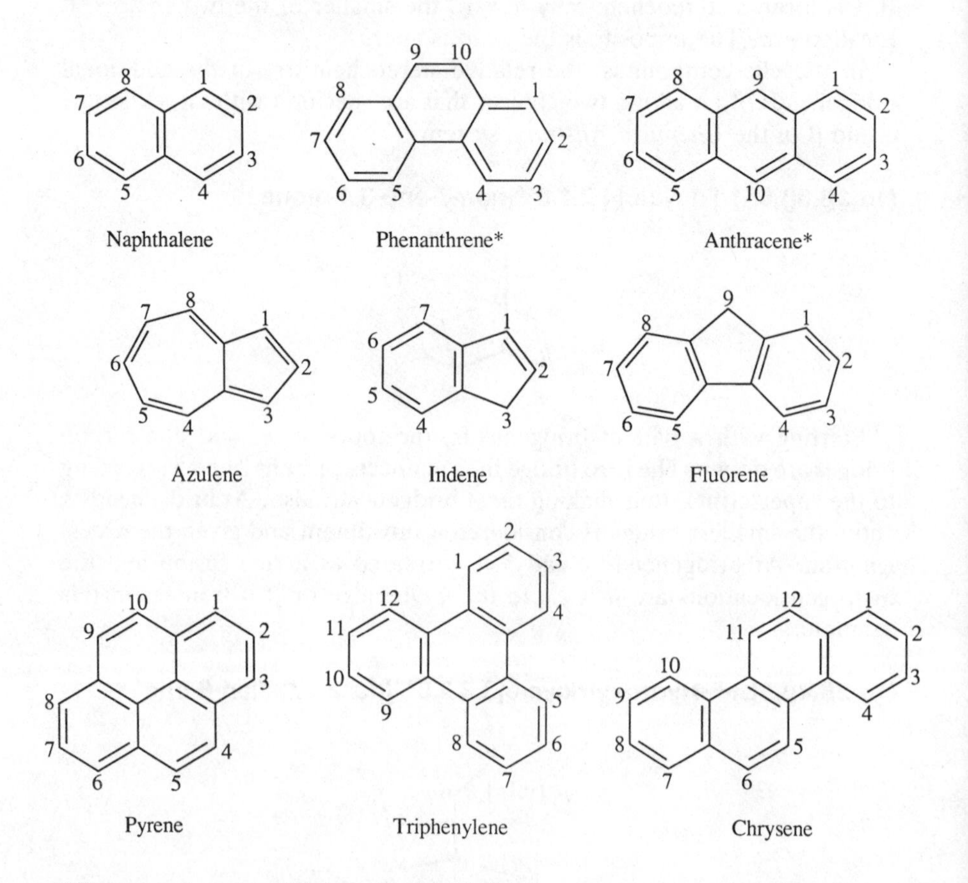

* Exceptions to systematic numbering (see p. 7).

Fusing more rings onto one of these basic systems may give another one with a trivial name. If not, a name with bracketed locants is used. For example,

Benzo[*a*]anthracene

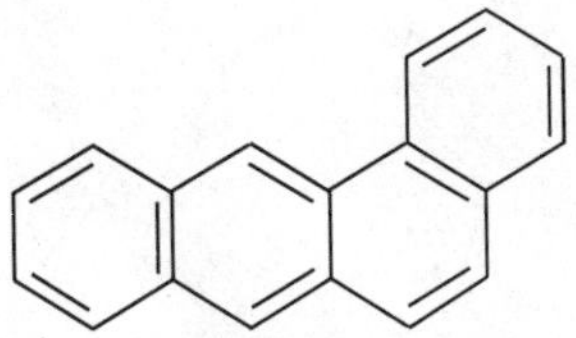

Since a side of a ring is shared, the sides are labeled *a*, *b*, *c*, and so on, where carbons numbered 1 and 2 constitute side *a* and 2 and 3 constitute side *b*, and so on.

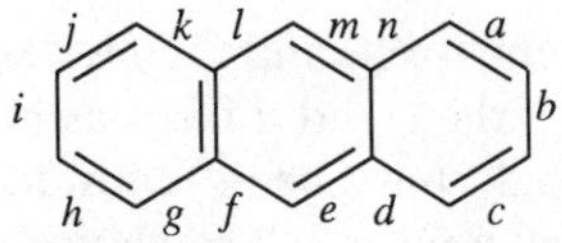

j k l m n a i b h g f e d c

Here the earliest letter of the anthracene is used to show the side fused, and the ring fused to it is given first with "o" ending.

The final combination is then renumbered to locate substituents or hydrogens added, reducing unsaturation. First orient the system so that a maximum number of fused rings are in a horizontal row. If there is still a choice, place the maximum number of rings to the upper right. Then number clockwise starting with the carbon not involved in fusion in the most counterclockwise position of the uppermost or uppermost-farthest right ring. See the numbering in the systems with trivial names on page 6. Atoms involved in the fusion that could carry a substituent only if the system were less unsaturated are given the previous position number plus an a, b, c, and so on. Where there is a choice, these too are minimized; for example, 4b < 5a:

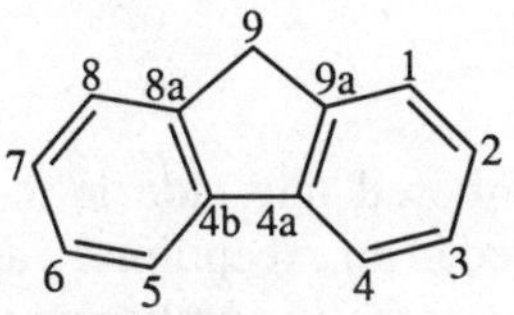

9 8 8a 9a 1 7 2 4b 4a 6 5 4 3

Correct

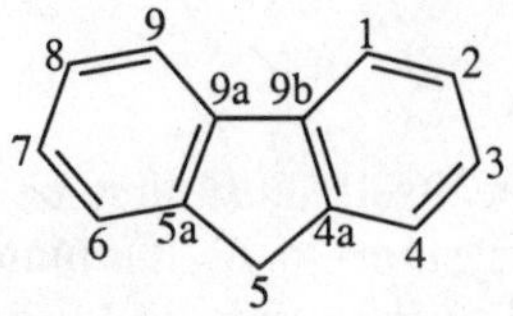

9 8 1 2 9a 9b 7 3 6 5a 4a 4 5

Incorrect

In those rings where one carbon remains without a double-bonding partner, it is denoted by (e.g., 9*H*) and given the lowest possible number within the earlier rules. This is called *indicated hydrogen* and is used even when an atom other than hydrogen is actually there in the molecule of interest.

1*H*-Benz[*cd*]azulene

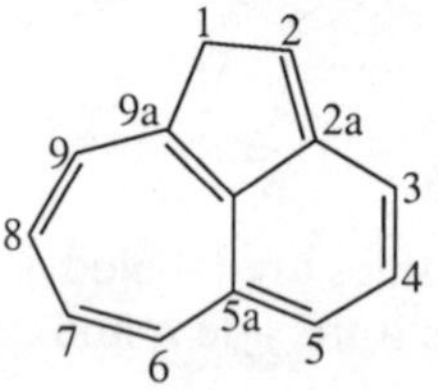

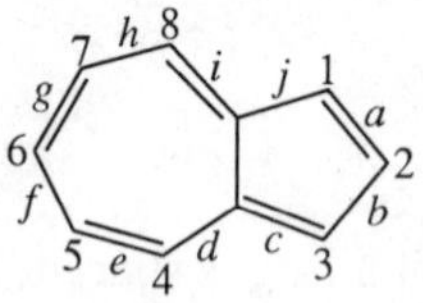

Azulene

First number around azulene itself, and then letter the sides as shown. The benzo ring is fused to both the *c* and *d* faces as indicated in brackets. The "o" is omitted when it would have been followed by a vowel. Now orient the three ring molecule as on page 7. The choice of which two rings go on the horizontal axis and which one on the right is determined by finding which orientation gives the smallest number for the fusion atoms; for instance, 2a instead of 3a or 4a. Since there is an odd number of carbons, one remains without a π bond. In this isomer it is the number 1 carbon as shown by the 1*H*.

***trans*-1,2,3,4-Tetrahydrobenzo[*c*]phenanthrene-3,4-diol diacetate**

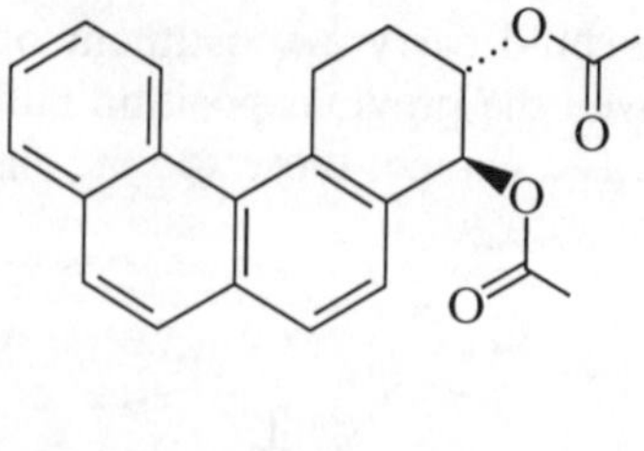

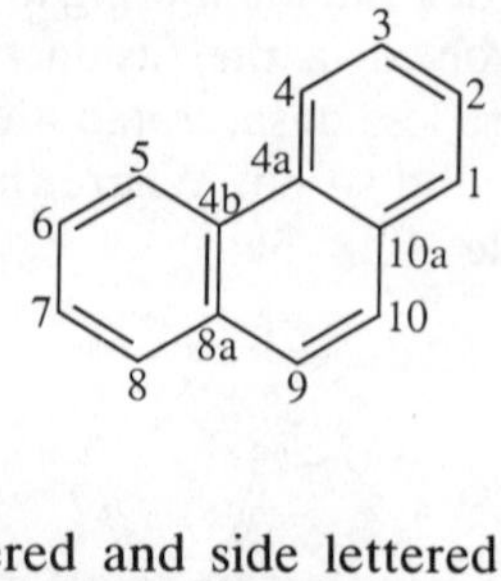

Phenanthrene itself must first be numbered and side lettered. As with several base systems (p. 6), it is numbered in an exceptional way. A benzene ring is fused to the *c* side and a new systematic numbering is made. The 1, 2, 3, and 4 carbons have hydrogens added to saturate the double bonds, and then the 3 and 4 carbons have hydroxy groups substituted for hydrogens in a trans arrangement. Finally, it is the acetate ester at both alcohols.

6,13-Dihydro-5*H*-indeno[2,1-*a*]anthracene-7,12-diol

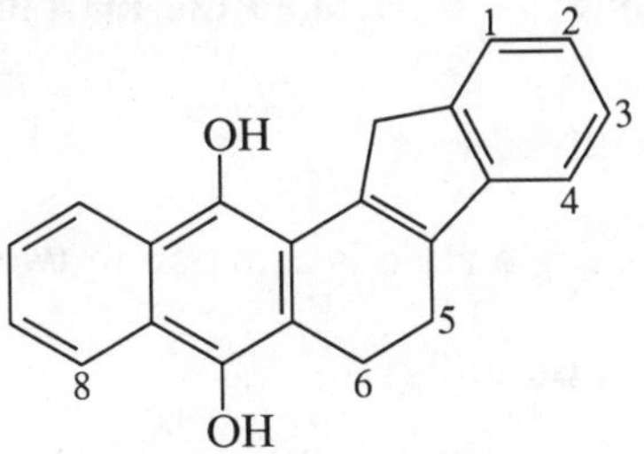

The sides of anthracene are lettered and the carbons of indene are numbered.

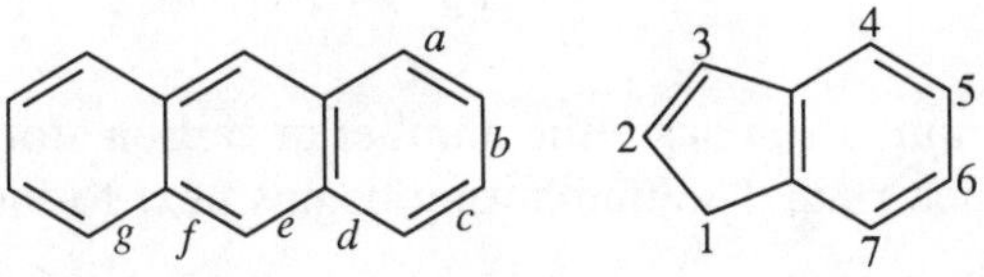

All sides of benzo were the same but not of indene; thus the bracket must include a designation for a side of it also. Furthermore, the direction of fusion must be shown since there are two choices for bringing the pair together. The numbers in the bracket are carbons of indene fused in order 2, then 1. These are fused to the side of anthracene, with the number 1 carbon constituting the number 2 carbon of the indene. The whole system is then renumbered according to the rules and the substituents and indicated hydrogen placed accordingly.

The other direction of fusion is in 1*H*-indeno[1,2-*a*]anthracene:

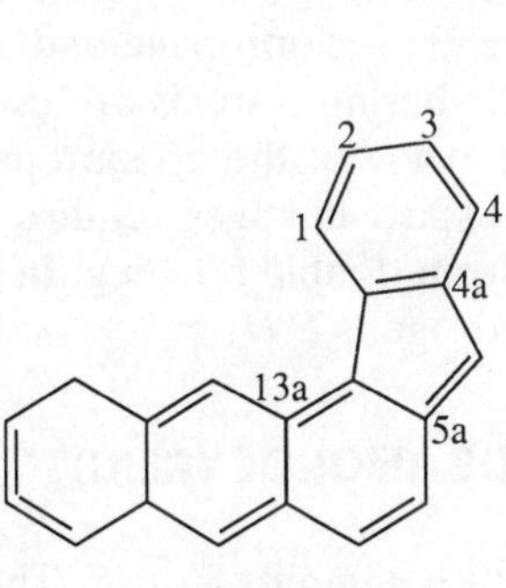

Wrong numbering

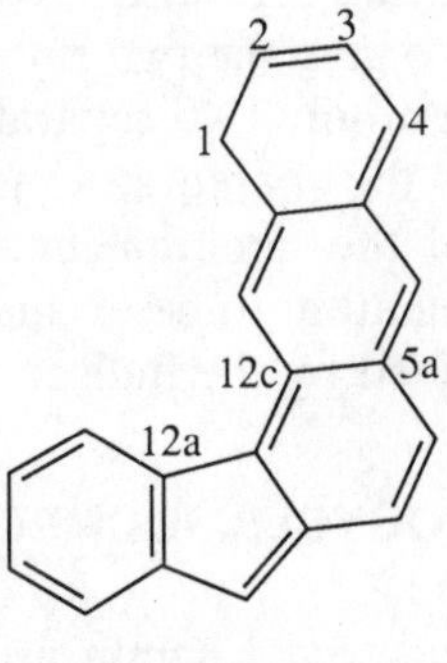

Correct numbering

Note that here a different orientation is used because it gives the lower fusion numbers:

Correct	4a,5a,7a,8a,12a,12b,12c,13a
Wrong	4a,5a,7a,8a,12a,13a,13b,13c

1.5 SPIRO COMPOUNDS

In spiro compounds a single atom is common to two rings:

Spiro[2.5]octane-1-carbonitrile

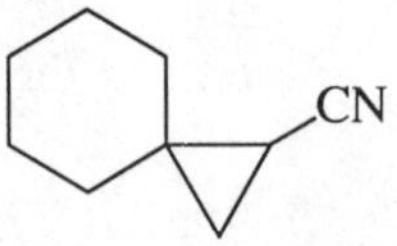

CN

The bracketed numbers indicate the number of carbon atoms linked to the spiro atom in each ring. The numbering begins next to the spiro atom in the smaller ring.

7′-Ethyl-3′,4′-dihydrospiro[cyclopropane 1,1′(2′*H*)-naphthalene]

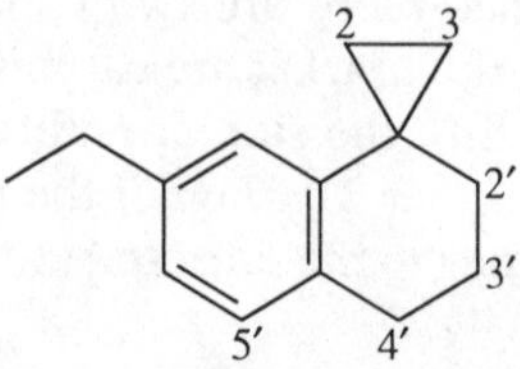

2 3 2′ 3′ 4′ 5′

Here a fused ring system is a component of the spiro pair. The brackets following spiro give the two ring systems, cyclopropane and naphthalene, that share an atom. Two separate numbering systems are used. The 1,1′ indicates that the shared atom is number 1 of the cyclopropane ring and number 1′ of the naphthalene. The spiro linkages require another naphthalene ring atom to be excluded from double bonding, in this case the 2′ as determined by the indicated hydrogen, 2′*H*.

1.6 HETEROCYCLIC COMPOUNDS (MONOCYCLIC)

Systematic and trivial names are both commonly in use. The systematic names consist of one or more prefixes from Table I and multipliers where needed designating the heteroatoms followed by suffix from Table II to

TABLE I. Prefixes in Order of Decreasing Priority[a]

Oxygen	ox-
Sulfur	thi-
Selenium	selen-
Nitrogen	az-
Phosphorus	phosph- (or phosphor- before -in or -ine)
Silicon	sil-
Boron	bor-

[a] An "a" is added after each prefix if followed by a consonant.

TABLE II. Suffixes Indicating Ring Size

	Containing Nitrogen		Containing No Nitrogen	
Atoms in the Ring	Maximally Unsaturated	Saturated	Maximally Unsaturated	Saturated
3	-irine	-iridine	-irine	-irane
4	-ete	-etidine	-ete	-etane
5	-ole	-olidine	-ole	-olane
6	-ine	—[a]	-in	-ane
7	-epine	—[a]	-epin	-epane
8	-ocine	—[a]	-ocin	-ocane
9	-onine	—[a]	-onin	-onane
10	-ecine	—[a]	-ecin	-ecane
>10[b]	—	—	—	—

[a] use the unsaturated name preceded by "perhydro."

[b] Use the carbocyclic ring name with heteroatom replacement prefixes: oxa-, thia-, and so forth.

give the ring size with an indication of the unsaturation, all preceded by substituents.

Numbering of ring atoms begins with the element highest in Table I and continues in the direction that gives the lowest locants to the next heteroatom, and so on.

2-(2-Bromoethyl)-2-methyl-1,3-dioxolane

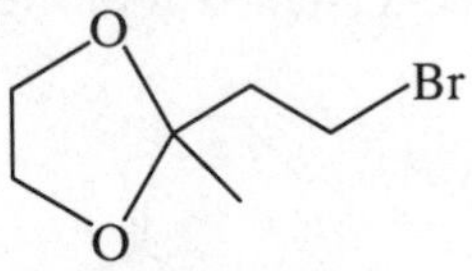

2-Butyl-3-ethyl-3-methyloxaziridine

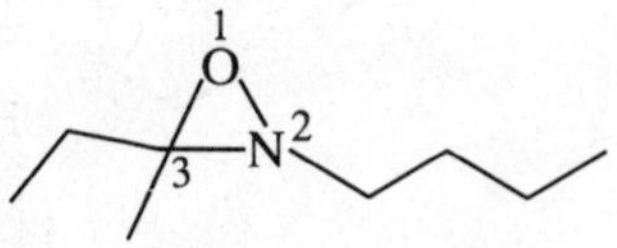

The ring numbering begins with the highest priority oxygen (ox) and continues with nitrogen (az), and then the ring size and saturation are indicated.

3,5-Di-*tert*-butyl-2-phenyl-1,4,2-dioxazolidine

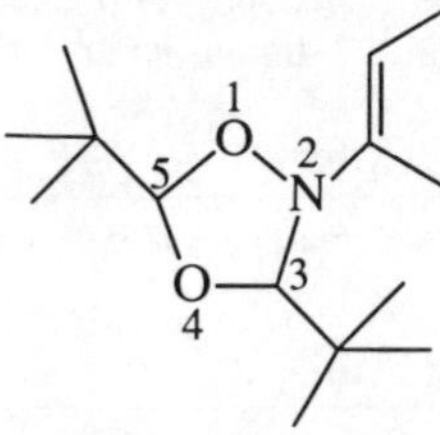

Here the numbering starts with the highest priority oxygen and proceeds to the nearest heteroatom, minimizing the numbers. In the name the two oxygens come first with their numbers first. The "-olidine" ending specifies a saturated five-membered ring.

4,7-Dihydro-2-methoxy-1-methyl-1*H*-azepine

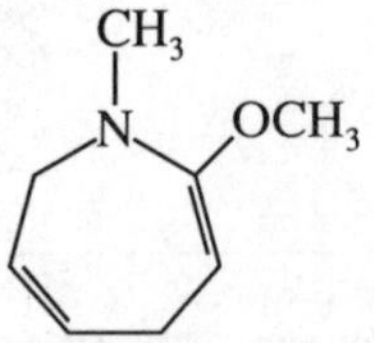

Azepine indicates an unsaturated seven-membered ring containing one nitrogen. One atom of the seven must have indicated *H*. Two others have additional hydrogen; thus the dihydro shows that there is one bond less than fully unsaturated. Notice that the indicated hydrogen is assigned the lowest numbered atom not in double bonding (the nitrogen) and *then* replaced by the substituent.

(6*R*,14*R*)-6,14-Dimethyl-1,7-dioxa-4-(1-propylthio)cyclotetradec-11-yne-2,8-dione

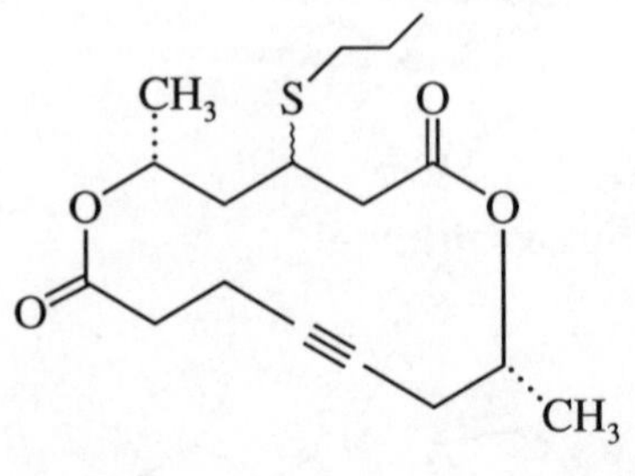

Here the ring is larger than 10 members; therefore, the hydrocarbon ring name cyclotetradecyne is used, modified by 1,7-dioxa, which is a replacement of the 1 and 7 carbons with oxygens. The "a" ending on "oxa" indicates replacement. The numbering starts with a heteroatom and proceeds to the other heteroatom by the shortest path.

Besides the systematic names developed from the table, many five- and six-membered rings have trivial names that are commonly used. A selection of the more common ones is as follows:

Unsaturated

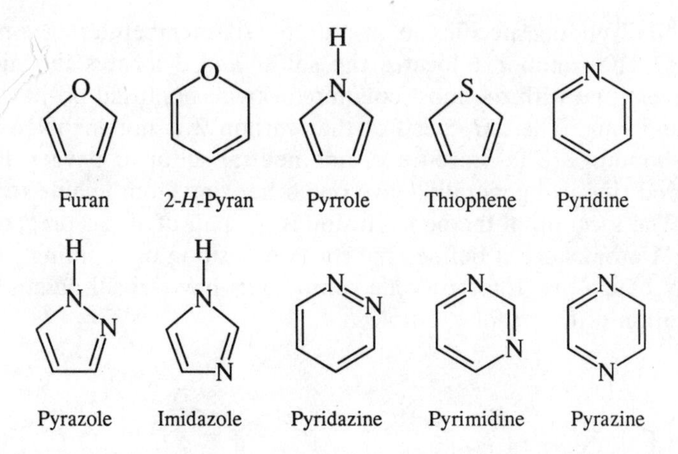

Saturated

Pyrrolidine Pyrazolidine Piperidine Morpholine

1.7 FUSED RING HETEROCYCLIC COMPOUNDS

The systematic heterocyclic names and the trivial names for the unsaturated heterocycles are the basis for fusion names.

4-(4-Chlorophenyl)-6,7-dimethoxy-2*H*-1,3-benzothiazine

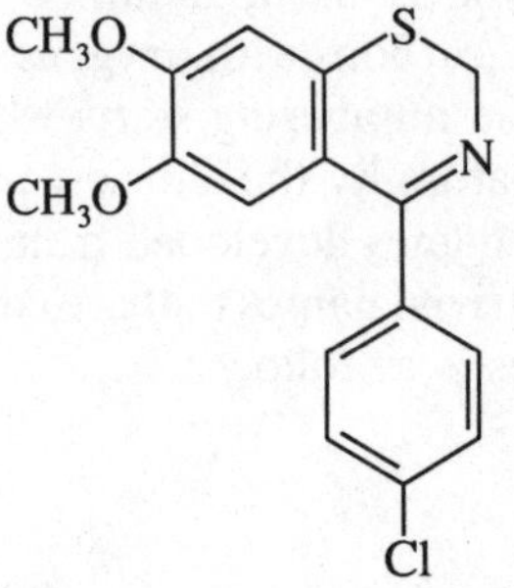

The "ine" ending specifies an unsaturated six-membered ring containing nitrogen. The number 1 locates the sulfur and 3 locates the nitrogen, starting as usual with the most counterclockwise nonfused position of the right-hand ring. The 2*H* specifies that carbon 2 is not involved in ring double bonding. (The presence of one neutral sulfur or oxygen in a six-membered ring will generally leave one other ring atom unable to double bond.) The location of the benzo fusion is the pair of atoms preceding the number 1 atom since it defines the starting position of counting.

Many benzo-fused heterocyclic compounds have trivial names for the combination of rings, for example,

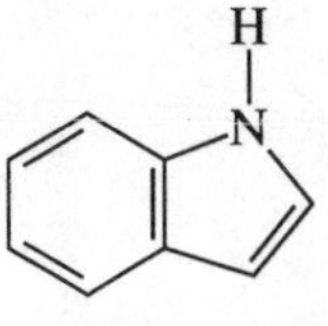

Indole 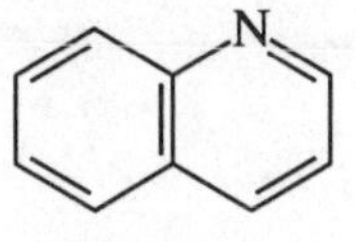 Quinoline 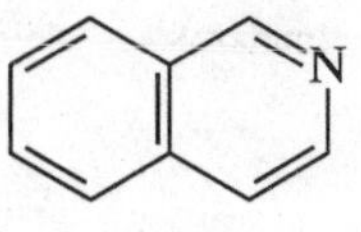 Isoquinoline 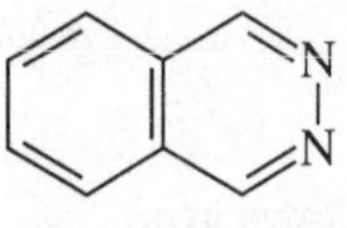 Phthalazine

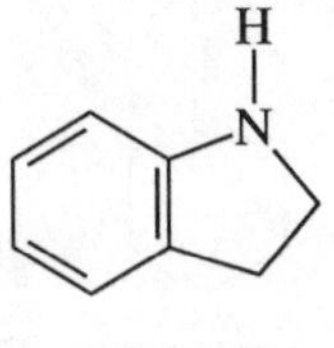

Indoline 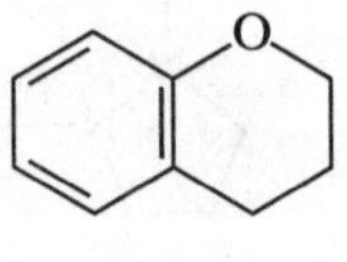 Chroman

1,2-Dihyro-3-methylbenzo[*f*]quinoline

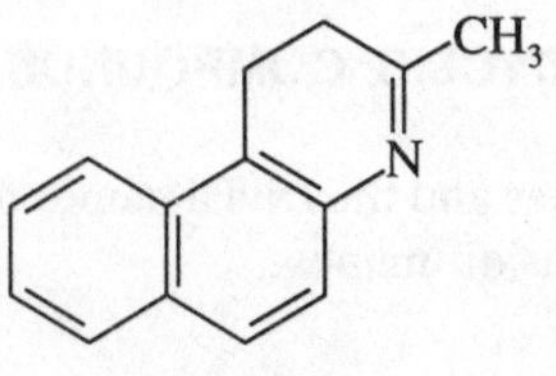

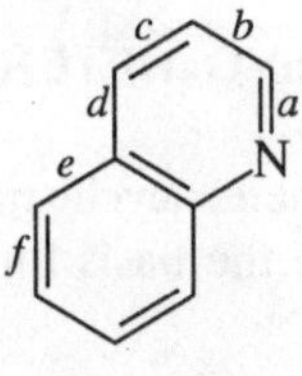

The sides of quinoline are lettered following the numbering system, and the benzo is fused to side *f*. The whole system is renumbered orienting as directed on page 7, that is, maximum number of rings in a horizontal row maximum in upper right, giving the heteroatom the lowest number and numbering from the most counterclockwise nonfused atom in the upper right ring.

When two heterocyclic rings are fused, we must indicate sides and direction of fusion in brackets as we did for the carbocyclic case on page 9. Examples of the six possible fusions between pyridine and furan are shown.

1. 6-Methylfuro[2,3-*b*]pyridine
2. 2,3-Dihydro-2,7-dimethylfuro[2,3-*c*]pyridine
3. 2,3-Dihydro-2-methylfuro[3,2-*b*]pyridine
4. 4-Bromofuro[3,2-*c*]pyridine
5. 5,7-Dihydrofuro[3,4-*b*]pyridine
6. Furo[3,4-*c*]pyridine-1,3-dione

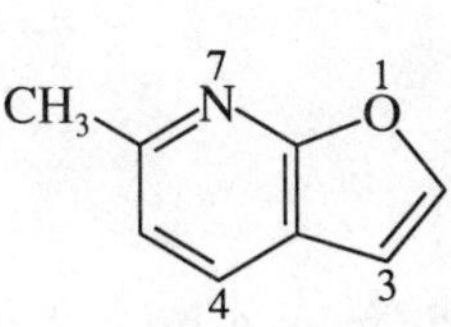

CH_3 7 N 1 O 4 3 **1**

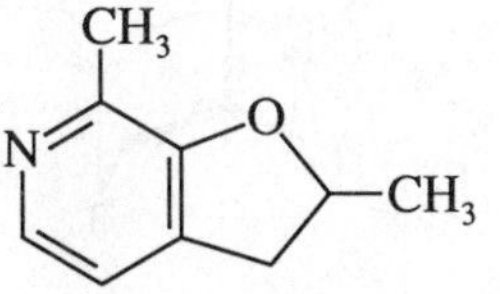

CH_3 N O CH_3 **2**

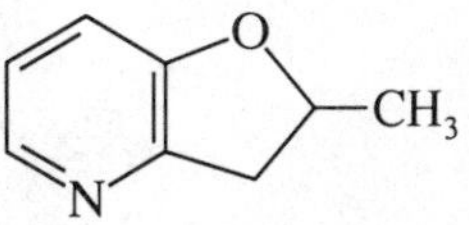

O CH_3 N **3**

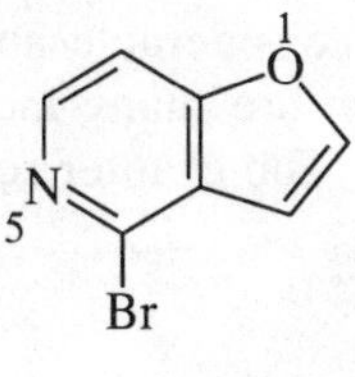

1 O N 5 Br **4**

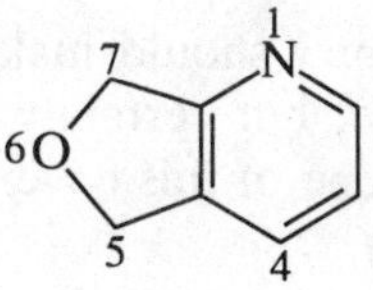

1 N 7 6O 5 4 **5**

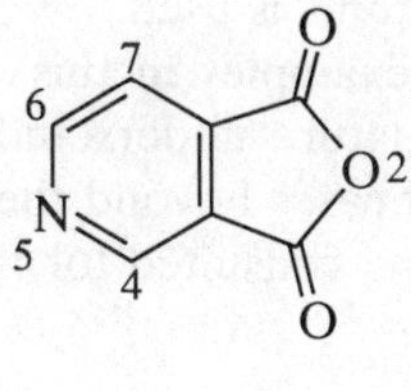

O 7 6 N 5 4 O2 O **6**

Furo[3,2-*b*] indicates that the 3 and 2 carbons of the furan are the 2 and 3 carbons of the pyridine, respectively. The two-ring system is renumbered, giving the two heteroatoms the lowest numbers possible, or if this leaves two choices (as in 1), giving the highest priority heteroatom the lowest number.

If the heteroatom is one of the fusion atoms, it appears in the name of both rings.

1*H*-Pyrido[1,2-*c*]pyrimidine

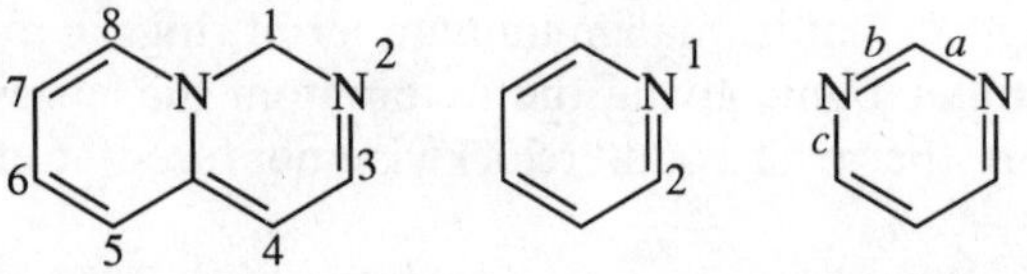

1.8 BRIDGED HETEROCYCLIC COMPOUNDS

Bridged heterocyclic compounds are named according to the replacement nomenclature; that is, the hydrocarbon name is used with oxa, aza, and so on to substitute heteroatoms for carbons as was seen in large ring monocyclic compounds (p. 12).

***exo*-4-Chloro-6-methyl-6-azabicyclo[3.2.1]octane**

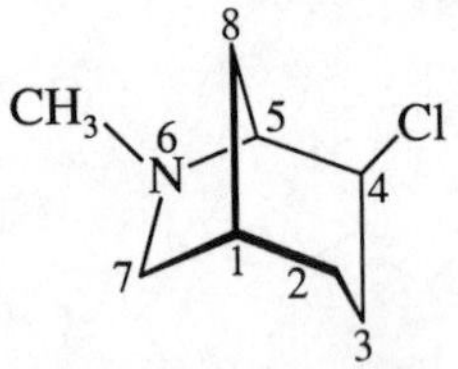

As in the hydrocarbons, the numbering proceeds from a bridgehead across the bridges in the order largest, medium, and smallest. Of the two choices of starting points, the one that gives the lowest numbers to the heteroatoms is used.

The examples in this chapter should make a considerable amount of nomenclature understandable, but certainly there are many more complicated cases beyond the scope of this coverage. The chapter references should be consulted for them.

PROBLEMS

Draw complete structures for each of the following compounds:

1. 2,4-Dimethylbenzo[*g*]quinoline

2. 2-(Bromomethyl)-4,7-dimethoxyfuro[2,3-*d*]pyridazine

3. Spiro[cyclopentane-1,3′-bicylo[4.1.0]heptane]

4. 6-Methoxyspiro[4,5]decane

5. 7-Methyl-7*H*-benzo[*c*]fluorene-7-carboxylic acid

6. *endo*-8-Hydroxybicyclo[5.3.1]undecan-11-one

7. (1α, 2α, 5α, 6β)-Tricyclo[4.3.1.$1^{2,5}$]undecan-1-amine

8. (1α, 2β, 3α, 5α)-6,6-Dimethylbicyclo[3.1.1]heptane-2,3-diol

REFERENCES

1. Rigaudy, J.; Klesney, S. P. (preparers). *Nomenclature of Organic Chemistry,* (International Union of Pure and Applied Chemistry), Pergamon, Oxford, 1979.
2. *Index Guide, Chemical Abstracts,* American Chemical Society, Columbus, OH, 1984, Appendix IV.
3. McNaught, A. D. The Nomenclature of Heterocycles, *Adv. Heterocyclic Chem.* **1976,** *20,* 176–319.
4. Cahn, R. S.; Dermer, O. C. *Introduction to Chemical Nomenclature,* 5th ed. Butterworths, London, 1979.

2

Searching the Literature

A truly vast and every-growing body of organic chemical information is recorded in the chemical literature. In a research library practically all of it can be searched[1–4] surprisingly quickly with the use of *Chemical Abstracts, Beilstein,* and the review literature.

Chemical Abstracts covers over 12,000 periodicals, patents, and other sources and produces brief summaries of the information in each article with a bibliographic heading. These appear in weekly issues. Six months of these issues constitute a volume that is accompanied by six indexes: *General Subject Index, Chemical Substance Index, Formula Index, Index of Ring Systems, Author Index,* and *Patent Index*. January to June of 1985 is volume 102. Ten volumes (5 years' coverage) are combined in the *Collective Indexes;* the tenth collective volumes are 86–95 (1977–1981). The tenth collective chemical substance index alone comprises 32 bound books. The indexes are based on the entire original document and will include compounds that are not specifically mentioned in the abstract.

For most purposes it is best to begin searching in the latest available volume indexes and continue to earlier volumes and into the collective indexes. Often the earlier literature on a subject or a substance is referenced or summarized in the more recent papers, and it will simplify your search. Before the ninth collective index, there were decennial indexes back to volume 1, 1907. There are several other changes that you will need to be aware of to search early literature, and these will be brought out in due course.

2.1 GENERAL SUBJECT SEARCH

The procedure for searching by subject is illustrated here by answering the question, "Is the structure of the mating pheromone of the Japanese beetle *Papillia japonica* known, and if so, has it been made synthetically?" The *General Subject Index* entries are made up of a controlled vocabulary to prevent scattering of references under several entries. Of the several words in this question, only certain ones are of value in the index. The first step is to examine the *Index Guide,* which refers the many alternative words and nonsystematic chemical names to the names used in the indexes. In the 1984 *Index Guide* under "Pheromones" we find that studies of these as a class and of new pheromones are given under this term but studies of known ones are at those specific headings. Under *Papillia japonica* we find "see Japanese beetle." Turning now to the latest *General Subject Index,* volume 101, under Japanese beetle we find attractant references but not the pheromone. In volume 100 we find "pheromone of, asym. synthesis of, 102990y." The number refers to an abstract in this volume. Consulting that abstract we find the structure pictured, the name (*R,Z*)-(-)-5-(1-decenyl)oxacyclopentan-2-one, a reference to the journal *Agric. Biol. Chem.,* the fact that it was synthesized, and a key step in that process. The original document should next be consulted for details and references to earlier syntheses. A few more volumes should be searched to catch any recent papers missed by those authors.

You should prepare a checklist record of all your literature searches, even if you find nothing on the subject, in order to determine how far you went with a search and to save having to repeat later.

If you need to continue into earlier literature with your subject search, check the earlier index guides in case changes in vocabulary were made. In the eighth collective index and earlier, the *Subject Index* contains both chemical substances and subjects alphabetized together. (Earlier than the eighth collective index, there are no index guides. Appropriate "see" and "see also" references are given in the subject index itself.)

In 1966 and earlier the indexes give column numbers instead of abstract numbers. For example, *65*:18621e refers to a column in the abstracts and the letter "e" refers to a distance down the column on the page.

For a more general subject such as the synthesis of all insect pheromones, it may be more profitable to start with a review article. These may be quickly located by using *Index of Reviews in Organic Chemistry* published by the Royal Society of Chemistry (London). This consists of annual and collective issues covering material back to the early 1960s. The recent issues have two sections: Section 1 covers compounds and Section 2 covers processes and phenomena. The reviews from 1940 to 1964 are found in three volumes by Kharasch and Wolf, *Index to Reviews, Symposia and Monographs in Organic Chemistry*. A number of reviews are found under insect in Section 2 and under pheromones in Section 1. The reviews are also found in *Chemical Abstracts* (CA), where they are recognized by a capital R preceding the abstract number: for example, *95*:R79987v leads to a 183-page review with 486 references on the synthesis of insect pheromones.

2.2 CHEMICAL SUBSTANCE NAME SEARCH

There are many possible names, common and systematic, that may be used for most compounds. Each compound appears at only one place in a volume of the *Chemical Substance Index*; therefore, you must find the CA name. That name will consist of a parent followed by modifiers, each listed alphabetically. Suppose you were interested in 1-benzyl-2-naphthoic acid. The parent name would be naphthoic acid. Check the *Index Guide* to see what parent name CA uses. In the 1984 issue we find under "Naphthoic acid:" "See Naphthalenecarboxylic acid [1320–04-3]." In the tenth collective *Chemical Substance Index* we find 1- and 2-naphthalenecarboxylic acid, each with long lists of modifiers. In the list of modifiers we find "———, 1-(phenylmethyl)-[73194-80-6], *92*:21518v;" and the next entry is "prepn. and cyclocodensation of, *95*:219898u." The abstracts themselves should then be consulted, and from these the original papers may be found. Other volumes of the index should also be checked starting with the most recent. If in proceeding to earlier literature you lose track and suspect a nomenclature change, consult the *Index Guide* of that period or the *Formula Index*.

Clues to the modifier names can be found in the *Index Guide* also; for example, under benzyl hydroperoxide we find "See hydroperoxide, phenylmethyl." *tert*-Butyl is also not now used by CA, but an idea can be found, for example, under "*tert*-Butyl hydroperoxide: See hydroperoxide, 1,1-dimethylethyl."

Since the *Chemical Substance Index* is alphabetized by the parent name first, it can be used to find examples of a class of compounds where you are unsure as to what cases might be known. For instance, if you were looking for an example of the structure

$$(CH_3)_3C\underset{\displaystyle OH}{\underset{|}{N}}-\overset{\displaystyle O}{\overset{\|}{C}}-OR$$

where the R group could be any size, the formula index would require selection of a particular case that might not happen to be known. However, under Carbamic acid (1,1-dimethylethyl)hydroxy-" we find in the tenth collective *Chemical Substance Index* ethyl ester, methyl ester, and phenyl ester.

The *Index Guide* is also useful for finding the true identity of trade-named materials such as the butylated hydroxytoluene on cereal box labels. The guide gives a complete name: "Phenol, 2,6-bis(1,1-dimethylethyl)-4-methyl [128-37-0]." The numbers in brackets are *registry numbers,* which are useful in computer searching; see page 25.

If you plan to search for a rare compound of fair complexity, it may be difficult for an infrequent user to devise the CA systematic name in order to use the *Chemical Substance Index*. In these cases it may be preferable to start with the *Formula Index*.

2.3 MOLECULAR FORMULA SEARCH

The elemental composition of each substance is specified in the order carbon, hydrogen, and other elements in alphabetical order. These formulas are arranged in order of increasing numbers of carbons and for a given number of carbons, in order of increasing numbers of hydrogens, and so on for the other elements. If the compound is a salt such as an amine hydrochloride, acetate, or sulfate, it will be found under the formula for the free amine. Quaternary ammonium salts are under the formula for the cation, omitting the anion.

For example, let us try to find the preparation and properties of the enol acetate of chloroacetone:

$$ClCH_2C(OC(=O)CH_3)=CH_2$$

The molecular formula is $C_5H_7ClO_2$. Turning to the latest *Formula Index*, volume 101 shows 28 substances with this formula. Simply look at the names and quickly reject almost all of them; for example, 2-butenoic acid, modified, is not relevant. Careful examination of the propenols shows that this compound is not listed here. Proceeding to earlier volumes, we eventually locate one reference as tabulated here:

101	No entry
100	No entry
99	No entry

98	No entry
97	No entry
96	No entry
10th coll. (86–95)	No entry
9th coll.	No entry
8th coll.	No entry
7th coll.	1-Propen-2-ol, 3-chloro, acetate *56*:7123d
6th coll.	No entry
41–50	No entry
14–40	No entry

The CA *Formula Index* does not extend earlier than 1920 but *Beilstein* has a formula index that does; see page 29.

Consulting the abstract to the only entry found, we find the reference is Euranto, E.; Kujanpaa, T. *Acta Chem. Scand.* **1961**, *15*, 1209. The article gives no references to other papers for this compound, but it includes the preparation and physical properties of 3-chloropropen-(1)-yl-(2) acetate.

The *Formula Index* gives only the abstract and registry numbers (and indicates patents with a *P*) but no indication of the paper content. For substances where many references exist, it may be better to use the *Formula Index* to get the CA name and then to go to the *Chemical Substance Index* where modifiers are given, to select which abstracts you want to read.

In searching the *Formula Index* it is important to watch for nomenclature changes. For example, in the tenth collective *Formula Index* under $C_{10}H_{18}O_2$ we find the following structure named 2(3*H*)-furanone, dihydro-3-methyl-5-pentyl, *95*:P115267w:

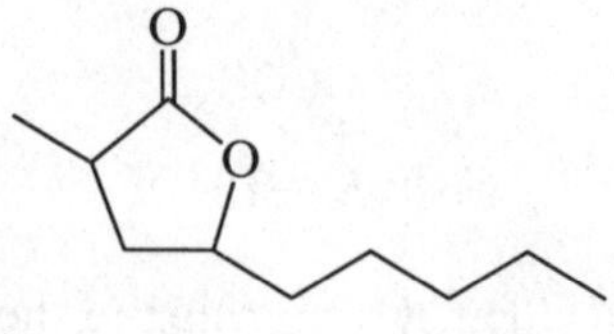

In the abstract itself it is referred to as an α-methyl-γ-alkyl-γ-butyrolactone. In the seventh collective *Formula Index* the name is nonanoic acid, 4-hydroxy-2-methyl-, γ-lactone *59*:11205d. The change in the locant for the methyl group should not be overlooked. Once again, it is not necessary to know all these nomenclature changes, but when the furanones disappear suddenly at the seventh collective index, it is time to look at the whole list under the formula and recognize the older terminology.

2.4 RING SYSTEM SEARCH

Sometimes the formula you are searching leads to names of many complex cyclic compounds and combing through them for the one you want is tedious. You can locate the particular parent name for the ring system first and then search in the *Formula* or *Chemical Substance Index* with it. This is done by using the *Ring Systems Handbook*.

Another occasion to use the *Handbook* is to find a compound containing the ring system of interest where you do not have a particular set of substituents in mind. For example, if an ethyl-substituted compound would be as suitable as a methyl, you would have to search too many formula guesses in the formula index. With the *Ring Systems Handbook* you can find all the substituted cases together regardless of their formulas. You find the parent name and then see the *Chemical Substance Index*. The *Ring System Handbook* includes monocyclic and fused, bridged, and spiro polycyclic compounds.

To demonstrate this, find an example compound containing the ring system

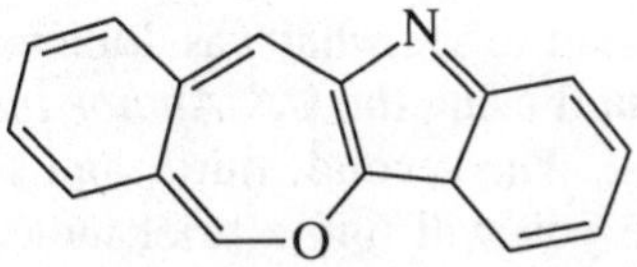

The hierarchy of index headings and subheadings is

1. Elemental formula of ring system: $C_{16}NO$
2. Number of rings: 4
3. Size of rings in increasing order: 5,6,6,7
4. Elemental formula of each ring in formula index order:

$$C_4N–C_6–C_6–C_6O$$

Consulting the *Ring Formula Index* under these four subheadings we find

$C_{16}NO$: 2 RINGS
⋮
$C_{16}NO$: 4 RINGS
$C_3NO–C_4N–C_6–C_7$
⋮
$C_4N–C_6–C_6–C_6O$

1*H*–[1]Benzoxepino[5,4–*b*]indole [*RF 36982*]
1*H*–[2]Benzoxepino[4,3–*b*]indole [*RF 36983*]
1*H*–Dibenz[2,3:6,7]oxepino[4,5–*c*]pyrrole [*RF 36984*]
Oxepino[3,2–*d*]carbazole [*RF 36985*]
C_4NO–C_5N–C_6–C_6*

We may recognize the name of the one sought or limit it to a few that we then view structurally in *Ring Systems File II* filed by the Registry File (*RF*) numbers. In this case all four compounds under C_4N–C_6–C_6–C_6O are found together as numbers *RF 36982–RF 36985*. We easily recognize the target ring system as *RF 36983*. The name found there is now searched in the *Chemical Substance Index*. An example CA citation is also given in the *File,* here CA *88*:6775v. Bridged polycyclic and spirocyclic compounds are indexed here also in the same way. Cumulative supplements to the *Ring Systems Handbook* are added semiannually.

2.5 AUTHOR NAME SEARCH

Occasionally you will want to see what was done next in a certain research group. This may be found using the CA *Author Index*. All the authors on each paper are indexed. The second, third, and so on will refer back to the first author, where you will find a brief indication of the content of each paper and an abstract number. The alphabetization is not the same as a telephone directory because many papers give only the last name and initials. The order is alphabetized by last name and then first initial, and then second initial. An author index accompanies each CA volume and collective index. (CA does maintain a name authority file and will use the full name of the author if it is known to them. Thus in the author index, you may find the author entered as Smith, James W., even though it is given on the paper as J. Smith.)

You may find out what other workers have since done with information from a particular paper by using the *Science Citation Index* published by the Institute for Scientific Information, Philadelphia, and covering the years 1961, 1964 to present. The first authors of all papers referred to are alphabetized, and the papers by the first author that were referred to are listed chronologically. After each paper is given a list of new papers that cite that original paper. Temporary 2-month issues are available for the current year and are replaced with the complete year index.

2.6 COMPUTER SEARCHING OF CHEMICAL ABSTRACTS

The Chemical Abstracts Service produces a computer readable file called "CA SEARCH." For each original paper the file includes the CA citation, abstract heading information, CA section code, *General Subject Index* headings and modifiers (called *descriptors*), entries from the individual issue keyword subject indexes (called *identifiers*), and registry numbers with modifiers from the *Chemical Substance Index*, which include uncontrolled vocabulary. This covers the literature from 1967 and is updated biweekly.

The current and earlier files are available for on-line searching from several vendors, including the DIALOG system by Lockheed Missiles and Space Company, Inc.; the BRS/SEARCH system by Bibliographic Retrieval Services; ORBIT Search System by System Development Corporation, and STN, which is a cooperative of database producers. The user pays a fee through an account with the vendor (about $100 per hour plus a charge for each reference typed or printed) and to the telecommunications system through which the user connects by telephone using the appropriate computer terminal. These vendors carry many other databases besides chemistry.

The computer searching should not be used as a substitute for the normal searching of printed CA indexes, but rather for concurrent searching of a combination of terms that would not come, or come with long labor, from the printed indexes. This is best seen in an example such as the following case done in the DIALOG system.

In DIALOG the CA SEARCH data are available in five files organized according to the CA collective index periods and also as a single file.

File	Collective Index Period	Dates
308	8	1967–1971
309	9	1972–1976
320	10	1977–1979
310	10	1980–1981
311	11	1982–
399	All	1967–

For instance, a researcher found an early reference in a monograph about thiazoles stating that thioformamides react with chloroacetone to give thiazolium salts: "Karimullah, *J. Chem. Soc.* **1937**, 961." The researcher now wanted to find any newer examples of this reaction. We are not looking for a particular thiazolium salt; therefore, the *Formula Index* is not useful.

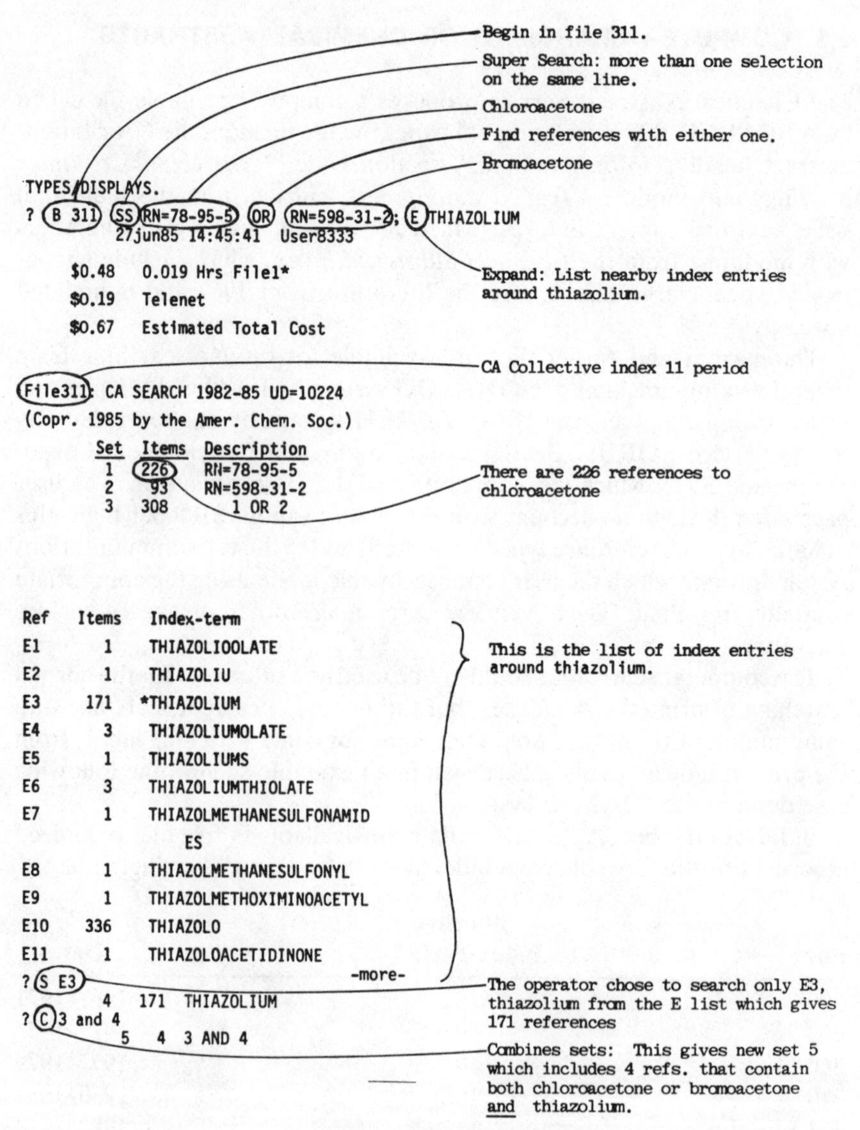

Begin in file 311.
Super Search: more than one selection on the same line.
Chloroacetone
Find references with either one.
Bromoacetone
TYPES/DISPLAYS.
? B 311 SS RN=78-95-5 OR RN=598-31-2; E THIAZOLIUM
27jun85 14:45:41 User8333
$0.48 0.019 Hrs File1*
$0.19 Telenet
$0.67 Estimated Total Cost
Expand: List nearby index entries around thiazolium.
CA Collective index 11 period
File311: CA SEARCH 1982-85 UD=10224
(Copr. 1985 by the Amer. Chem. Soc.)
Set Items Description
1 226 RN=78-95-5
2 93 RN=598-31-2
3 308 1 OR 2
There are 226 references to chloroacetone
Ref Items Index-term
E1 1 THIAZOLIOOLATE
E2 1 THIAZOLIU
E3 171 *THIAZOLIUM
E4 3 THIAZOLIUMOLATE
E5 1 THIAZOLIUMS
E6 3 THIAZOLIUMTHIOLATE
E7 1 THIAZOLMETHANESULFONAMID ES
E8 1 THIAZOLMETHANESULFONYL
E9 1 THIAZOLMETHOXIMINOACETYL
E10 336 THIAZOLO
E11 1 THIAZOLOACETIDINONE
-more-
This is the list of index entries around thiazolium.
? S E3
4 171 THIAZOLIUM
? C 3 and 4
5 4 3 AND 4
The operator chose to search only E3, thiazolium from the E list which gives 171 references
Combines sets: This gives new set 5 which includes 4 refs. that contain both chloroacetone or bromoacetone and thiazolium.

Figure I. A search in DIALOG files for examples of the reaction of thioformamides with chloroacetone to give thiazolium salts. Reprinted by permission of DIALOG® Information Services, Inc. This information is copyrighted by the American Chemical Society and is reprinted by permission. No further copying is allowed.

```
                              Type set 5 in format 3, all four references.
 ? T 5/3/1-4
5/3/1
   00068228     CA: 100(9)68228u       JOURNAL
   Cyclization of 3-(acylmethylthio)-4-cyano-2H-isothiazole-5-thiones
   AUTHOR(S): Gewald, Karl; Roellis, Hans
   LOCATION: Sekt. Chem., Tech. Univ. Dresdan, DDR-8027, Dresdan, Ger.  Dem. Rep.
   JOURNAL: Monatsh. Chem.    DATE: 1983  VOLUME: 114  NUMBER: 8-9  PAGES: 999-1003
   CODEN: MOCMB7  ISSN:  0026-9247  LANGUAGE: German
5/3/2
   99212491  CA:  99(25)212491e          JOURNAL
   Imidazo(1,2,-b)-1,2-benzisothiazoles and Pyrimidol(1,2-b)-1,2-benzisothiazolium
salts
   AUTHOR(S):  Chuiguk, V.A.; Komar, E.L.
   LOCATION: Kiev, Gos. Univ., Kiev, USSR
   JOURNAL: Khim. Geterotsikl. Soedin,  DATE: 1983  NUMBER: 8   PAGES: 1134-5
CODEN: KGSSAQ   ISSN:  0453-8234   LANGUAGE: Russian

5/3/3
   98107200  CA:  98(13)107200r    JOURNAL
   Cyanine dyes, new potent antitumor agents
   AUTHOR(S): Minami, Isoa;  Kozai, Yoshio; Nomura, Hiroaki; Tashiro, Tazuko
   LOCATION:  Cent. Res. Div., Takeda Chem. Ind. Ltd., Osaka, Japan, 582
   JOURNAL:  Chem. Pharm. Bull.  DATE: 1982  VOLUME: 30  NUMBER:  9  PAGES: 3106-20
CODEN:  CPBTAL  ISSN:  0009-2363  LANGUAGE:  English

5/3/4
   96085485        CA:  96(11)85485s    JOURNAL
   Studies on 1, 4-benzothiazine.  III.  Chemical reactions of 2-acylbenzathiazines
and related compounds
   AUTHOR(S):  Sugiyama, Kazuaki; Ogawa, Takako; Hirano, Hiroshi
   LOCATION: Ogaka Coll. Pharm., Matsubara, Japan
   JOURNAL: Yakugaku Zasshi  DATE:  1981  VOLUME: 101  NUMBER:  10 PAGES: 904-11
CODEN:  YKKZAN  ISSN: 0372-7750  LANGUAGE:  Janpanese

 ?END/SAVE  TEMP
Serial#TF11
      27jun85 14:48:06  User8333
   $3.19  0.042 Hrs File311  3 Descriptors
   $0.42  Telenet
   $1.44  8 Types
   $5.05  Estimated Total Cost

   DIALOG accession number, CA citation, It is a journal article.  Save
   the search instructions to use in the next file.
   This number was assigned to these instructions by DIALOG.
```

Figure I. (*Continued*)

There are 20 pages of references under thiazolium in the tenth collective *Chemical Subject Index,* and chloroacetone is not included in the modifiers there; thus we cannot make this search in the printed indexes.

The chloroacetone is best specified as a registry number, and at the same time the alternative bromoacetone is included as the corresponding registry number as found in the *Chemical Substance Index*. In order to examine adjacent terms to thiazolium in the file index, the command "Expand" gave a list of 11 entries, including 2 before and 8 after (see Fig. I). It was decided to keep only thiazolium, for which there are 171 references in file 311. The mechanics of the process are explained on the figure showing the search. The lines beginning with "?" were typed by the op-

```
 ┌─Begin in file 310 using the search strategy saved under this number.
 │
?(B  310; .EXS TF11)
      27jun85   14:54:10  User8333
    $0.11  0.007 Hrs File201
    $0.07  Telenet
    $0.18  Estimated Total Cost

File310:CA  Search - 1980-1981
(Copr. 1985 by the Amer. Chem. Soc.)
        Set    Items  Description
        ---    -----  -----------
          1     138    RN=78-95-5
          2      62    RN=598-31-2
          3     190      1 OR 2
          4      37     37 E3
          5       0  3  AND  4

?  E  THIAZOLIUM
  Ref Items   Index-term
  E1      2   THIAZOLINYLMETHYLHYDRAZI
                DE
  E2      1   THIAZOLINYLTHIOACETATE
  E3     96  *THIAZOLIUM
  E4      2   THIAZOLIUMOLATE
  E5      2   THIAZOLIUMOLATES
  E6      1   THIAZOLIUMYL
  E7      1   THIAZOLIZATION
  E8    199   THIAZOLO
  E9      1   THIAZOLOAZABENZOTHIAZOLI
               UM
  E10     1   THIAZOLOAZEPIN
                                   -more-
  ? S E3
           6  96  THIAZOLIUM
  ?(C 3 AND 6)
 ┌
 └Combining sets 3 and 6 gave only one reference, now called set 7.
        7   1  3  AND 6

? T  7/3/1
7/3/1
    94065258  CA: 94(9)65258a   JOURNAL
  ┌ Asymmetric benzoin condensation catalyzed by optically active
  │thiazolium salts in micellar two-phase media
┌─┤ AUTHOR(S): Tagaki, Waichiro; Tamura, Yoshiharu; Yano, Yumihiko
│ │ LOCATION: fac. Eng., Gunma Univ., Kiryu, Japan, 376
│ │ JOURNAL: Bull. Chem. Soc.  Jpn.  DATE: 1980  VOLUME: 53  NUMBER 2  PAGES:
│ └478-80  CODEN: BCSJA8  ISSN:  0009-2673  LANGUAGE:  English
└──This reference contains a good example of the reaction sought.
? (LOGOFF)
 ┌    27jun85 14:48:51  User8333
 └─Disconnect from DIALOG
```

Figure I. (*Continued*)

erator, and the rest came from DIALOG. File 311 gave four references. File 310 next gave one reference that included a good example of the reaction sought. File 320 gave two more references not pertinent to the thioformamide reaction, and file 309 gave seven and file 308 gave none.

A search of this sort is not thorough; that is, it may well not uncover all the cases of that particular kind of reaction, but it often gives enough good leads to be very worthwhile. This search missed a relevant paper in the ninth collective period because the registry number for chloroacetone was not indexed for this paper; however, it was referred to in the 1980 paper, so it was indirectly available from the search.

There are many other useful commands for searching not mentioned here. A manual should be consulted before undertaking a new search, and the actual process is economically run by an experienced operator with you standing by to make quick decisions.

Computer Searching by Substructure. Chemical Abstracts Service (CAS) offers a substructure searching system in its registry file that is part of CAS ONLINE (available on STN).

If you are interested in finding references to a particular substructure where other parts of the structure may vary, you can specify that substructure by typing commands that construct rings and chains and specify atoms and bonds between particular atoms. You may partially generalize by using "not" commands; for example, you could exclude bromine from a particular site. As you proceed, you may order the partial structure drawn. Alternatively, the substructure can be constructed on a graphics terminal, which uses a menu of commands, by manipulating the cursor with the ⇅ and → ← buttons or with a tablet and stylus or "mouse." The search is therefore not limited to a particular compound or structures grouped by nomenclature coventions. The search can be tested in a learning file of 100,000 structures or in a sample search of the file, and then used in the full CAS Registry File, which contains over 8 million substances from CA since 1965 and is updated weekly.

The results of the search can be printed on-line or off-line and include a structural picture for each compound retrieved plus the registry number, name, synonyms, and up to 10 recent CA references.

2.7 BEILSTEIN HANDBOOK OF ORGANIC CHEMISTRY

If you are interested in the preparation and/or properties of a particular organic compound, searching *Chemical Abstracts* may be frustrating because many of the references you uncover will give a use for the compound and not the data you are seeking. This particular information may be found quickly in *Beilstein's Handbuch der Organischen Chemie*. *Beilstein* is an organized collection of preparations and properties of organic compounds that were known before 1960. The fourth edition (*Vierte Auflage*) consists of a basic series (*Hauptwerk*) covering work up to 1909, and four supplementary series (*Erganzungswerk*) covering the literature to 1959. A fifth supplement will eventually cover 1960–1979.

The *Handbook* consists of 29 volumes. The main divisions are as follows: acyclic compounds, volumes 1–4; carbocyclic compounds, volumes 5–16; heterocyclic compounds, volumes 17–27; *General Subject Index,* volume 28; and *General Formula Index,* volume 29. The *General Formula Index* is complete in the second supplement and covers the *Hauptwerk* and first and second *Erganzungswerk*. Many of these volumes consist of sets of bound subvolumes. It is hardly a "handbook" now since it occupies about 45 feet of shelf space.

The compounds are arranged in the volumes according to the rules of the *Beilstein* system, which allow you to search directly in the volumes without using indexes, finding similar compounds located together. These rules are beyond the scope of this chapter but are available elsewhere.[5] The example search shown below begins instead with a formula index.

The illustrative search example is as follows. Find the melting point and preparations of *N,N'*-diisopropylurea, $C_7H_{16}N_2O$. Consulting the *General-Formelregister,* volume 29, we find 11 isomeric ureas listed, including "*N,N'*-Diisopropyl-harnstoff *4*, 155, II 631." This indicates that it is in volume 4 in the *Hauptwerk* on page 155, not covered in the first *Erganzungswerk,* but in the second *Erganzungswerk* on page 631. Consulting these pages, we find

Syst. No. 337.] DERIVATE DES ISOPROPYLAMINS. 155

ameisensäuremethylester und Isopropylamin in Wasser (THOMAS, *R.* **9**, 71). — Flüssig. Kp: 165,5°. D^{15}: 0,981.

N-Isopropyl-harnstoff $C_4H_{10}ON_2 = (CH_3)_2CH \cdot NH \cdot CO \cdot NH_2$. *B.* Durch Reduktion der Verbindung $(CH_3)_2C$—$N \cdot CO \cdot NH_2$ (O) (Syst. No. 4190) mit Aluminiumamalgam (CONDUCHÉ, *A. ch.* [8] **13**, 65). — Nadeln. F: 154°. Löslich in Wasser, Alkohol, Chloroform, siedendem Benzol und Aceton, weniger in Essigester, schwer in kaltem Äther und kaltem Benzol.

N.N'-Diisopropyl-harnstoff, symm. Diisopropylharnstoff $C_7H_{16}ON_2 = (CH_3)_2CH \cdot NH \cdot CO \cdot NH \cdot CH(CH_3)_2$. *B.* Aus N-Brom-isobutyramid durch Erhitzen mit Na_2CO_3, neben Isopropylisocyanat (A. W. HOFMANN, *B.* **15**, 756). — Nadeln (aus Alkohol). F: 192°. Unlöslich in Wasser und Äther.

Brom auf Malonsäure-bis-isopropylamid in heißem Eisessig (WEST, *Soc.* **127**, 751). — Nadeln (aus Alkohol). F: 204°. — Geschwindigkeit der Reaktion mit Jodwasserstoffsäure in 4% Wasser und 2% Essigsäure enthaltendem Methanol bei 25° und 30,2°: W.

N.N'-Diisopropyl-harnstoff $C_7H_{16}ON_2 = (CH_3)_2CH\cdot NH\cdot CO\cdot NH\cdot CH(CH_3)_2$ (H 155). *B.* Neben Spuren von N-Isopropyl-N'-isobutyryl-harnstoff bei der Einw. von 1 Mol 10%iger Natronlauge auf N-Chlor-isobutyramid ohne Kühlung (ROBERTS, *Soc.* **123**, 2782). Beim Erwärmen einer wäßr. Lösung des Kalium- oder Natriumsalzes des Isobutyrhydroxamsäure-benzoats (JONES, SCOTT, *Am. Soc.* **44**, 421). — Krystalle (aus verd. Alkohol).

We now locate it in E III and/or E IV, which are not included in that index. Knowing that it is in volume 4 from the first formula index, we now turn to the new *Gesamtregister,* volume 4 formula index, which covers all parts of volume 4, including the fourth supplement. Under the formula we find "Harnstoff, *N,N'*-Diisopropyl-, *4* 155b, II 631a, III 277a, IV 521." The letters indicate how far down the page to look; the first compound on the page is a, the second b, the eighth h, and so on. The melting point is 197°C, and quite a variety of syntheses are given with references.

In those volumes where the new formula index is not yet available, you can locate compounds in E II and E IV by examining the heading on the right hand pages in these supplements, looking for those pages that correspond to ones in the earlier issues. For example, in volume 13, on top of page 141 we find

E IV 13 Syst.—Nr. 1771/H 96-97; E III 169–172* 141

Material on this page is an extension of that on pages 96–97 of the *Hauptwerk* and pages 169–172 of the third supplement of this volume.

There is also a formula and subject index in each individual subvolume in the third and fourth supplements.

A few parts of the *Fifth Supplementary Series* have appeared, covering the literature from 1960 to 1979. This series is in English, while the earlier ones are all in German.

If the compound you are seeking does not appear before 1929, you will need to use the *Beilstein* system to determine in which volume it should appear in the third and later supplements. For a concise introduction, see the booklet *How to Use Beilstein,* Beilstein Institute, Springer-Verlag, Berlin, 1978.

2.8 GENERAL SOURCES

Thus far we have discussed finding original references to chemical information. A general idea of the reactions and the properties of classes of compounds is more readily obtained from compilations of organic chemistry. These are encyclopedic works spanning the entire field. One of the best is *Methoden der Organischen Chemie,* known as "Houben-Weyl" for the editors of the first edition. The fourth edition was completed in 1985 with 65 volumes and a general index. This is an organized, completely referenced, very detailed collection of methods of preparing essentially all classes of organic compounds plus their reactions. It includes selected experimental details and extensive tables of examples. The publisher is Georg Thieme Verlag, Stuttgart.

Comprehensive Organic Chemistry is a five-volume work plus a sixth volume of indexes again spanning the whole field and providing many leading references. The editorial board was chaired by D. Barton and W. D. Ollis. It was published complete and contemporaneous in 1979 by Pergamon Press, Oxford.

There are several other such compendia and also hundreds of individual monographs available on specialized topics within the field.

PROBLEMS

1. Find the best literature preparation of the following compound. Give the registry number, the journal reference, the reaction, the melting or boiling point, and the ^{1}H NMR chemical shift value of the α-methyl groups.

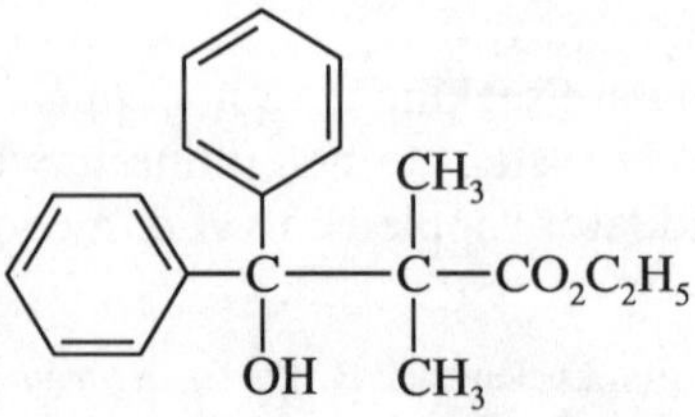

CH_3
—C—C—$CO_2C_2H_5$
OH CH_3

2. Find the best literature preparation of the following compound. Give the journal reference, the melting point, and the ^{1}H NMR values for the vinyl protons.

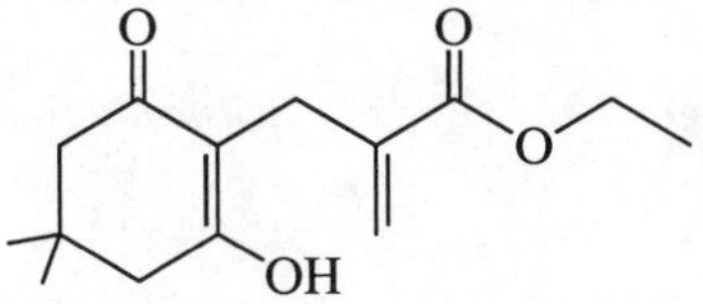

3. Locate and give the reference for a recent review on the use of carbon disulfide as a reactant in organic synthesis.
4. Locate a reference to the preparation of 3,5-dimethoxy-4-bromophenylacetonitrile.
5. Locate a reference to a compound that is an example of the following ring system and give the CA name for that ring system.

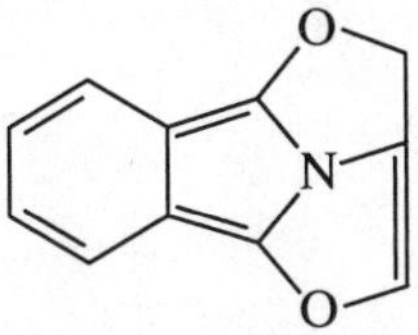

6. Locate a reference to the following compound where R is an alkyl group.

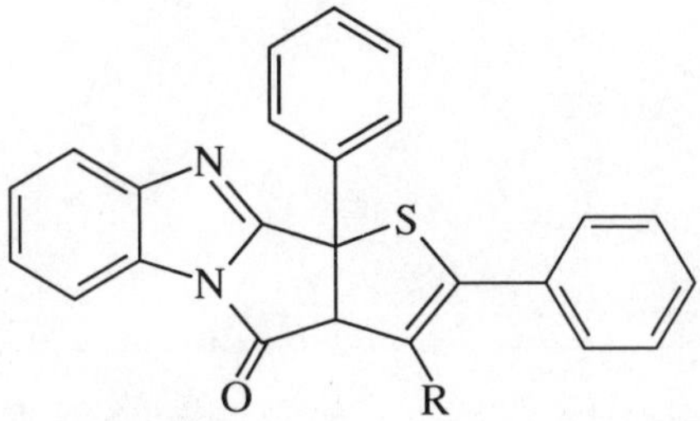

7. Using *Beilstein,* find the melting point of 4-bromo-3,4′-dinitrobiphenyl.

REFERENCES

1. Antony, A. *Guide to Basic Information Sources in Chemistry,* Wiley, New York, 1979.
2. Bottle, R. J. *Use of the Chemical Literature,* 3rd ed., Butterworths, London, 1979.

3. Maizell, R. E. *How to Find Chemical Information,* 2nd ed., Wiley-Interscience, New York, 1987.
4. Skolnik, H. *The Literature Matrix of Chemistry,* Wiley-Interscience, New York, 1982.
5. Runquist, O. *A Programmed Guide to Beilstein's Handbuch,* Burgess, Minneapolis, 1966.

3

Stereochemistry

The shapes and properties of molecules can depend not only on the order of connection of atoms but also on their arrangement in three-dimensional space. Molecules differing only in configuration are called stereoisomers, and are the principal subject of this chapter.[1]

3.1 REPRESENTATIONS

Some organic molecules such as benzene are planar as defined by the point locations of all nuclei present. These are easily represented on our planar printed page.

Most organic molecules are three-dimensional structures, best viewed and represented in solid molecular models. The necessity of using paper requires pictures that show depth as perspective does in artwork and photography. The mere projection onto the plane of the paper, as in the shadow of a molecular model, loses the real difference between left- and right-handed structures. The best alternative on paper is a stereo pair of pictures viewed through a stereopticon. Most commonly in journals and handwritten material we use representations where depth is portrayed via conventions instead of the pictorially obvious. These are exemplified in Figure I for 2-butanol.

In the pictorial representation your three-dimensional cues are the front and back emergence of bondsticks on the spherically shown atoms. In the

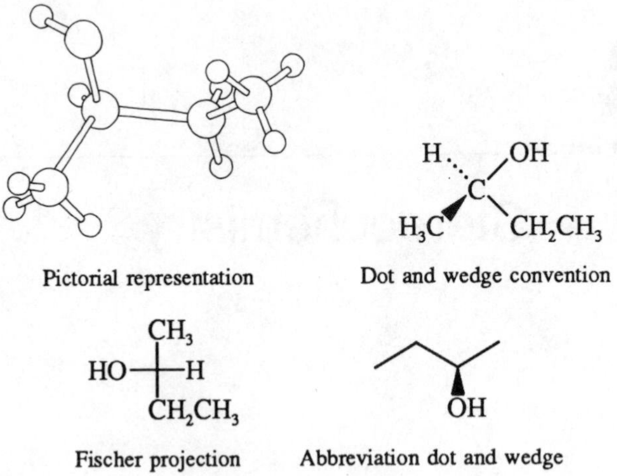

Pictorial representation

Dot and wedge convention

Fischer projection

Abbreviation dot and wedge

Figure I. (*R*)-2-Butanol.

dot–wedge convention the group on the broad end of the wedge is defined as being above the plane of the paper, the dotted bond extends below the plane of the paper, and the line bonds are in the plane of the paper. In the abbreviated form the hydrogens on carbon are not shown but defined as completing the tetravalency of carbon.

In the Fischer projection, the center of the crossed lines is a carbon atom and those bonds emanating from it to the side are defined as extending above the plane of the paper toward the viewer and those extending toward

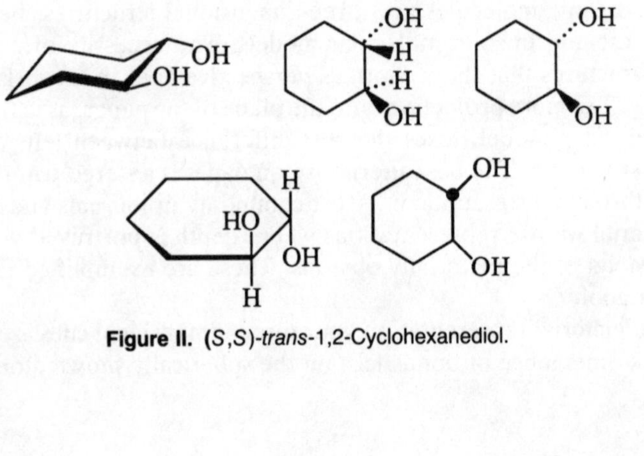

Figure II. (*S*,*S*)-*trans*-1,2-Cyclohexanediol.

the top and bottom of the page are defined as extending below the plane of the paper, away from the viewer.

Related conventions are used for portraying ring compounds as exemplified in Fig. II for (*S*,*S*)-*trans*-1,2-cyclohexanediol. Many authors will draw one enantiomer of each molecule in a reaction scheme when they are actually using racemic materials. Their text should indicate this meaning.

3.2 VOCABULARY

A molecule or other object that is different from its own mirror image (e.g., as is a shoe) is *chiral.* Molecules that are identical to their mirror image are *achiral.* The conformational flexibility of molecules allows many different representations; therefore, in testing of a pair of structures for identity or mirror image relationship, the models should be flexed or the drawings redrawn to attempt a match. A left-hand fist is not the mirror image of an open right hand, yet we will refer to left and right hands as mirror images generally.

A chiral molecule and its mirror image molecule are *enantiomers;* that is, their relationship is enantiomeric. A pair of shoes is an enantiomeric pair. A *racemic mixture* or *racemate* is a combination of equal amounts of enantiomers.

Most common chiral molecules contain one or more chiral centers. A *chiral center* is an atom in tetrahedral hybridization (usually a carbon) with four all-different groups bonded to it. In sulfoxides a nonbonding pair of electrons serves as the fourth group. Carbon 2 in 2-butanol and carbons 1 and 2 in 1,2-cyclohexanediol are chiral centers. The two possible spatial arrangements about a chiral center are called *configurations,* and each one is designated (*R*) or (*S*) according to the Cahn–Ingold–Prelog system.[2]

If a molecule contains more than one chiral center, there will usually be *diastereomeric* pairs. Diastereomers have the same order of connection of atoms in their structures, but one differs in spatial arrangement from the other and from the mirror image of the other in all reasonable conformations. Diastereomeric substances must therefore differ in many physical properties. (The term "diastereomer" is now used to relate cis and trans alkenes and cis and trans ring compounds even if they are achiral.) The term *stereoisomer* includes enantiomers and diastereomers. All three stereoisomers of 1,2-cyclohexanediol are shown in Fig. III.

The (*S*,*S*) and (*R*,*R*) isomers are mirror images and, therefore, enantiomers. The (*R*,*S*) isomer is different from either (*S*,*S*) or (*R*,*R*) and

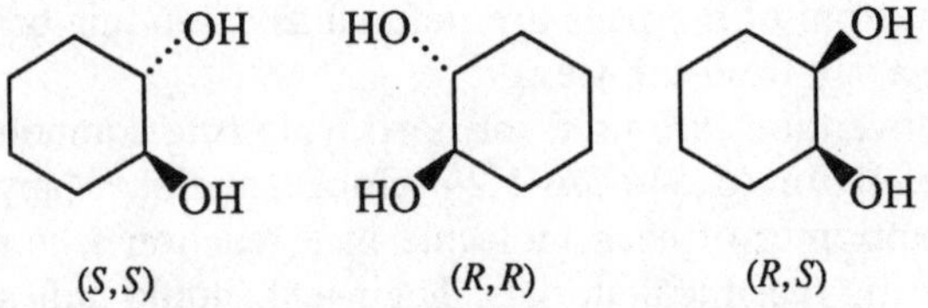

(S,S) (R,R) (R,S)

Figure III. 1,2-Cyclohexanediol stereoisomers with configurational designation.

is, therefore, a diastereomer of each. Note that the (*R*,*S*) isomer is achiral despite the presence of chiral centers. Achiral molecules containing chiral centers are termed *meso.* Meso structures may be recognized by the presence of a mirror plane within the molecule in certain conformations.

You can expect more stereoisomers for structures that contain more chiral centers. For example, the C_6 glycopyranoses contain five chiral centers (Fig. IV), and there are 32 stereoisomers. Half of them are enantiomers of the other half.

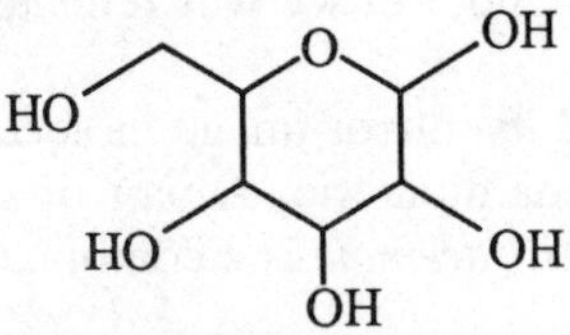

Figure IV. The C_6 glycopyranoses.

Each additional chiral center doubles the number of stereoisomers except where meso compounds occur, or where a polycylic ring system prohibits some configurations. Compound **1** contains four chiral centers, but there are only two stereoisomers.

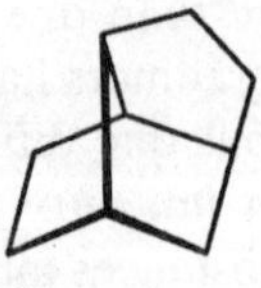

1

Certain diastereomeric relationships are designated by prefixes derived from two carbohydrates. D-Threose (**2**) and D-erythrose (**3**) have two chiral centers and each center has an —H and an —OH group. Other molecules that differ in the analogous fashion are prefixed *threo-* and *erythro-*, as generally represented in Fig. V.

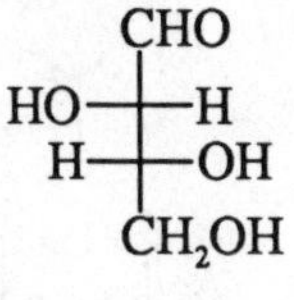

2 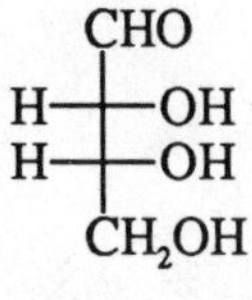 **3**

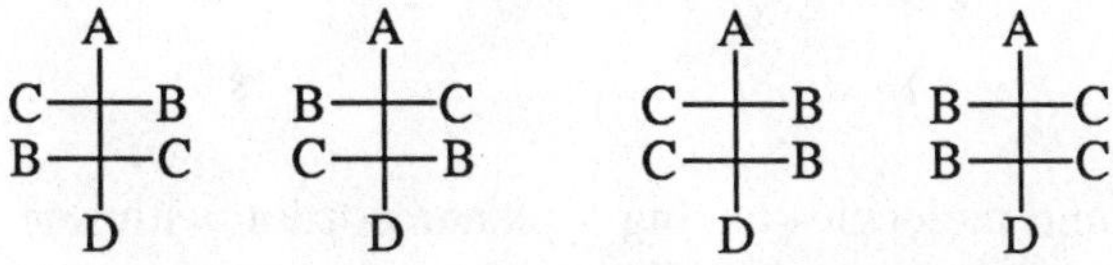

Threo *Erythro*

Figure V. *Threo* and *erythro* stereoisomers.

3.3 CHIRAL MOLECULES WITH NO CHIRAL CENTERS

Molecules with a twist along an axis such as allenes (**4**), spiro compounds (**5**), and exocyclic double-bonded compounds (**6**) can be chiral.

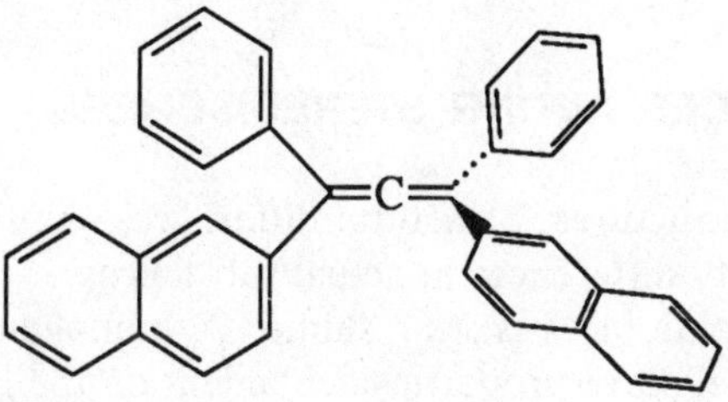

4 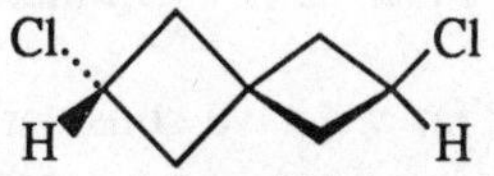 **5**

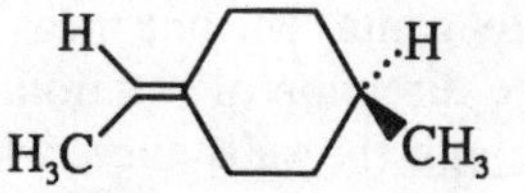

6

Crowding of groups in a molecule may restrict rotation about single bonds or prevent planarity, again generating a twist, giving conformational enantiomerism (**7, 8**).

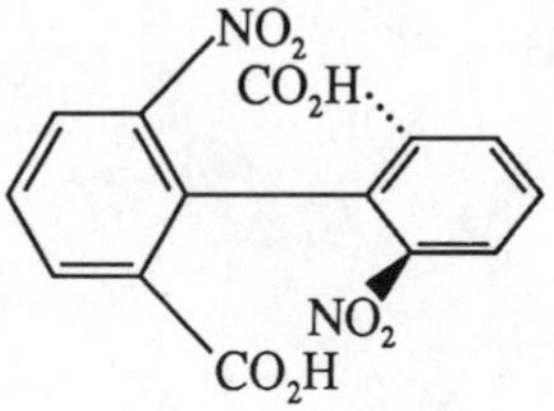

Finally, some molecules having a planar portion with one face distinguished from the other and lacking a plane of symmetry are chiral. Examples include paracyclophane **9** and *trans*-cyclooctene **10**.[3]

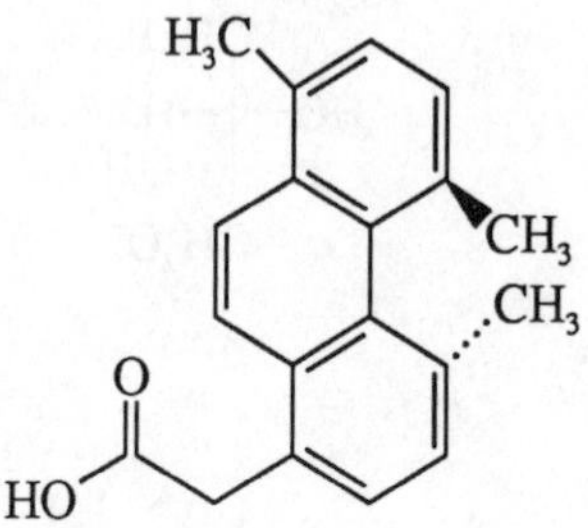

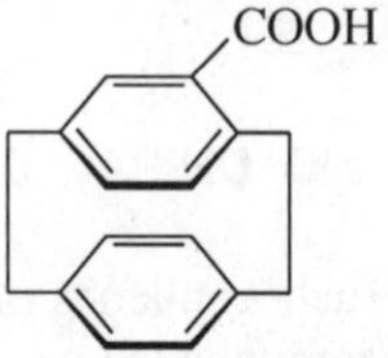

3.4 PROPERTY DIFFERENCES AMONG STEREOISOMERS

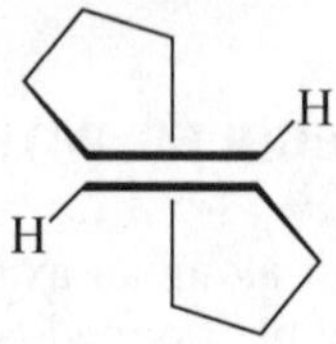

Thus far we have considered structures. Structural differences should be manifest in measurable property differences in actual substances.

The difference between enantiomers is very subtle. A homogeneous sample of a pure enantiomer will have properties dependent on the intermolecular attractions in the sample, and the mirror image will be no different in that regard. Thus the melting point and boiling point of each will be identical. So also will be their refractive index density, spectra, and rate of reaction toward achiral reagents. The one measurable physical difference between enantiomers is the direction of rotation of the plane of polarized light when it is passed through the substance, which is opposite for each.

The amount of rotation of the plane of polarized light depends on the path length through the sample and the density of the material, or the concentration in grams per milliliter in solution. The specific rotation is defined as the number of degrees of rotation for a sample of density 1 g/mL and 10 cm long and is a fundamental characteristic of an enantiomer. Measurements made under different conditions may be proportioned to the specific value. Samples that rotate the plane of polarized light are called *optically active.* Racemic material is not optically active. If a sample is 75% of one enantiomer and 25% of the other, the rotation will be one-half of the maximum and that sample will be labeled 50% optically pure or 50% enantiomeric excess. Rotation measurements are commonly reported for light from a sodium lamp of wavelength 5890 Å. If a white light source is used, you will see various colors as one polarizer is rotated because the rotation of the plane by a sample varies with the wavelength. A graph of the rotation versus wavelength is called an *optical rotary dispersion curve.*

As mentioned before, the melting point of one enantiomer of a chiral substance will be identical to that of the other enantiomer. However, if both enantiomers are together in a sample, the melting point is likely to be different from that of a pure enantiomer. Consider rows of people shaking hands using all right hands (or all left hands). A certain fit will exist. Consider instead random shaking, including right with left. This produces a different fit. By analogy, interactions between (+) and (−) enantiomers will be different from (+) and (+), and we can expect a different melting point for the mixture. If solutions or melts of various ratios of enantiomers are cooled to produce solid, we find one of three possible behaviors:

1. The enantiomers may crystallize separately, giving a mixture of (+) and (−) crystals called a *conglomerate.* The melting point of a conglomerate is then a mixed melting point and is lower than that of pure enantiomer. Various ratios of (+) and (−) give melting points graphed in Fig. VI*a*.
2. In other cases a stronger attraction exists between opposites and a racemic compound is formed. The crystals will contain both enantiomers in equal amounts. The melting point of the racemate is depressed by adding a small amount of one enantiomer (Fig. VI*b*, VI*c*).

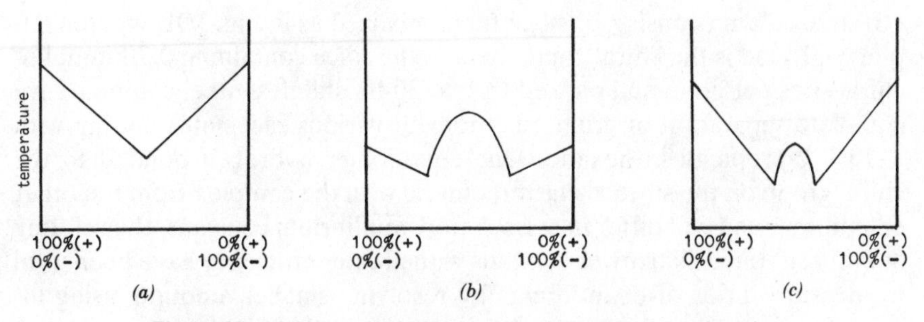

Figure VI. Melting point behavior of various enantiomeric pairs.

The racemate may melt either lower or higher than a pure enantiomer.

3. If the difference between enantiomers is small, the substance may crystallize as an ideal solid solution, where the ratio of (+) to (−) has no effect on the melting point, and the graph is flat. A few give nonideal solutions with maxima or minima.

An example of Fig. VI*b* behavior is found with *trans*-2-*tert*-butylcyclohexanol.[4] The pure (−) rotating enantiomer and the pure (+) rotating enantiomer melt at 50–52°C, while the racemic material melts at 84.4–85°C.

The structural difference between diasteromers is greater than that of enantiomers. They are not mirror images; therefore, all their physical properties will differ to some extent. For example, racemic *cis*-2-*tert*-butylcyclohexanol melts at 56.8–57.7°C (compare above). Furthermore, the cis and trans isomers are separable by thin-layer chromatography. Similarly, *meso*-2,3-diphenylbutane melts at 126.4–127°C, while the racemic form is liquid at room temperature and shows a different retention time on gas chromatography.

3.5 RESOLUTION OF ENANTIOMERS

Enantiomers behave identically toward achiral materials. However, a chiral solvent, adsorbent, catalyst, or reagent that is present as only one enantiomer itself will differentiate between enantiomers.[5] A left foot will fit well in a left shoe but not in a right shoe. The left-foot–left-shoe combination is diastereomeric with the left-foot–right-shoe combination and thus has different properties.

Such differences have been used to separate one enantiomer from the other, a process called *resolution*. Chromatographic separation of racemates on various chiral substrates has been demonstrated.[6] A particularly effective column consists of silica functionalized as in Fig. VII, where (*R*)-phenylglycine is the chiral component.[7] The silica containing 0.70 mmol of chiral sites per gram and packed in 2 × 30-in. stainless-steel columns gave complete separation of gram quantities of various racemates eluting with 5–10% 2-propanol in hexane. One enantiomer hydrogen bonded to the chiral group on the silica is diastereomeric with the complex from the other enantiomer and will differ in dissociation equilibrium constant, thus eluting at different rates. Narrower columns with greater efficiency have been used to measure ratios of enantiomers by resolving smaller amounts using ultraviolet absorption detection of the isomers in the elluate.

Ratios of enantiomers may sometimes be measured without separation,

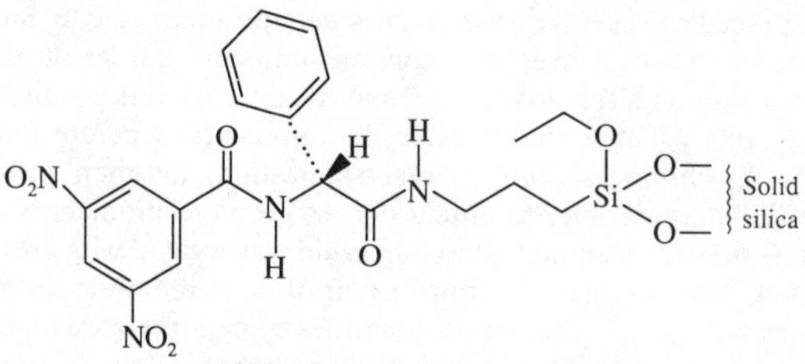

Solid silica

Figure VII. A chromatographic resolving agent.

using a chiral solvating agent to make them spectroscopically differentiable.[8] For example, the 100-MHz ^1H NMR spectrum of **11** in CCl_4 solution with 0.4 *M* **12** added gave separate signals for the methyl groups in the *R* and *S* isomers of **11** (7.6 Hz apart).

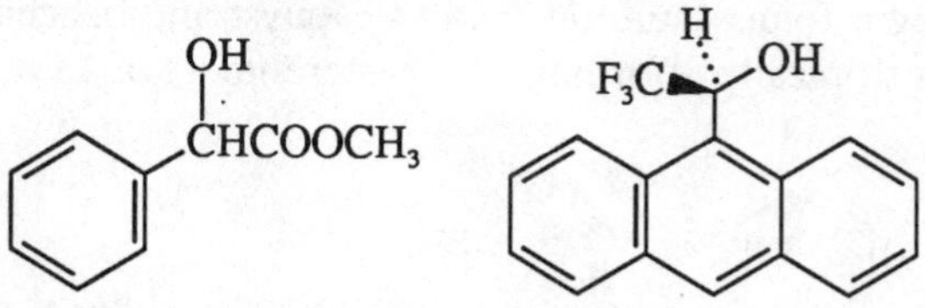

(*R*) and (*S*)
11

12

Enzymes are chiral and will often form a reactive complex with one enantiomer of a substance to the complete exclusion of the other. Thus the enzyme can give resolution by catalyzing, for example, hydrolysis of an ester leaving one enantiomer unchanged. Hog liver esterase gives pure (−)-*trans*-2-phenylcyclohexanol from the racemic acetate. The (+) enantiomer remains as the acetate and can be readily separated and hydrolyzed by acid or base (Eq. 1) to give the (+)-alcohol.[9]

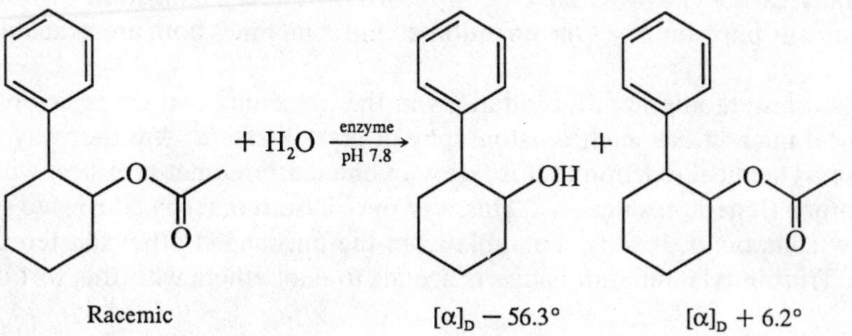

$$ + H_2O \xrightarrow[\text{pH 7.8}]{\text{enzyme}} \quad + \qquad (1)$$

Racemic $\qquad [\alpha]_D - 56.3°$ $\qquad [\alpha]_D + 6.2°$

In the preceding cases the enantiomers were differentiated by formation of diastereomeric complexes with column stationary phases or enzymes. The more common alternative is to bond the enantiomers covalently to a chiral resolving agent to make stable diastereomers, separate those diastereomers by chromatography or recrystallization, and then disassemble each purified diastereomer to obtain the resolved enantiomers.

Racemic *trans*-2-cyanomethylcyclohexanol was treated with (*R*)-(−)-1-(1-naphthyl)ethyl isocynate to afford a pair of diastereomeric carbamates. These were separable in multigram quantities by high-pressure liquid chromatography on silica gel eluting with benzene–ether.[10] Each separate diastereomer was treated with trichlorosilane to regenerate the isocyanate resolving agent and release a pure single enantiomer of *trans*-2-cyanomethylcyclohexanol. This was hydrolyzed in a subsequent step to give the enantiomerically pure lactone.

(−)-Mandelic acid is a suitable resolving agent for isolating one enantiomer of various chiral alcohols by recrystallization of a mixture of diastereomeric esters.[11] For example, 2-octanol was esterified with (−)-mandelic acid using *p*-toluenesulfonic acid as catalyst and benzene to remove the water azeotropically affording the diastereomers of **13** in 97% yield.

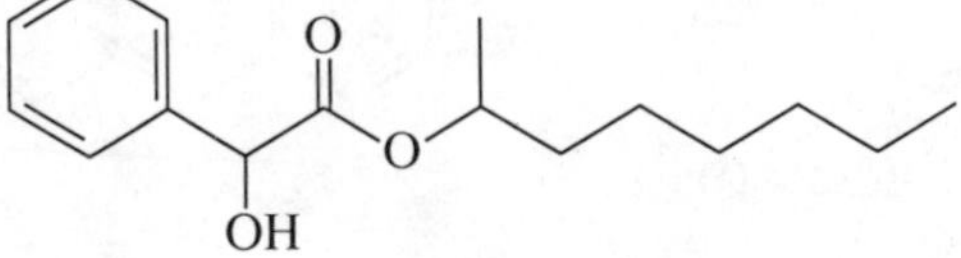

O
O
OH

13

Recrystallization gave a 59% yield of one diastereomer. Saponification then afforded one optically pure enantiomer of 2-octanol. The diastereomeric purity of the ester could be determined readily by ^{1}H NMR or ^{13}C NMR where the α-methyl group of the alcohol gave distinctly different signals in the two diastereomers.

Many carboxylic acids have been resolved by recrystallization of salts of optically pure amines. One enantiomer and sometimes both are available this way.

The diastereomeric differentiations in the preceding two cases involve physical interactions in chromatography or crystallization. Another way is to use a chemical reaction that is fast with one diastereomer and slow with the other (kinetic resolution). This way one diastereomer is converted to a new compound, readily separable from the unchanged other diastereomer. Triisobutylaluminum converts acetals to enol ethers with this sort of selectivity. This reaction was used to resolve many unsymmetrically substituted cyclic ketones as exemplified in Eq. 2.[12]

$$\text{2-methylcyclohexanone} \longrightarrow \text{acetal} \xrightarrow[0°C]{Al(CH_2CH(CH_3)_2)_3} \text{enol ether} + \text{acetal} \quad (2)$$

The readily separable enol ether and acetal can each be hydrolyzed in acid to give the optically active ketone. If the reaction is carried to 35% completion, the enol ether furnishes (*R*)-2-methylcyclohexane with greater than 95% enantiomeric excess. If carried to 70% completion, the remaining acetal gave (*S*)-2-methylcyclohexanone with greater than 95% enantiomeric excess.

Some racemates that give conglomerates may be resolved with no external diastereomeric influences.[13] Direct crystallization of individual enantiomers from saturated solutions of racemates may be localized by seeding with pure enantiomers, especially if large crystals will grow as, for example, with hydrobenzoin. A practical variation on this process, called *entrainment,* begins by enriching a solution of racemate with one enantiomer, cooling to saturation, and seeding with the one in excess. In favorable cases a crop about twice the size of the original excess is obtained. The solution then contains an excess of the other enantiomer. More racemate is added, cooled to saturation, and seeded with the other enantiomer to gain a crop of it as large as was obtained for the first. This is then repeated indefinitely. In this way 13,000 tons of L-glutamic acid was produced annually from synthetic racemate. Many other amino acids have been resolved on a smaller scale; however, most organic racemates give racemic compounds on crystallization and are therefore unsuitable for resolutions by entrainment.

If one of the preceding resolution procedures has given material of less than 100% optical purity, simple recrystallization or sublimation[14] will often give the pure enantiomer. This sort of fractionation must be avoided if you are isolating a product, intending to measure the optical purity as it comes from the reaction.

3.6 ASYMMETRIC SYNTHESIS

In Section 3.5 pure enantiomers were obtained by separation of racemic material. Another alternative is to begin with an achiral compound and generate a single enantiomer of a chiral compound from it.[15,16] This requires a chiral influence from another component in a chemical reaction.

A chiral "template" may be temporarily attached to an achiral molecule and then a new chiral center made under that influence, and finally the original "template" is removed.[17] To illustrate this, consider first the alkylation of simple ketone enolates such as cyclohexanone enolate, where a chiral center is formed. Without a chiral influence the products are always racemic. If a chiral imine enolate is employed as in Eq. 3, the incoming alkylating agent is guided with high specificity to a particular face of the enolate.[18]

Ph H. O + H_2N OCH_3 → Ph H. N OCH_3 $\xrightarrow[\text{THF, }-20°\text{C}]{\text{LDA}}$

23% yield

Ph H. N Li O CH_3 $\xrightarrow{\text{I}}$ Ph H. N OCH_3 $\xrightarrow[\text{HOAc}]{H_2O}$ O

99% (*S*), 76% yield

(3)

By this procedure 12 cases on various rings and with various alkylating agents all gave the same choice of attacking face and the analogous enantiomer product. Reduction of (*S*)-phenylalanine followed by methylation provides the chiral primary amine.

The chiral influence may be a catalyst, in which case a small amount of chiral material can lead to large amounts of enantiomerically pure (or enriched) product. Several *N*-acyl (*R*)-amino acids were prepared by hydrogenation of α,β-unsaturated precursors using the soluble rhodium catalyst with the (*S*,*S*)-phosphine ligand shown in Eq. 4.[19]

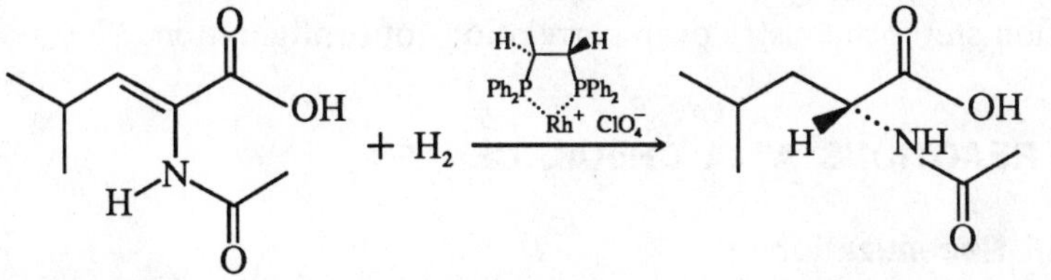

$$+ H_2 \xrightarrow{Rh^+\ ClO_4^-} \qquad (4)$$

The illustrated synthesis of *N*-acetylleucine gave pure (100%) (*R*) enantiomer in tetrahydrofuran solvent. In ethanol the product was 93% optically pure. Other amino acid derivatives were prepared similarly in 74–100% optical purity.

Bakers yeast can provide the chiral enzyme catalyst and the reducing agent as in the reduction of ethyl acetoacetate to ethyl (*S*)-(+)-3-hydroxybutyrate in 86% enantiomeric excess.[20]

The reagent itself may be chiral. A chiral tridentate ligand prepared from (*S*)-aspartic acid was added to lithium aluminum hydride (Eq. 5). This modified hydride was then used to reduce α,β-unsaturated ketones, consistently giving the (*S*) alcohols in 57–95% isolated yield with 28–100% enantiomeric excess.[21]

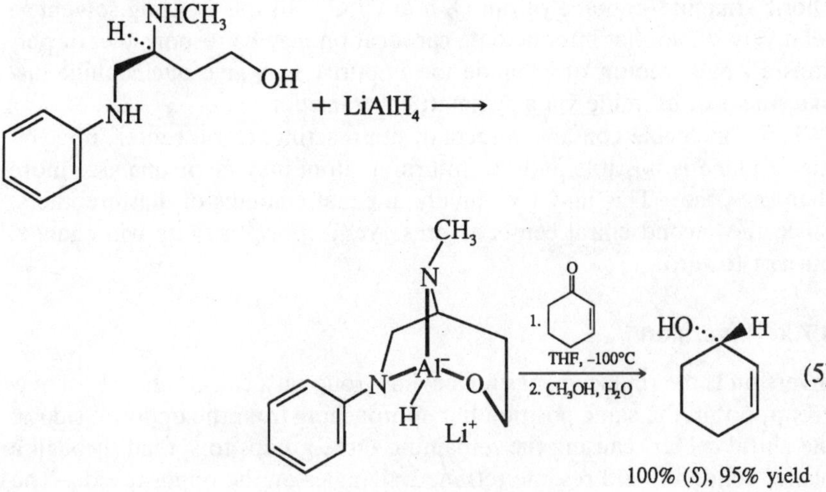

$$+ LiAlH_4 \longrightarrow$$

$$\xrightarrow[2.\ CH_3OH,\ H_2O]{1.\ \text{cyclohexenone, THF, } -100^\circ C} \qquad (5)$$

100% (*S*), 95% yield

Resolution and asymmetric synthesis depend on the energy difference between a pair of diastereomeric complexes or between crystals of diastereomers. These differences are very small compared to reaction enthalpies and are, therefore, not as generalizable. The examples selected here are some of the best in each category; there are many examples that give far less selectivity. You should not expect routine application of these techniques to new cases, but rather much trial-and-error development. On the other hand, partially resolved materials can serve very well in studies of reaction stereochemistry or in correlations of configuration.

3.7 REACTIONS AT A CHIRAL CENTER

3.7.1 Racemization

In molecules that contain only one chiral center, certain conditions will lead to a loss of optical activity, eventually giving racemic material. These conditions lead to the formation of an intermediate structure where a plane of symmetry passes through the former chiral center atom. For instance, removal of a proton from the chiral center to give a carbanion allows a planar or rapidly inverting pyramidal structure to exist at that atom. The return of a proton will then be equally probable on either face leading to either enantiomer. After sufficient time, equal amounts of enantiomers will be present, and the result is called *racemization.*

Other circumstances where a plane of symmetry may occur at an intermediate stage include nucleophilic substitution, neighboring group participation, rearrangements, and carbocation formation. 1-Bromoethylbenzene shows an optical half-life in solution in hexamethylphosphoric triamide–pentane of only 8 h at 27°C.[22] In this ionizing solvent, a relatively stable, flat intermediate carbocation may be responsible, or perhaps a small amount of bromide ion impurity may give nucleophilic displacement of bromide via a symmetrical transition state.

If the molecule contains a second, nonreacting, chiral center, no symmetry plane is possible and the returning atom may favor one side more than the other. This may give finally unequal amounts of diastereomers. Since the second chiral center is preserved, optical activity will change, but not to zero.

3.7.2 Inversion

Inversion is the replacement of a leaving group on a chiral center by a new group, not in the same position but approaching from the opposite side of the chiral center, causing the remaining three groups to spread through a planar condition and resume tetrahedral angles on the opposite side. The first example in which inversion was known to occur in a particular step is shown in Eq. 6.[23]

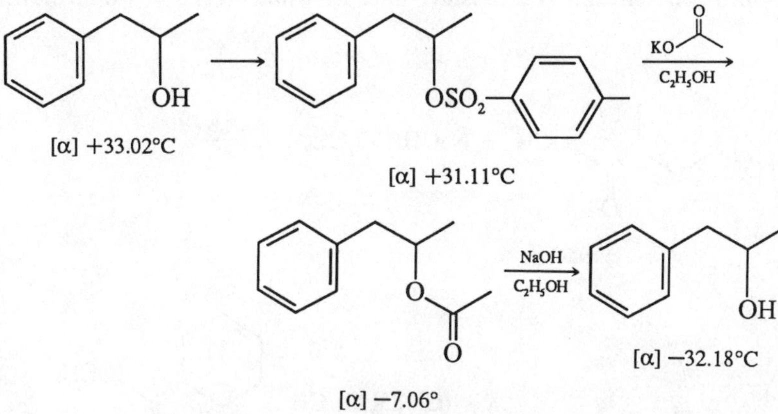

OH

[α] +33.02°C

OSO_2

[α] +31.11°C

KO (acetate), C_2H_5OH

NaOH, C_2H_5OH

O

O

OH

[α] −7.06°

[α] −32.18°C

(6)

The overall three-step process gave alcohol of inverted configuration as found by rotation measurements. The first and third step did not involve bonds to the chiral center and could not give inversion. The second step must thus have given inversion. Notice that this conclusion was made without knowing the absolute configurations (Section 3.8). Nucleophilic substitution reactions that follow clean second-order kinetics generally give complete inversion of configuration.

3.7.3 Retention

Retention of configuration occurs when an incoming group replaces a leaving group directly (front-side) without inversion. Retention is also found when a two-step substitution occurs, that is, a temporary group arrives with inversion and is, in turn, replaced by a final group with a second inversion. The front-side substitution occurs when the incoming group is attached to the leaving group and thus held to the front side. In other words, rearrangements such as the Beckmann (Eq. 7),[24] Hofmann, Curtius, Schmidt, Lossen, and Baeyer–Viliger give retention.

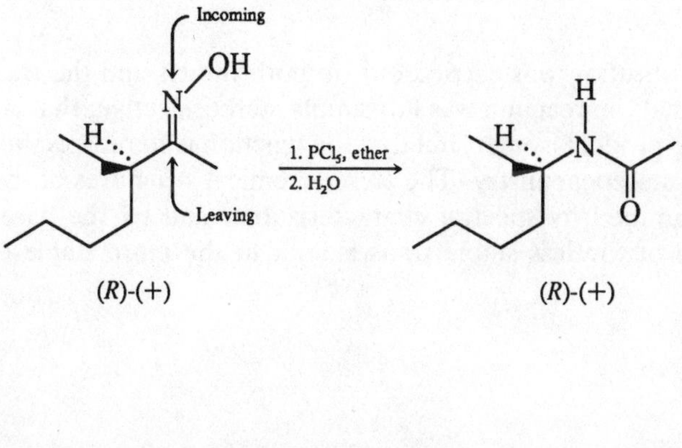

Incoming

OH

N

H

Leaving

1. PCl_5, ether

2. H_2O

H

H

N

O

(7)

(*R*)-(+)

(*R*)-(+)

Retention was proved without use of optically active compounds in the case shown in Eq. 8.[25] Here a second chiral center is present so that retention at one site gives a diastereomer of what inversion would have given.

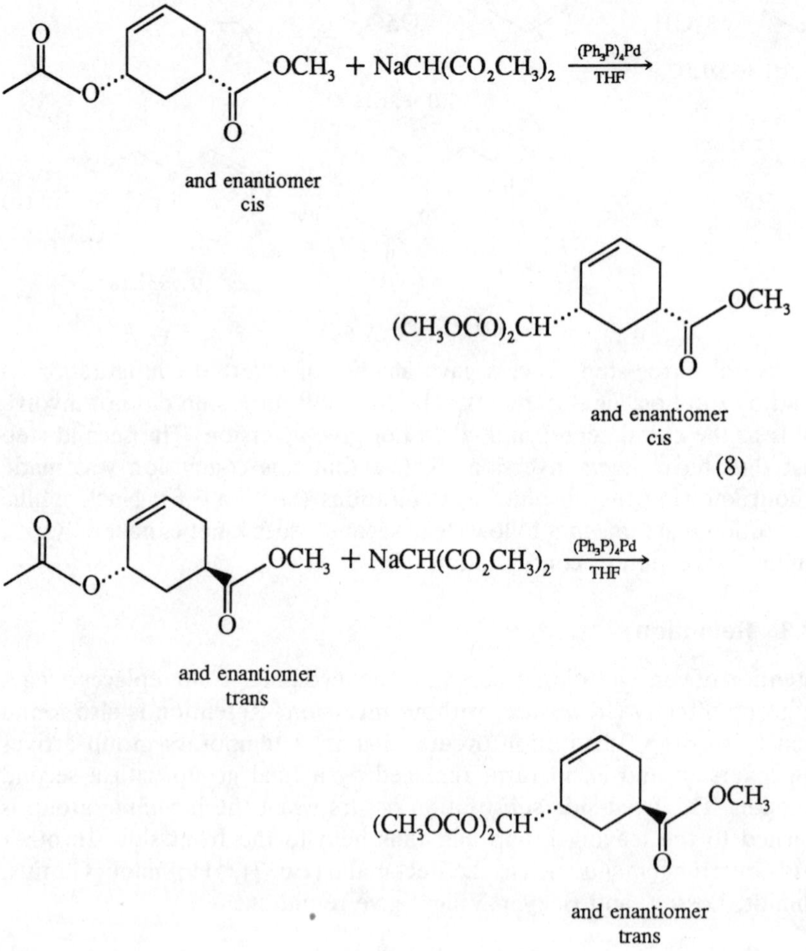

$OCH_3 + NaCH(CO_2CH_3)_2 \xrightarrow[THF]{(Ph_3P)_4Pd}$

and enantiomer
cis

$(CH_3OCO)_2CH$ OCH_3

and enantiomer
cis

(8)

$OCH_3 + NaCH(CO_2CH_3)_2 \xrightarrow[THF]{(Ph_3P)_4Pd}$

and enantiomer
trans

$(CH_3OCO)_2CH$ OCH_3

and enantiomer
trans

The substitution was carried out on both the cis and the trans isomers to assure that the reaction was not simply stereoselective, that is giving the most stable product isomer, but that the reaction is stereospecific according to starting stereochemistry. The stereochemical structures of the products were determined by spectral characterization and by the base-catalyzed conversion of the less stable trans isomer to the more stable cis isomer.

The substitution reaction presumably proceeds via π-allylpalladium complexes.

3.7.4 Transfer

A new chiral center may be formed stereospecifically while an original chiral center flattens. The Claisen rearrangement of optically active allylic alcohols shows such a transfer (Eq. 9).[26] The (*R*)-*cis*-alcohol gave the (*S*)-*trans* ester with 97–99% chiral transmission via a chair–six-membered ring transition state. By close examination of the transition state, you can see that the (*S*)-*trans* gives the same stereoisomeric product.

$$+ (C_2H_5O)_3CCH_3 \xrightarrow[142°C]{CH_3CH_2COOH}$$

(9)

3.8 RELATIVE AND ABSOLUTE CONFIGURATIONS

If models or drawings of an enantiomeric pair are made, we can label each with an *R* or an *S* for each chiral center. If two actual samples of the enantomeric materials are on hand, we can make a measurement of rotation direction and label each as (+) or (−). Now, which go together? Does the (+) sample have the *R* or *S* structure? Until the work of J. M. Bijvoet in 1951,[27] there was no way of determining this. Now, for crystalline samples, the analysis of anomalous scattering in X-ray crystallography leads to such determinations, a great many of which have been done. For a particular substance, correlating the sign of rotation with the configurational designation of structure gives the *absolute configuration*. For example, in 1972[28] the cobalt(II) salt of (−)-malic acid was examined by the anomalous dispersion method and determined to have the structure **14,** which we label *S* by the Cahn–Ingold–Prelog system.

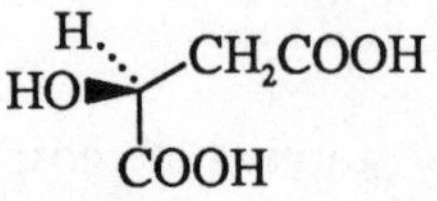

14: (*S*)-(−)-Malic acid

Now that this is known, the absolute configuration of many other compounds immediately becomes known because they have been related by synthesis at an earlier time.

For example, in 1963 (+)-2-hydroxy-3-phenylpropionic acid was ozonolyzed to the (+)-enantiomer of malic acid. This now requires that the starting acid had the *R* configuration:

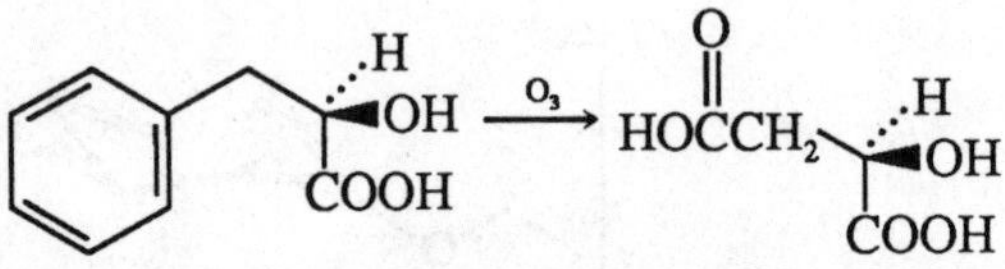

$$\xrightarrow{O_3} \quad (10)$$

(*R*)-(+)-2-Hydroxy-3-phenylpropionic acid (*R*)-(+)-Malic acid

In 1921 (+)-2,4-dihydroxybutyric acid was oxidized to (+)-malic acid, which now requires this also to be labeled *R*.

Before the absolute determination, the relationship of these three compounds was known and useful, even though an enantiomeric picture could not be drawn with certainty. This relationship is called *relative configuration.* The early statement that (+)-2,4-dihyroxybutyric acid and (+)-malic acid and also (+)-2-hydroxy-3-phenylpropionic acid all have the "same" configuration is a statement of relative configuration. Since they do not have the identically same four groups around the chiral center, the term "same" could be ambiguous, especially if several groups were modified in reactions. Therefore, a statement of relative configuration should be accompanied by a description of the reactions to be sure of the meaning of "same."

It is interesting to note that the uses of configurational information such as proof of inversion in S_N2 reactions was made with relative configurations before absolute ones were available. In fact, the absolute configurations are not useful in themselves, except as another means of obtaining more relative configurations.[29]

Keep in mind that two compounds with the "same" configuration may have different configurational designations and also that they may well have opposite signs of rotation as in

$$\text{(R)-(+)-1-Phenylethylamine} + H_2 \xrightarrow{\text{catalyst}} \text{(R)-(−)-1-Cyclohexylethylamine} \quad (11)$$

(*R*)-(+)-1-Phenylethylamine (*R*)-(−)-1-Cyclohexylethylamine

The preceding reactions demonstrating configurational correlation did not involve bonding changes at the chiral carbon and are quite reliable. Many other correlations involve reactions at the chiral center but with known stereochemistry such as S_N2 with inversion. These are reasonably reliable also. Other correlations have been made by observing a constancy of direction of rotation in a family of compounds such as *n*-alkyl secondary alcohols where the *S* isomer is generally (+). Yet other correlations and statements of absolute configuration have been made on the basis of order of elution in chromatography[30] and on generalizations on stereospecificity in asymmetric syntheses.

In the earlier part of this chapter, absolute configurations were used in the descriptions and illustrations. All of these were established by correlations similar to those mentioned here.

An extensive, referenced, illustrated list of stereochemical correlations and absolute configurations is available,[31] and also a list of 6000 selected absolute configurations.[32]

3.9 TOPISM

Up to this point we have considered whole molecules differing as stereoisomers. We now turn to an atom or group within a molecule and examine the three-dimensional shape of the environment of that atom or group, within the molecule.[33] For example, bromochloromethane is not chiral and has no stereoisomers, but the environment of one of the hydrogen atoms is the mirror image of the environment of the other one (Fig. VIII). These hydrogen atoms are called *enantiotopic,* which means that they reside in mirror-image environments.

In ordinary circumstances enantiotopic atoms or groups exhibit identical character, but in chiral media their environments become more different than mere mirror images. This differentiation is demonstrated graphically for the general case in Figure IX. The tetrahedral molecule containing

Environment of H^a Environment of H^b

Figure VIII. Environments in bromochloromethane.

enantiotopic groups can attach to a chiral substrate with all complimentary groups and sites matched by using one of the enantiotopic groups; however, attempting to use the other enantiotopic group fails to give a match. They are thus differentiable. Observable differentiations of this sort occur in enzyme-catalyzed reactions. For example, isobutyric acid can be hydroxylated to give optically pure (+)-3-hydroxy-2-methylpropionic acid by bacterial fermentation:[34]

$$(CH_3)_2CHCOOH + O_2 \xrightarrow[H_2O]{\textit{Pseudomonas putridas},} HO{-}CH_2CH(CH_3)COOH \qquad (12)$$

$[\alpha]^{25}_{578}$ + 7.6°
48% conversion

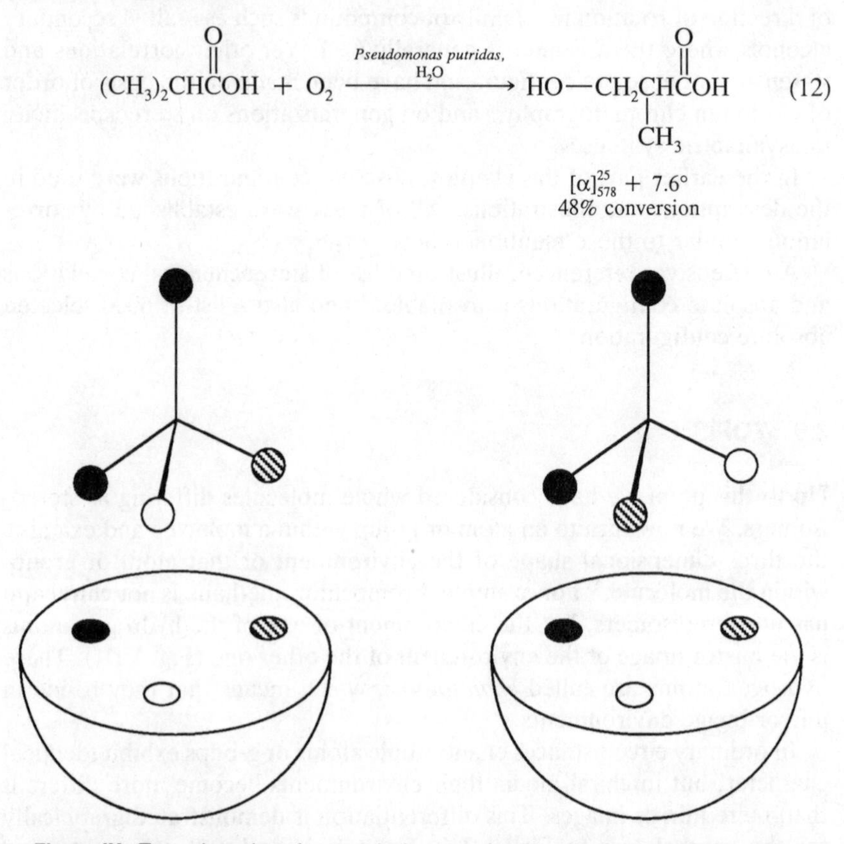

Figure IX. Two orientations for complexing enantiotopic groups on a chiral surface.

Figure X. Identifying enantiotopic pairs.

The two methyl groups in isobutyric acid are enantiotopic, and the chiral enzyme has selectively oxidized one and not the other.

There is a simple thought test to determine whether certain groups in a molecule are enantiotopic. Simply imagine replacing one of the two groups with an atom X. Then do likewise with the other one instead. (Fig. X). If this produces a pair of enantiomers, the groups were enantiotopic. These distinguishable enantiotopic groups may each be labeled. If the X group has a higher priority (Cahn–Ingold–Prelog) than the group it replaced but not higher than the next higher original group, and the resulting chiral center is of *R* configuration, the original enantiotopic group replaced is designated *pro-R*. If the other enantiotopic group had been replaced, the chiral center would have been *S;* therefore, the other enantiotopic group is designated *pro-S*.[33] Thus H^a in Fig. VIII is *pro-S* and H^b is *pro-R*. The central atom (carbon carrying the enantiotopic groups; Caabc) is called a *prochiral center.*

Pairs of atoms or groups whose replacement with X does not generate a new chiral center are called *homotopic,* as, for example, the hydrogens in dichloromethane. They have identical environments and can be interchanged by rotation of the molecule and are absolutely indistinguishable.

If replacing each of two identical groups with X produces a pair of

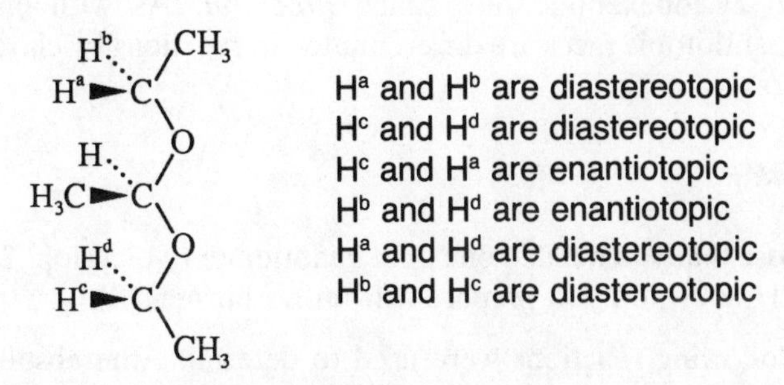

Hᵃ and Hᵇ are diastereotopic
Hᶜ and Hᵈ are diastereotopic
Hᶜ and Hᵃ are enantiotopic
Hᵇ and Hᵈ are enantiotopic
Hᵃ and Hᵈ are diastereotopic
Hᵇ and Hᶜ are diastereotopic

Figure XI. Stereotopic relationships in acetaldehyde diethyl acetal.

diastereomers, those original groups are *diastereotopic*. The H^a and H^b in acetaldehyde diethylacetal (Fig. XI) have this relationship. Replacement of H^a with X generates two new chiral centers. Replacement of H^b gives a structure diastereomeric with the first one; thus H^a and H^b are diastereotopic. The environments within the molecule of H^a and H^b are diastereomeric, and H^a and H^b give separate signals in the ^1H NMR spectrum (Chapter 10). In Fig. XI the two ethyl groups are enantiotopic. You should build molecular models to assure yourself of these relationships and also those listed in Fig. XI. In contrast, the diethyl acetal of formaldehyde contains no diastereotopic groups. Molecules that contain one chiral center will contain no enantiotopic groups.

The two faces of a flat molecular site may be enantiotopic. If an sp^2 carbon has three different groups bonded to it, the faces are enantiotopic as exemplified by acetaldehyde (Fig. XII). Looking directly at one face, if the three groups give a clockwise decrease in priority, the face toward you is designated *re*. If it is turned about, the face now toward you will give counterclockwise decrease and be designated *si*.[35]

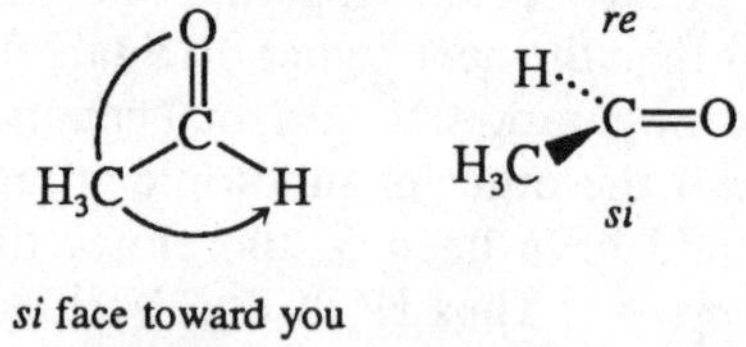

O
C
H_3C H

si face toward you

re
H
C=O
H_3C
si

Figure XII. Acetaldehyde faces.

The chiral hydride reducing agent in Eq. 5 selectively adds a hydride to the *re* face of α,β-unsaturated ketones. A complimentary hydride reducing agent with a different tridentate ligand was also prepared for selective addition of a hydride to the *si* face of these ketones.[21]

Molecules with enantiotopic faces, including acetaldehyde, benzaldehyde, and cyclohexenone, are called *prochiral*. As with enantiotopic groups, enantiotopic faces are differentiated in reactions by chiral reagents or catalysts.

PROBLEMS

1. Suppose that you needed one pure enantiomer of bicyclo[2.2.1]heptan-2-ol. How would you prepare it from norbornene?[11]

2. The following reactions were used to determine the absolute config-

uration of 3-hydroxy-2-methylpropanoic acid. (+)-(*S*)-Alanine was esterified and reduced with $LiAlH_4$ to give 2-amino-1-propanol. Phosgene then gave the 4-methyl-1,3-oxazolidin-2-one, melting point (mp) 53–54°C, $[\alpha]^{20}_{578}$ − 7.8°. A sample of racemic 3-hydroxy-2-methylpropanoic acid was resolved to give material of rotation $[\alpha]^{20}_{578}$ − 7.6°. The resolved acid was esterified and then treated with hydrazine to give the hydrazide. Treatment of the hydrazide with nitrous acid gave the 4-methyl-1,3-oxazolidin-2-one, mp 53–54°C, $[\alpha]^{20}_{578}$ + 7.9°. Assign the absolute configuration of the acid.[36]

3. Using the information in problem 2, determine whether the *pro-R* or *pro-S* methyl group is oxidized in Eq. 12 of this chapter.

4. (+)-3-Tetradecanol is known to have the *S* configuration. Treatment of (+)-1,2-epoxytridecane with methyllithium gives (+)-3-tetradecanol. What is the absolute configuration of (+)-1,2-epoxytridecane? Draw a three-dimensional representation of it. How would you determine the absolute configuration of (−)-2-tridecanol?[37]

5. The specific rotation of the (+)-3-tetradecanol prepared in problem 4 was $[\alpha]^{25}_{D}$ +6.7°. The previously reported rotation for this compound was $[\alpha]^{25}_{D}$ +5.1°. What is the maximum possible optical purity of the 5.1° rotating sample? What is the maximum percent of (+)-enantiomer in the 5.1° rotating sample?

6. Draw a three-dimensional representation of chlorocyclopropane. Identify a pair of enantiotopic atoms. Identify a pair of diastereotopic atoms.

7. What is the relationship between the two methoxy groups in a molecule of 3-bromobutanal dimethyl acetal?

8. What is the relationship between the two benzylic hydrogens in *meso*-azobis-α-phenylethane? What is the relationship between the two benzylic hydrogens in (*R*,*R*)-azobis-α-phenylethane?

9. The following steps were used to convert the (*R*)-alkynol to the (*R*)-γ,δ-unsaturated ester. How would you convert the (*R*)-alkynol into the (*S*)-γ,δ-unsaturated ester?[38]

OH
R
1. $NaAl(OCH_2CH_2OCH_3)_2H_2$
2. $CH_3C(OC_2H_5)_3$, propionic acid, heat
R
O
O

10. Photochemical chlorination of (+)-(*S*)-2-bromobutane with *t*-butyl hypochlorite at −78°C gave the following, among other products:

	Yield (%)	Enantiomeric Purity
erythro-2-Bromo-3-chlorobutane	20	Optically pure
threo-2-Bromo-3-chlorobutane	6	Optically inactive
1-Chloro-3-bromobutane	4	Optically pure
1-Bromo-2-chlorobutane	3.2	Optically pure

Draw three-dimensional representations of each of these products, and explain in terms of intermediates why three of these are optically pure and one is not.[39]

11. The two stereoisomers of 3-methyl-2,4-dibromopentane shown below were cyclized by treatment with zinc in 1-propanol–water. The products were analyzed for the ratio of *cis*- to *trans*-trimethylcyclopropanes; the results are

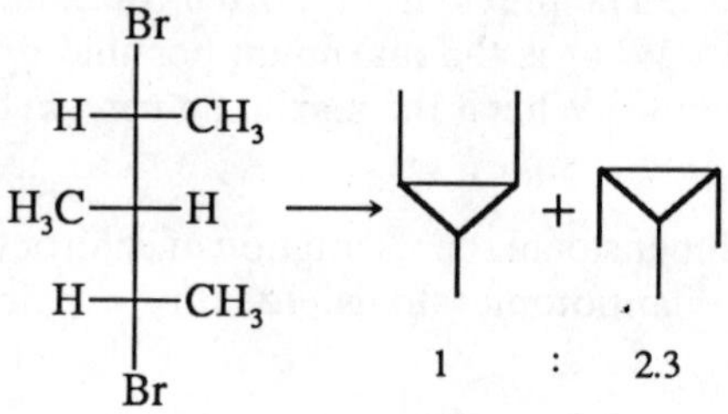

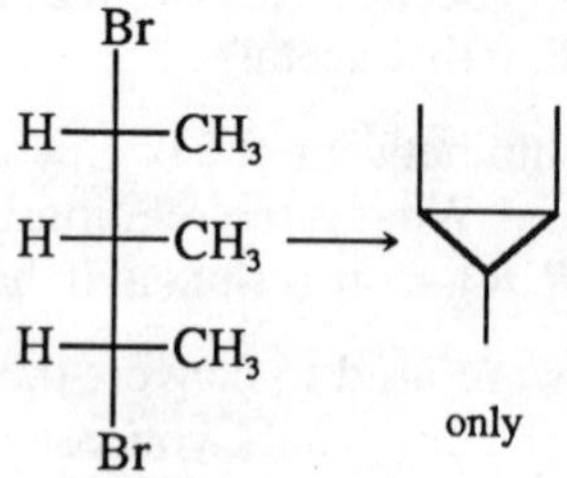

Br
H—+—CH_3
H_3C—+—H → (cis) + (trans)
H—+—CH_3
Br
1 : 2.3

Br
H—+—CH_3
H—+—CH_3 → (cis)
H—+—CH_3
Br
only

The overall stereochemical possibilities are retention at both sites, retention at one and inversion at the other site, and inversion at both sites. Considering all the results, what is the overall stereochemistry for the process or processes that give(s) the major product from the first isomer? What is the overall stereochemistry for the formation of the minor product from the first isomer? Explain with drawings how you reached your conclusions.[40]

12. The following two sequences of reactions were carried out. Why does one give material with deuterium α to the benzene ring and the other no deuterium α to the benzene ring?[41]

D(Ph)C=O → (D)(Ph)C=C oxazolone (N=C–Ph) → acid → D(Ph)C=C(COOH)(NHCOPh) → H_2, Pd → PhCHDCH(CO_2H)(NHCOPh) → acid → resolution → L-PhCHDCH(COO^-)(NH_3^+) → L-Phenylalanine ammonia lyase → PhCD=CHCOOH

H(Ph)C=O → (H)(Ph)C=C oxazolone (N=C–Ph) → acid → H(Ph)C=C(COOH)(NHCOPh) → D_2, Pd → PhCHDCD(CO_2H)(NHCOPh) → acid → resolution → L-PhCHDCD(COO^-)(NH_3^+) → L-Phenylalanine ammonia lyase → PhCH=CDCOOH

REFERENCES

1. Eliel, E. L. *Stereochemistry of Carbon Compounds,* McGraw-Hill, New York, 1962.
2. Cahn, R. S.; Ingold, C. K.; Prelog, V. *Angew. Chem. Internatl. Ed.* **1966,** *5,* 385.
3. Cope, A. C.; Banholzer, K.; Jones, F. N.; Keller, H. *J. Am. Chem. Soc.* **1966,** *88,* 4700.

4. Djerassi, C.; Hart, P. A.; Warawa, E. J. *J. Am. Chem. Soc.* **1964,** *86,* 78.
5. Jacques, J.; Collet, A.; Wilen, S. H. *Enantiomers, Racemates and Resolutions,* Wiley, New York, 1981.
6. Blaschke, G. *Angew. Chem. Internatl. Ed.* (review) **1980,** *19,* 13.
7. Pirkle, W. H.; Welch, C. J. *J. Org. Chem.* **1984,** *49,* 138; Pirkle, W. H.; Finn, J. M. *J. Org. Chem.* **1982,** *47,* 4037.
8. Pirkle, W. H.; Hoover, D. J. *Topics Stereochem.* **1982,** *13,* 263.
9. Whitesell, J. K.; Chen, H.-H.; Lawrence, R. M. *J. Org. Chem.* **1985,** *50,* 4663.
10. Pirkle, W. H.; Adams, P. E. *J. Org. Chem.* **1980,** *45,* 4111.
11. Whitesell, J. K.; Reynolds, D. *J. Org. Chem.* **1983,** *48,* 3548.
12. Mori, A.; Yamamoto, H. *J. Org. Chem.* **1985,** *50,* 5444.
13. Collet, A.; Brienne, M. J.; Jacques, J. *Chem. Rev.* **1980,** *80,* 216.
14. Garin, D. L.; Greco, D. J. C.; Kelley, L. *J. Org. Chem.* **1977,** *42,* 1249.
15. Valentine, D., Jr.; Scott, J. W. *Synthesis* (review) **1978,** 329.
16. Morrison, J. D., Ed. *Asymmetric Synthesis,* Vols. 1–3, Academic Press, New York, 1983–1984.
17. Meyers, A. I. In *Asymmetric Reactions and Processes in Chemistry,* E. L. Eliel and S. Otsuka, Eds. (ACS Symposium Series, Vol. 185) American Chemical Society, Washington, DC, 1982, p. 83.
18. Meyers, A. I.; Williams, D. R.; Erickson, G. W.; White, S.; Druelinger, M. *J. Am. Chem. Soc.* **1981,** *103,* 3081.
19. Fryzuk, M, D.; Bosnich, B. *J. Am. Chem. Soc.* **1977,** *99,* 6262.
20. Seebach, D.; Sutter, M. A.; Weber, R. H.; Zuger, M. F. *Org. Syn.* **1984,** *63,* 1.
21. Sato, T.; Goth, Y.; Wakabayashi, Y.; Fujisawa, T. *Tetrahedron Lett.* **1983,** *24,* 4123.
22. Hutchins, R. O.; Masilamani, D.; Maryanoff, C. A. *J. Org. Chem.* **1976,** *41,* 1071.
23. Phillips, H. *J. Chem. Soc.* **1923,** *123,* 44.
24. Kenyon, J.; Young, D. P. *J. Chem. Soc.* **1941,** 263.
25. Trost, B. M.; Verhoeven, T. R. *J. Org. Chem.* **1976,** *41,* 3215.
26. Chan, K.-K.; Cohen, N.; De Noble, J. P.; Specian, A. C., Jr.; Saucy, G. *J. Org. Chem.* **1976,** *41,* 3497.
27. Bijvoet, J. M.; Peerdeman, A. F.; Van Bommel, A. J. *Nature* **1951,** *168,* 271.
28. Kryger, L.; Rasmussen, S. E. *Acta Chem. Scand.* **1972,** *26,* 2349.
29. Fiaud, J. C.; Kagan, H. B. *Determination of Configurations by Chemical Methods,* Georg Thieme, Stuttgart, 1977, p. 2.
30. Doolittle, R. E.; Heath, R. R. *J. Org. Chem.* **1984,** *49,* 5041.
31. Klyne, W.; Buckingham, J. *Atlas of Stereochemistry,* 2nd ed., Vols. 1, 2, Oxford University Press, New York, 1978.

32. Jacques, J. *Absolute Configuration of 6000 Selected Compounds with One Asymmetric Carbon Atom,* Georg Thieme, Stuttgart, 1977.
33. Mislow, K.; Raban, M. *Topics Stereochem.* **1967,** *1,* 1.
34. Goodhue, C. T.; Schaeffer, J. R. *Biotechnol. Bioeng.* **1971,** *13,* 203.
35. Hanson, K. R. *J. Am. Chem. Soc.* **1966,** *88,* 2731.
36. Retey, J.; Lynen, F. *Biochem. Biophys. Res. Commun.* **1964,** *16,* 358.
37. Coke, J. L.; Richon, A. B. *J. Org. Chem.* **1976,** *41,* 3516.
38. Chan, K.-K.; Specian, A. C., Jr.; Saucy, G. *J. Org. Chem.* **1978,** *43,* 3435.
39. Skell, P. S.; Pavlis, R. R.; Lewis, D. C.; Shea, K. J. *J. Am. Chem. Soc.* **1973,** *95,* 6735.
40. Applequist, D. E.; Pfohl, W. F. *J. Org. Chem.* **1978,** *43,* 867.
41. Kirby, G. W.; Michael, J. *Chem. Commun.* **1971,** 415.

4

Synthesis of Functional Groups

A chemist who undertakes the synthesis of an organic compound of some complexity must consider three aspects: (1) the synthesis of the functional groups in the final molecule plus those needed at intermediate stages,[1] (2) the formation of carbon–carbon bonds to develop larger molecules, and (3) the strategy of selecting starting materials and intermediate goals. A chapter is devoted to each; the first is concerned with functional groups.

The current practical alternatives for preparing each functional group include many classical reactions with relatively known mechanisms, plus many modern ones with complex or often unknown mechanisms. The introductory texts favor conceptually simple methods applied to small monofunctional molecules. Most of those synthetic products are commercially available; therefore, more generalizable and selective methods are chosen here. The functional group syntheses that involve joining carbon atoms are presented in Chapter 5.

4.1 CARBOXYLIC ACIDS AND RELATED DERIVATIVES

The high oxidation state of carbon in which there are three bonds to electronegative atoms is the characteristic of carboxylic acids and the related acid chlorides, anhydrides, esters, orthoesters, amides, and nitriles. The transformations may involve oxidation from hydrocarbons or other partially oxidized substrates or exchange among the various electronegative atoms on the carbon.

4.1.1 Carboxylic Acids

Benzylic sites containing at least one hydrogen in hydrocarbons may be oxidized to the carboxylic acid state by using strong agents, including dichromate, permanganate, or nitric acid. For example, *p*-cymene was converted to *p*-toluic acid:[2]

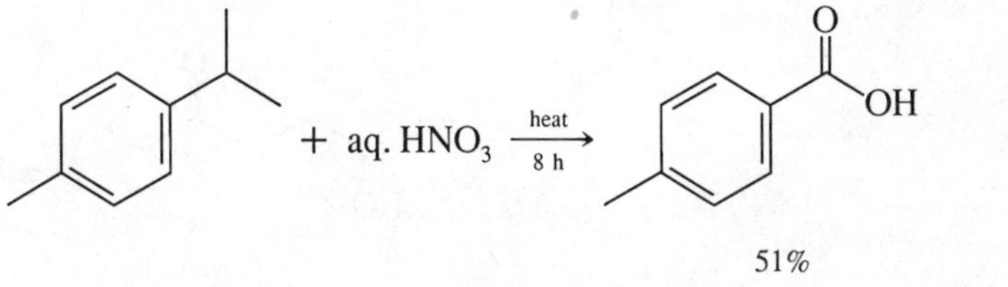

$$+ \text{ aq. } HNO_3 \xrightarrow[\text{8 h}]{\text{heat}} \quad (1)$$

51%

Carbons that are already partially oxidized such as alkenes, primary alcohols, aldehydes, and methyl ketones are commonly raised to the carboxylic acid oxidation state also. Appropriately substituted alkenes may be cleaved by using ozone followed by hydrogen peroxide to give carboxylic acids. A convenient alternative is the combination of sodium periodate and a catalytic amount of permanganate (Eq. 2).[3]

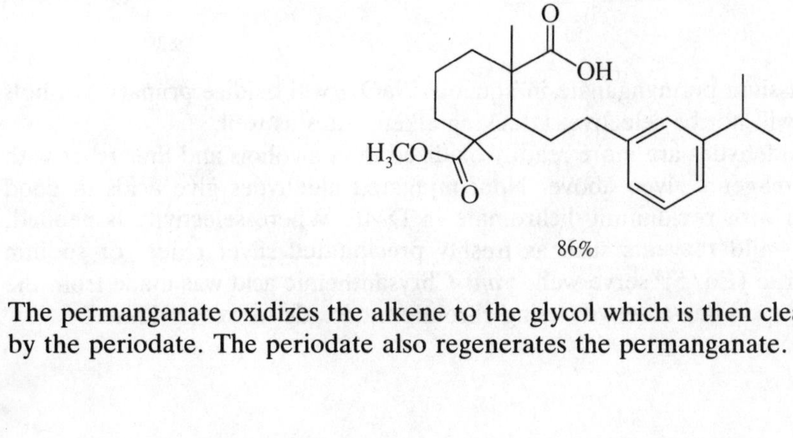

$$+ \ NaIO_4 \xrightarrow[t\text{-BuOH–water, rt}]{KMnO_4}$$

H$_3$CO, O, OH, H$_3$CO, O

(2)

86%

The permanganate oxidizes the alkene to the glycol which is then cleaved by the periodate. The periodate also regenerates the permanganate.

Aqueous potassium permanganate will oxidize alkenes rapidly to the acids if a small amount of tricaprylmethylammonium chloride is present as a phase-transfer catalyst (Chapter 9). In this way 1-decene was converted to nonanoic acid in 91% yield in 3 min.[4]

Primary alcohols are oxidized by the easily prepared pyridinium dichromate in DMF (Eq. 3)[5] or by Jones reagent in acetone (Eq. 4).[6]

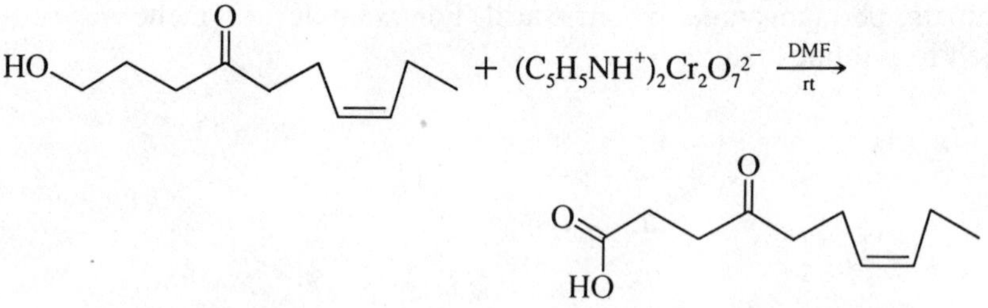

$$\text{HO(CH}_2)_3\text{CO(CH}_2)_2\text{CH=CHCH}_2\text{CH}_3 + (C_5H_5NH^+)_2Cr_2O_7^{2-} \xrightarrow[\text{rt}]{\text{DMF}} \text{HOOC(CH}_2)_2\text{CO(CH}_2)_2\text{CH=CHCH}_2\text{CH}_3 \quad (3)$$

93%

$$\text{HOCH}_2\text{-[OCH}_3\text{ furanoside, O-isopropylidene]} + CrO_3 + H_2SO_4 \xrightarrow[0°C]{H_2O,\ \text{acetone}} \text{HOOC-CH}_2\text{-[OCH}_3\text{ furanoside, O-isopropylidene]} \quad (4)$$

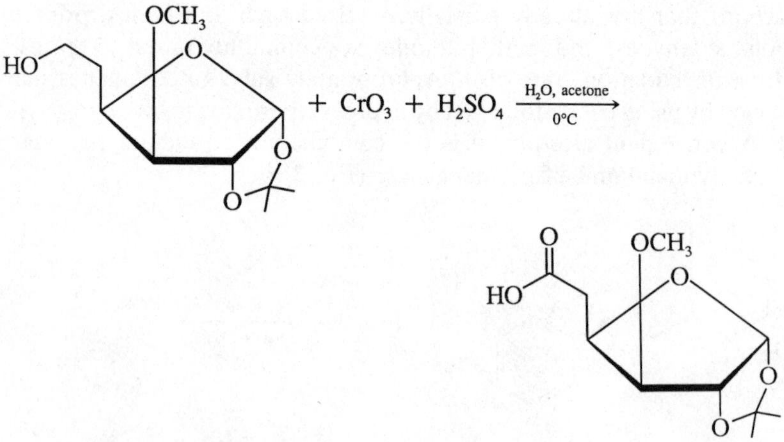

82%

Potassium permanganate in aqueous NaOH will oxidize primary alcohols but will not be selective, attacking alkene sites as well.

Aldehydes are more readily oxidized than alcohols and thus react with the reagents given above. Nonconjugated aldehydes give acids in good yield with pyridinium dichromate in DMF. Where selectivity is needed, very mild reagents such as freshly precipitated silver oxide[7] or sodium chlorite (Eq. 5)[8] serve well. *trans*-Chrysanthemic acid was made from the corresponding aldehyde using chromium trioxide in wet pyridine.[9]

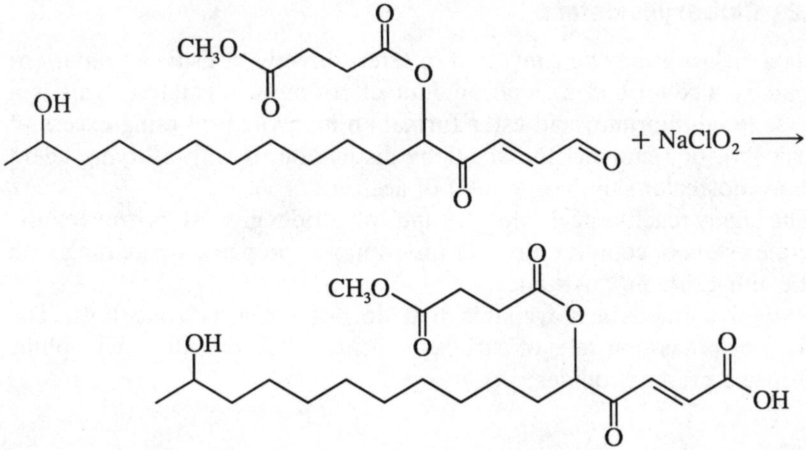

CH_3O O OH O O O + $NaClO_2$ ⟶

CH_3O O OH O O O OH (5)

69%

Methyl ketones are oxidized by alkaline hypochlorite to acid salts plus chloroform.[10]

Within the same oxidation level, heating any of the acid derivatives with aqueous acid or base leads ultimately to the acid or the salt thereof. Nitrile hydrolysis is particularly difficult, requiring prolonged heating in water-ethylene glycol:[11]

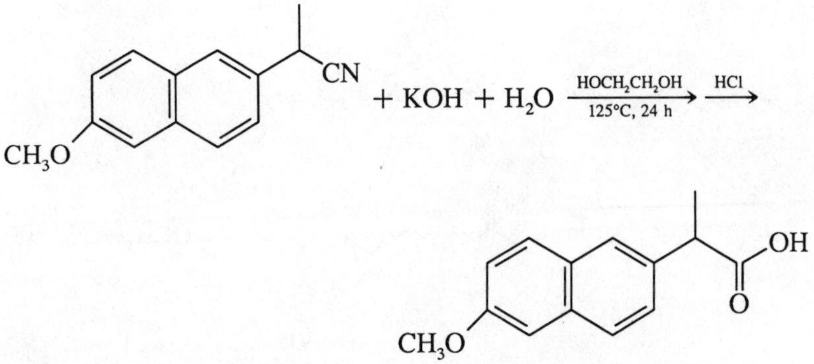

CN CH_3O + KOH + H_2O $\xrightarrow[125°C,\ 24\ h]{HOCH_2CH_2OH}$ $\xrightarrow{HCl}$

OH O CH_3O (6)

85%

There are many syntheses of acids where a carbon–carbon bond is formed such as carbonation of Grignard reagents, malonic ester alkylation, and Reformatzki reactions. Some are covered in Chapter 5, and you should review others in your introductory text.

4.1.2 Carboxylic Esters

Carboxylic acids may be converted to esters directly by using a primary or secondary alcohol and a small amount of strong acid catalyst. This is a reversible equilibrium, and ester formation is favored by using excess of the alcohol or removing the water by distillation or with a drying agent such as molecular sieves or a ketal of acetone.

The highly reactive acid chlorides and anhydrides give esters irreversibly. Acetate esters of complex alcohols are routinely prepared by treating with acetic anhydride and pyridine.

Several methods are available that do not begin with alcohols. The sodium or potassium salts of carboxylic acids are sufficiently nucleophilic to displace primary iodides:[12]

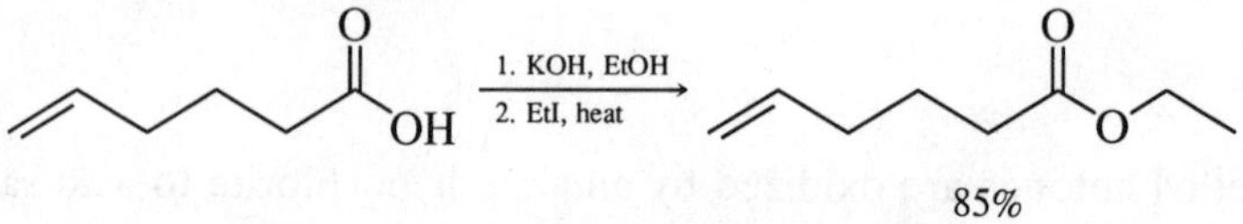

1. KOH, EtOH
2. EtI, heat

85%

(7)

Under neutral conditions a carboxylic acid will react with diazomethane in ether to give nitrogen gas plus the methyl ester in high yield and purity (Eq. 8).[13] This is ordinarily used on a small scale because diazomethane is volatile, toxic and explosive.

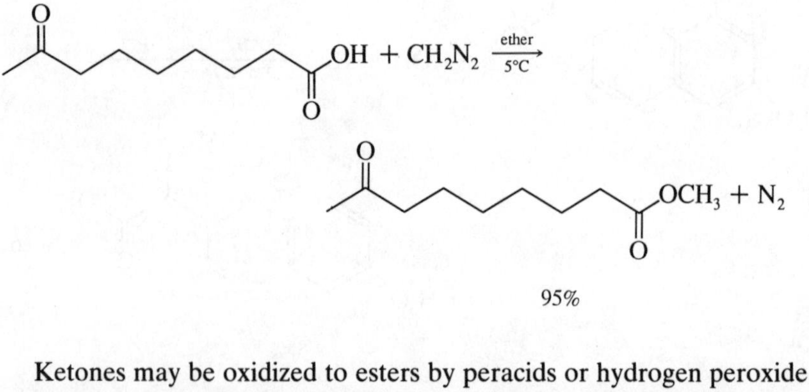

OH + CH_2N_2 $\xrightarrow[5°C]{ether}$

OCH_3 + N_2 (8)

95%

Ketones may be oxidized to esters by peracids or hydrogen peroxide, a process known as the Baeyer–Villiger oxidation. Unsymmetrical ketones are oxidized selectively at the more substituted α carbon, and that carbon migrates to oxygen with retention of configuration. Trifluoroperacetic acid generated *in situ* gave the double example in Eq. 9.[14] Cyclic ketones afford lactones:[15]

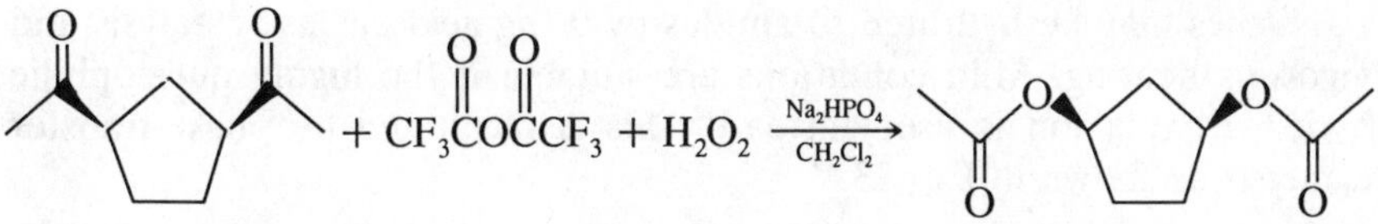

$$+ CF_3COCCF_3 + H_2O_2 \xrightarrow[CH_2Cl_2]{Na_2HPO_4} \quad (9)$$

$$+ H_2O_2 \xrightarrow[50°C]{CH_3CO_2H} \quad (10)$$

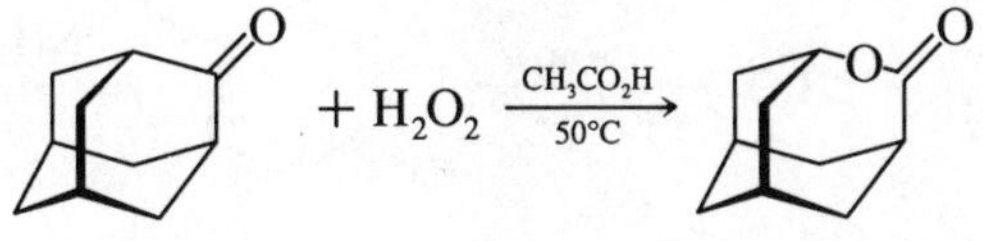

91%

Enol esters may be prepared from ketones by reaction with an anhydride or by exchange with isopropenyl acetate under acidic catalysis:[16]

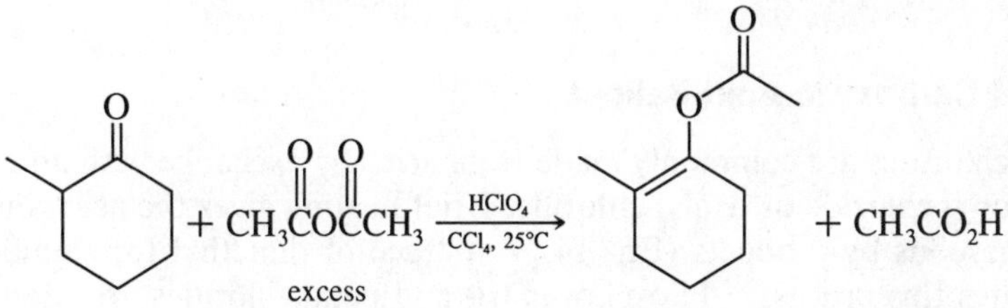

$$+ CH_3COCCH_3 \xrightarrow[CCl_4,\ 25°C]{HClO_4} \quad + CH_3CO_2H \quad (11)$$

excess

87–92%

4.1.3 Carboxylic Amides

Heating a carboxylic acid with ammonia or urea gives a carboxamide. For example, heptanoic acid and urea at 140–180° gives heptanamide in 75% yield plus CO_2 and H_2O.[17] The highly reactive acid halides and anhydrides combine with ammonia or primary or secondary amines to give amides at ordinary temperatures. Esters will react slowly with ammonia at room temperature (Eq. 12).[18] Higher boiling amines may be used if the alcohol is removed continuously by distillation.

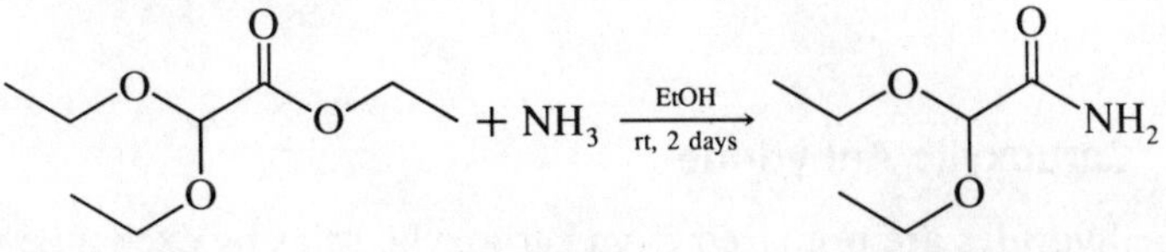

$$+ NH_3 \xrightarrow[rt,\ 2\ days]{EtOH} \quad NH_2 \quad (12)$$

100%

Nitriles may be hydrated to amides by using acid or base catalysis and vigorous heating. Mild conditions are suitable if the highly nucleophilic hydroperoxide ion is used instead. This is facilitated by phase-transfer catalysis as shown in Eq. 13.[19]

$$p\text{-}ClC_6H_4CN + 2H_2O_2 \xrightarrow[n\text{-}Bu_4N^+HSO_4^-,\ CH_2Cl_2,\ 25°C]{NaOH,\ H_2O} p\text{-}ClC_6H_4C(O)NH_2\ (97\%) + O_2 + H_2O \quad (13)$$

It can also be done under neutral conditions using colloidal black copper catalyst (from $NaBH_4$ reduction of $CuSO_4$) at 90°C.[20] With this catalyst, the sensitive acrylonitrile was converted to acrylamide in 89% yield.

4.1.4 Carboxylic Acid Halides

Acid chlorides are commonly made from acids by exchange with an excess of thionyl chloride or oxalyl chloride. Brief heating gives the acid chloride plus gaseous by-products (Eq. 14).[21] A trace of dimethyl formamide accelerates this process.[22] Phosphorus tri- and pentachlorides are used similarly. The acid bromides are made with phosphorus tribromide or oxalyl bromide.[21]

$$C_6H_5CH_2CH_2C(O)OH + ClC(O){-}C(O)Cl \xrightarrow{warm} C_6H_5CH_2CH_2C(O)Cl\ (98\%) + CO_2 + CO + HCl \quad (14)$$

4.1.5 Carboxylic Anhydrides

Most anhydrides are prepared from carboxylic acids by exchange with the readily available acetic anhydride. Heating these and then distilling the acetic acid and excess acetic anhydride shifts the equilibrium toward the higher boiling product (Eq. 15).[23] Five- and six-membered cyclic anhydrides usually form simply upon heating the dicarboxylic acid to about 120°C.

$$Ph_2CHCOH + CH_3COCCH_3 \rightleftharpoons Ph_2CHCOCCHPh_2 + CH_3COH \quad (15)$$

90–92%

A few unsymmetrical anhydrides are useful. Although formic anhydride cannot be made, acetic formic anhydride can be prepared by stirring sodium formate with acetyl chloride in ether (64% yield, bp 27–28°C).[24] It is useful for the formylation of alcohols and amines. Ethyl chloroformate gives a mixed anhydride with crotonic acid, and this was used to make crotonamide.[25]

4.1.6 Nitriles

Nitriles may be prepared by dehydration of amides. Phosphorus pentoxide and various acid chloride–base combinations have been used at elevated temperatures, but it can be done readily at 0°C to room temperature with a Vilsmeier reagent.[26,27] Oxalyl chloride and dimethylformamide in acetonitrile gives a precipitate of the reagent, an iminium salt, which is used as shown:[26]

H_3CO ... NH_2 + $ClCH{=}N^+(CH_3)_2\ Cl^-$ —(1. 0°C, CH_3CN, 5 min; 2. pyridine)→ H_3CO ... CN (16)

90%

Aldoximes may likewise be dehydrated by using chlorosulfonyl isocyanate–triethylamine[28] or other combinations.[29] Nitriles are also commonly made by displacements with cyanide ion (Chapter 5).

4.1.7 Ortho Esters

Ortho esters are acid derivatives in which the carboxyl carbon is sp^3-hybridized;[30] however most cannot be made from carboxylic acids. They are usually made by a two-stage alcoholysis of nitriles. Treatment of a nitrile with anhydrous hydrogen chloride in an alcohol gives the hydrochloride of an imidic ester. Treatment of this with an alcohol in a separate step (Eq. 17)[31] leads to the ortho esters.

$$C_2H_5CN + C_2H_5OH + HCl \xrightarrow{0°C} C_2H_5C(=\overset{+}{N}H_2\ Cl^-)-OC_2H_5 \xrightarrow[Et_2O]{C_2H_5OH} C_2H_5C(OC_2H_5)_3$$

85–95% 75–78%

(17)

The orthoformates and orthobenzoates are made from chloroform or trichloromethyl compounds by reaction with sodium alkoxides.

The alcohol parts of ortho esters may be exchanged under acidic conditions to give new orthoesters, especially where the incoming alcohol is a diol. This reaction is important in the Claisen rearrangement (Section 6.7). In contrast, the acid portion cannot be exchanged; that is, an acid cannot be converted to an ortho ester directly by transesterification. A few acids such as chloroacetic acid may be converted to bicyclic orthoesters by reaction with a triol with azeotropic removal of water. A general route to bicyclic ortho esters begins with 3-methyl-3-hydroxymethyloxetane as shown in Eq. 18.[32] The oxetane is prepared from neopentanetriol and diethyl carbonate.

Cl + HO O + N $\xrightarrow{CH_2Cl_2}$

O O $\xrightarrow[CH_2Cl_2,\ -15°]{BF_3 \cdot Et_2O}$ O O O

84% 91%

(18)

4.2 ALDEHYDES, KETONES, AND DERIVATIVES

The intermediate oxidation state of carbon in which there are two bonds to electronegative atoms is attained by reduction of acid derivatives or oxidation of alcohols and hydrocarbons. Interconversions at the same oxidation level such as hydration of alkynes are also valuable.

4.2.1 Aldehydes

The reduction of acid derivatives to aldehydes requires control because aldehydes are reduced to alcohols with greater ease. The palladium catalyzed hydrogenation of acid chlorides in the presence of 2,6-dimethylpyridine,[22] a modification of the Rosenmund reduction, shows this selectivity. If alkene sites are present, palladium on barium sulfate will leave them unchanged (Eq. 19), otherwise palladium on carbon is suitable. Aroyl halides require higher temperatures and quinoline-S-poisoned catalyst. The reaction is carried out at 1 to 4 atm of hydrogen pressure, monitoring gas uptake. Acid chlorides may also be reduced with sodium borohydride and $CdCl_2 \cdot 1.5$ DMF.[33]

Pd–$BaSO_4$, THF

96%

(19)

Nitriles and esters, especially lactones, may be reduced to aldehydes or hemiacetals by using diisobutylaluminum hydride (Eqs. 20, 21)[34,35] or various alkoxyaluminum hydrides such as $NaAlH_2(OC_2H_4OCH_3)_2$.[36] Any free aldehyde function already present will be reduced to the alcohol even faster.

toluene, −78°C; CH_3OH

80%

(20)

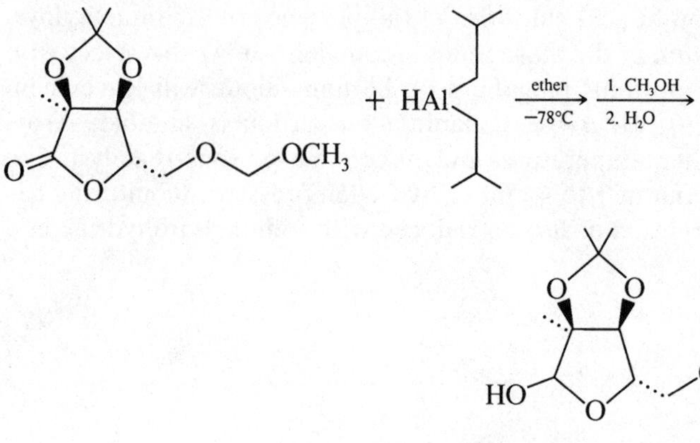

$$+ \ HAl(CH_2CH(CH_3)_2)_2 \xrightarrow[-78°C]{ether} \xrightarrow[2.\ H_2O]{1.\ CH_3OH}$$

O O O O O OCH$_3$

O O HO O O OCH$_3$ (21)

99.7%

Oxidation of primary alcohols can give aldehydes, and again control is necessary because simple oxidizing agents such as chromic acid will carry on to the carboxylic acid stage. A great many selective reagents have been used for this transformation including DMSO–acetic anhydride, oxygen over platinum, and many Cr^{6+} complexes. Of the chromium complexes, pyridinum chlorochromate is frequently the best choice (Eq. 22).[37] It is prepared by dissolving CrO_3 in 6 *M* aqueous HCl and adding pyridine, which results in a yellow filterable, air-stable solid that is not appreciably hygroscopic.

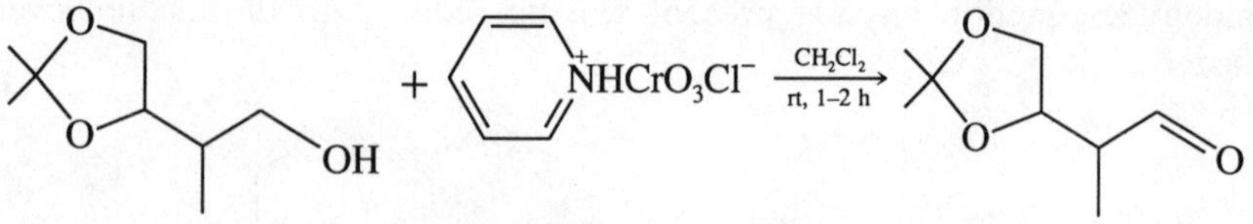

O O OH

$$+ \ C_5H_5\overset{+}{N}HCrO_3Cl^- \xrightarrow[rt,\ 1\text{–}2\ h]{CH_2Cl_2}$$

O O O

85%

(22)

Another frequent choice with sensitive substrates is the Swern oxidation where dimethylsulfoxide is the oxidizing agent:[38]

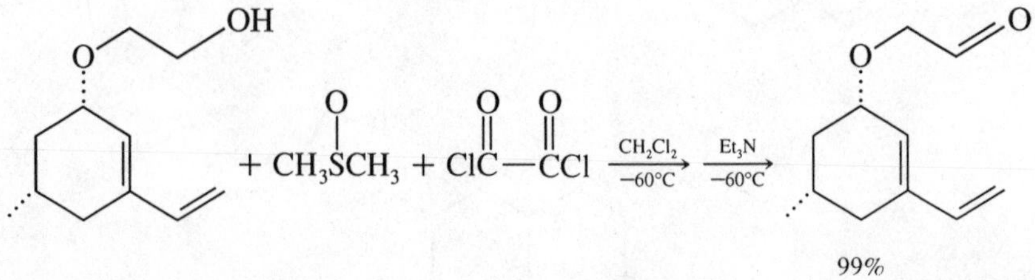

O OH

$$+ \ CH_3\overset{O}{\overset{\|}{S}}CH_3 + Cl\overset{O}{\overset{\|}{C}}-\overset{O}{\overset{\|}{C}}Cl \xrightarrow[-60°C]{CH_2Cl_2} \xrightarrow[-60°C]{Et_3N}$$

O O

99%

(23)

Allylic and benzylic primary and secondary alcohols are more easily oxidized, and a number of reagents selective for these are in use including freshly precipitated manganese dioxide, silver carbonate, dichlorodicyanoquinone, and potassium ferrate. 4-(Dimethylamino)pyridinium chlorochromate is mild and selective:[39]

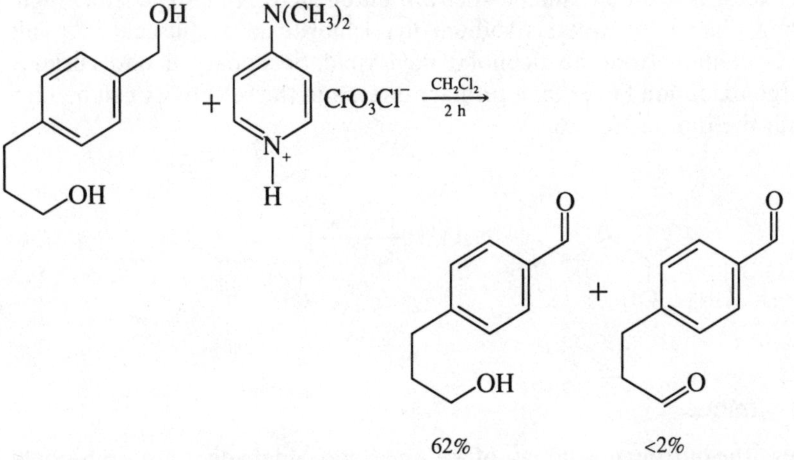

CrO_3Cl^- $\xrightarrow[2\ h]{CH_2Cl_2}$ (24)

62% <2%

Oxidative cleavage of appropriate alkenes can give aldehydes. Where ozone is used, the intermediate ozonides have more oxidizing power that can oxidize the desired aldehydes to carboxylic acids during hydrolysis. To avoid this interference, dimethyl sulfide is added as a reducing agent;[40] for example,[41]

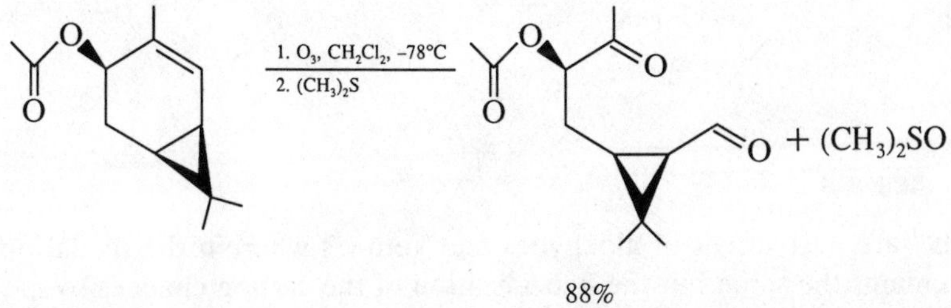

1. O_3, CH_2Cl_2, –78°C
2. $(CH_3)_2S$

$+ (CH_3)_2SO$ (25)

88%

The same overall result can be attained by using sodium periodate with a catalytic amount of osmium tetroxide. The OsO_4 gives the glycol, which the periodate cleaves to the aldehyde. The periodate also regenerates the OsO_4.

4.2.2 Ketones

Ketones are far less suceptible to oxidation than aldehydes and are readily prepared by oxidation of appropriately substituted alkenes and secondary alcohols. The conditions given in Sections 4.1.1 and 4.2.1 for oxidation of alkenes to acids or aldehydes are applicable for ketones as well, as already shown in Eq. 25.

The oxidation of secondary alcohols is often done with CrO_3 and H_2SO_4 in water (Jones reagent)[42] added to a solution of the alcohol in acetone. If an excess is avoided, alkene sites are untouched. An inexpensive, high-yielding reagent is aqueous sodium hypochlorite in acetic acid.[43,44] This gave 2-octanone from the alcohol in 96% yield. Secondary alcohols usually undergo oxidation faster than primary ones, and the selectivity can be high as with the diol in Eq. 26.

OH OH + NaOCl $\xrightarrow[25°C]{CH_3CO_2H}$ OH O (26)

85%

4.2.3 Imines

Imines, the nitrogen analogs of ketones and aldehydes, are commonly prepared by using primary amines and dehydrating conditions[45] as exemplified in Eq. 27.[46] These are useful as intermediates in the α-alkylation of ketones and aldehydes (Section 5.1.1).

NH_2 + O $\xrightleftharpoons{TsOH}$ N + H_2O (27)

85%

4.2.4 Acetals

Acetals[47] are derivatives of aldehydes and ketones wherein the oxidation level remains the same but the hybridization of the carbon changes to sp^3. This renders the former carbonyl carbon unattractive to nucleophiles and is therefore a temporary protecting device.

Aldehydes are converted to acetals by treating with excess alcohol and an acid catalyst. The reaction is reversible and the equilibrium is driven toward the acetal by the excess alcohol or by removal of the water as it is produced. If a 1,2- or 1,3-diol is used, the cyclic acetal forms readily, even exothermally, but is sometimes difficult to remove.

The greater steric hindrance in ketones makes acetal formation more difficult. Ethylene glycol or other diols are commonly used and the water is removed azeotropically with solvents such as toluene:[48]

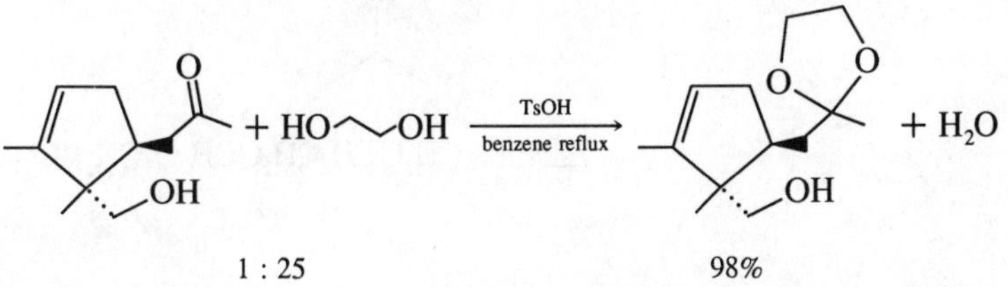

TsOH, benzene reflux; + H_2O (28)

1 : 25 98%

The water may alternatively be consumed *in situ* by including an orthoester such as triethyl orthoformate, which hydrolyzes to ester and alcohol during the acetalization. Trans acetalization is also useful. Treatment of a ketone with excess 2-ethyl-2-methyl-1,3-dioxolane in acid gives 2-butanone and the new ketal (Eq. 29).[49] In some cases a smaller excess of dioxolane is used and the reaction is driven by distilling the 2-butanone as it is formed.

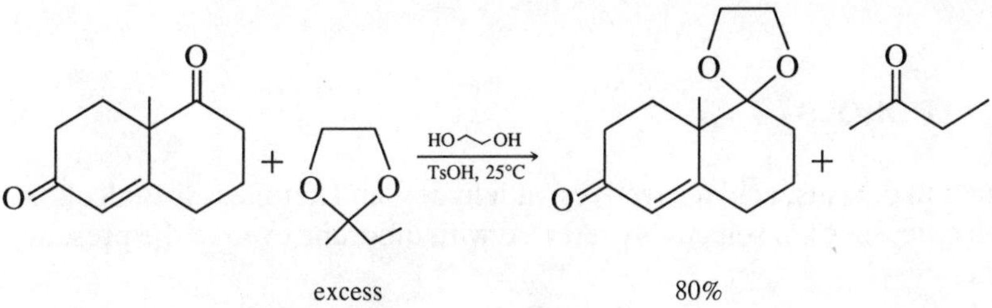

HO⁀OH, TsOH, 25°C (29)

excess 80%

Acetals can be used to protect alcohols also. In these cases a vinyl ether is used in place of the aldehyde as in Eq. 30.[34] Under basic conditions an α-haloether will convert an alcohol to an acetal as in Eq. 31.[50]

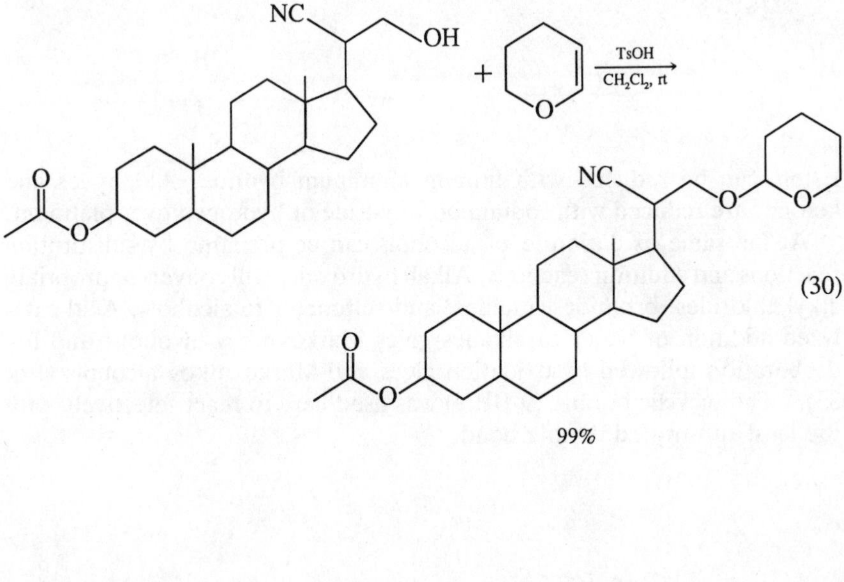

NC, OH; TsOH, CH_2Cl_2, rt; NC (30)

99%

$$CH_3OCH_2CH_2OCH_2Cl + HO(CH_2)_8Br + (i\text{-}Pr)_2NEt \xrightarrow[0°C\ \text{to rt}]{CH_2Cl_2} CH_3OCH_2CH_2OCH_2O(CH_2)_8Br \quad (31)$$

76%

Where the acetals were used as temporary protecting groups, they may be removed with aqueous acid to recover the ketone or aldehyde. Deprotection of alcohols is done in aqueous or alcoholic acid. The methoxyethoxymethyl ethers (Eq. 31) can be removed specifically by anhydrous zinc bromide followed by aqueous bicarbonate, conditions that leave other acetals intact.[51]

4.3 ALCOHOLS

Reduction of acids, acid derivatives, aldehydes, and ketones gives alcohols. Carboxylic acids are selectively reduced with diborane even in the presence of esters:[52]

$$\text{NC–acetoxy acid} + BH_3\text{•}THF \xrightarrow[-50°\ \text{to rt}]{THF} \xrightarrow{H_2O} \text{NC–acetoxy alcohol} \quad (32)$$

63%

Esters can be reduced with lithium aluminum hydride. Aldehydes and ketones are reduced with sodium borohydride or hydrogen over platinum.

At the same oxidation level, alcohols can be prepared by substitution reactions and addition reactions. Alkali hydroxides will convert appropriate alkyl chlorides, bromides, iodides, and sulfonates to alcohols. Acid-catalyzed addition of water to alkenes gives Markovnikov alcohols, and hydroboration followed by oxidation gives anti-Markovnikov alcohols (Eq. 33).[53] The bicyclic borane (9-BBN) was used here to react selectively with the least substituted double bond.

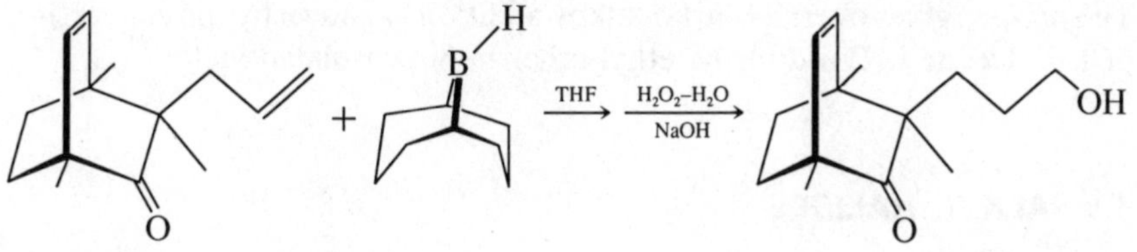

THF; H_2O_2–H_2O, NaOH

>77%

(33)

4.4 ETHERS

Like alcohols, ethers are commonly prepared by nucleophilic substitution or by addition to alkenes. In the Williamson method an alkoxide will displace a halide or sulfate group from a primary carbon. Suitable concentrations of alkoxides are available by using sodium hydroxide as a slurry in DMSO,[54] or in aqueous solution with a phase-transfer catalyst (Chapter 9). In the example in Eq. 34, an excess of dibromobutane was used to minimize formation of a diether.[55]

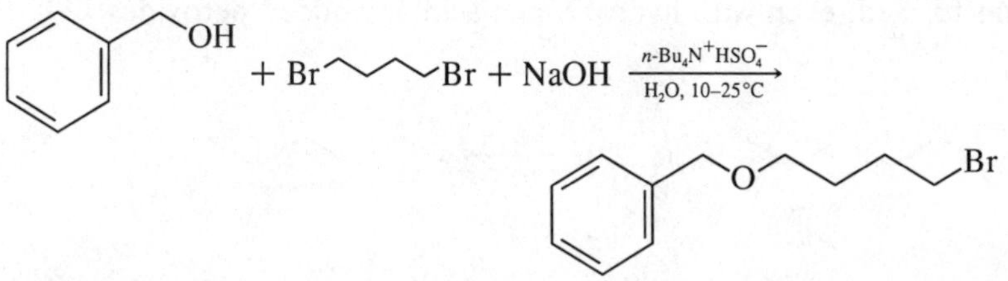

+ Br–(CH$_2$)$_4$–Br + NaOH $\xrightarrow[H_2O,\ 10\text{–}25°C]{n\text{-}Bu_4N^+HSO_4^-}$

88%

(34)

Overall addition of alcohols to alkenes is accomplished by alkoxymercuration followed by reduction as shown in Eq. 35 for the preparation of cyclohexyl isopropyl ether.[56]

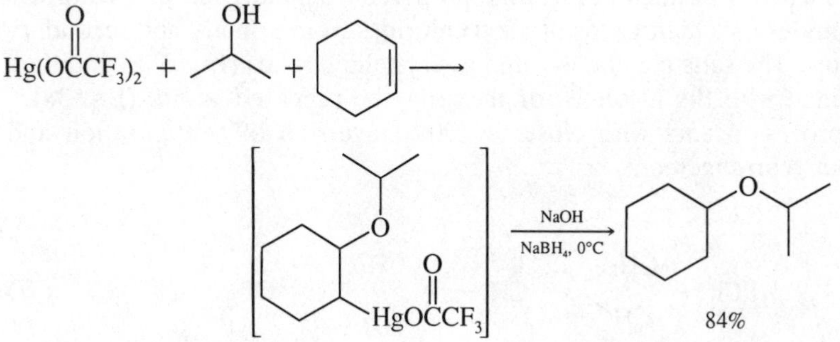

$Hg(OCCF_3)_2$ + OH + ⟶

$HgOCCF_3$ $\xrightarrow[NaBH_4,\ 0°C]{NaOH}$

84%

(35)

The process gives overall Markovnikov addition as shown by the conversion of 1-dodecene to 2-*n*-dodecyl ethyl ether in 80% isolated yield.

4.5 ALKYL HALIDES

Chlorine or bromine may be incorporated by substitution for hydrogen under free-radical conditions. This is useful where the substrate is a highly symmetrical compound or contains a site especially prone to free-radical formation. Otherwise complex mixtures of isomers are obtained.

Anhydrous hydrogen chloride, bromide, or iodide will add to alkenes by a carbocationic mechanism to give Markovnikov products. The products may have rearranged structures if the first intermediate carbocation can improve in stability by a 1,2-hydride or alkyl shift. In the presence of radical initiators, hydrogen bromide will add in the reverse direction to give anti-Markovnikov products.

These same addition reactions may be carried out by using the more easily handled aqueous acids if a phase-transfer catalyst such as hexadecyl tributylphosphonium bromide is used (Eq. 36).[57] Only Markovnikov products are formed, even with hydrobromic acid and added peroxides.

$$\text{1-octene} + \text{aq. HBr} \xrightarrow[115°C]{\text{catalyst}} \text{2-bromooctane} \quad (36)$$

1 : 5 88%

The most versatile and specific routes begin with alcohols. These and some other methods are detailed in the following sections.

4.5.1 Alkyl Chlorides

Iminium salts (Vilsmeier reagents), prepared from inorganic acid chlorides and amides give high yields of alkyl chlorides from primary and secondary alcohols. The salts may be isolated as crystalline solids (Eq. 37)[58] and then combined with the alcohols, or they may be prepared *in situ* (Eq. 38).[59] The process occurs with close to 100% inversion of configuration and without rearrangement.

$$PCl_5 + (CH_3)_2N{-}CH{=}O \longrightarrow (CH_3)_2\overset{+}{N}{=}C(Cl)H \; Cl^- \quad (37)$$

$$\text{2-octanol} + POCl_3 + DMF \xrightarrow[\text{rt}]{CHCl_3} \text{2-chlorooctane (90\%)} \tag{38}$$

A very mild, neutral procedure is treatment of the alcohol with triphenylphosphine in carbon tetrachloride. This gives chloroform and a salt that rapidly decomposes to the alkyl chloride (Eq. 39).[60] Rearrangement is absent even in the conversion of neopentyl alcohol to chloroneopentane.

$$\text{HOCH}_2\text{-dioxolane} + Ph_3P + \underset{\text{excess}}{CCl_4} \xrightarrow{\text{reflux}} \left[Ph_3\overset{+}{P}\text{—O—CH}_2\text{-dioxolane} \quad Cl^- \right] + CHCl_3 \longrightarrow \underset{80\%}{\text{ClCH}_2\text{-dioxolane}} + Ph_3PO \tag{39}$$

Some tertiary alcohols may be converted to the chlorides by simply shaking with concentrated aqueous HCl. This involves carbocations and may lead to rearrangements.

4.5.2 Alkyl Bromides

As with chlorides, the iminium salts are particulary useful for converting primary and secondary alcohols to bromides.[61] Phosphorus tribromide and DMF give the analogous reagent which, in dioxane, converted (−)-2-octanol to (+)-2-bromooctane in 85% yield with some loss of optical purity but no rearrangement.[58]

Triphenylphosphine with carbon tetrabromide at ice bath temperatures gives good yields of primary alkyl bromides but is poor with most secondary alcohols.[62]

Primary, secondary and tertiary alcohols are converted in high yield to the bromides by treatment with trimethylsilyl bromide in chloroform at 25–50°C.[63] Again the stereochemistry is mostly inversion (93.8% with 2-octanol).

Allylic hydrogens may be replaced with bromine by using *N*-bromosuccinimide under sunlamp irradiation. This is a free-radical chain mechanism, and two structural isomers may result from attachment at either end of the conjugated radical system. Benzylic hydrogens may be replaced similarly with *N*-bromosuccinimide, or simply with Br_2 in carbon tetrachloride in the presence of a catalytic amount of solid $La(OCOCH_3)_3$. With this catalyst, room fluorescent lighting is sufficient to initiate the reaction (Eq. 40). Yields are 50–90%.[64]

$$p\text{-}BrC_6H_4CH_3 + Br_2 \xrightarrow[La(OCOCH_3)_3,\ lamp]{CCl_4} p\text{-}BrC_6H_4CH_2Br \quad (65\%) \qquad (40)$$

4.5.3 Alkyl Iodides

Primary, secondary, and tertiary alkyl iodides may be prepared from alcohols by treatment with a solution of P_2I_4 in carbon disulfide at 0°C in yields of 45–90%, again without rearrangement.[65] Trimethylsilyl iodide or a combination of trichloromethylsilane and sodium iodide will also convert alcohols to iodides, but it is highly electrophilic and will cleave ethers, esters, and lactones also.[66]

The classic Finkelstein synthesis of primary iodides begins with the corresponding chloride, bromide, or tosylate. Treatment with a dry acetone solution of sodium iodide gives a precipitate of the corresponding sodium salt plus the alkyl iodide:[67]

$$\text{HO(CH}_2)_2\text{CH=CH(CH}_2)_2\text{O-THP} + \text{TsCl} \xrightarrow[0^\circ C]{pyridine} \xrightarrow[acetone,\ rt,\ 3\ h]{NaI} \text{I(CH}_2)_2\text{CH=CH(CH}_2)_2\text{O-THP} \quad (83\%) \qquad (41)$$

4.5.4 Alkyl Fluorides

Primary, secondary, and tertiary alcohols can be converted directly to fluorides, usually in high yield, by treatment with diethylaminosulfur trifluoride[68] (Eq. 42).[67] Most cases proceed without rearrangement.

$$Br(CH_2)_8OH + Et_2NSF_3 \xrightarrow[-78\ to\ 25^\circ C]{CH_2Cl_2} Br(CH_2)_8F \quad (61\%) \qquad (42)$$

The reagent is prepared by combining diethylaminotrimethylsilane and SF_4 at low temperature, and it is also commercially available.

Primary and secondary chlorides, bromides, and tosylates may be converted to fluorides in isolated yields of 35–66% by stirring with almost anhydrous tetrabutylammonium fluoride at 25–40°C without solvent.[69] The by-products include alcohols from the small amount of water present and alkenes arising because of the high basicity of unsolvated fluoride.

4.6 AMINES

Amines are commonly prepared by treating primary or secondary halides or sulfonates with excess ammonia. The primary amine produced may compete with ammonia to give some secondary and even tertiary amine. The excess is used to minimize this. A secondary or tertiary amine may be prepared by treating a primary or secondary amine with an alkyl halide or sulfonate in the same way. Primary amines can be made cleanly using potassium phthalimide instead of ammonia (Chapter 9, Eq. 9). The free amine may be released from the alkylated phthalimide by shaking briefly with aqueous methylamine at room temperature.[70] This gives *N,N′*-dimethylphthalamide and the desired amine.

Reductive amination of ketones and aldehydes gives primary, secondary, and tertiary amines. Sodium cyanoborohydride[71] or platinum-catalyzed hydrogenation are used as shown:[72,73]

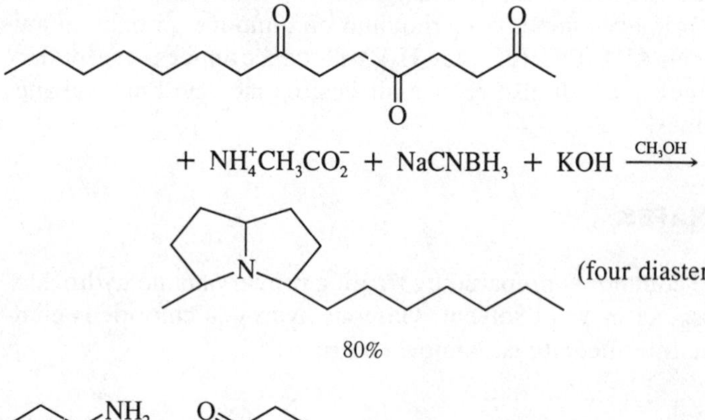

$$+ NH_4^+CH_3CO_2^- + NaCNBH_3 + KOH \xrightarrow{CH_3OH}$$

(four diastereomers) (43)

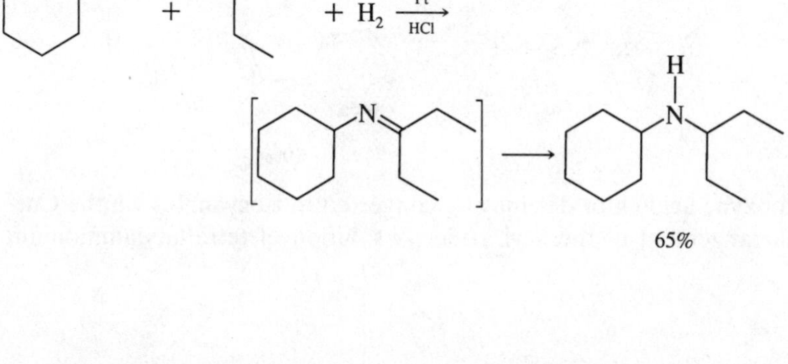

80%

$$NH_2 \quad + \quad O \quad + H_2 \xrightarrow[HCl]{Pt}$$

H

N

N

(44)

65%

Tertiary alkyl groups cannot be attached to nitrogen by the preceding reactions, and one must resort to carbocation methods. In the Ritter reaction an alcohol is treated with a strong acid in the presence of a nitrile. The alcohol is converted to a carbocation which attacks the nitrogen of the nitrile to give, after hydration, a hydrolyzable amide:[74]

$$+\ CH_3CN \xrightarrow[CH_3CO_2H]{H_2SO_4} \xrightarrow{H_2O}$$

$$\xrightarrow[H_2O,\ 200°C]{KOH} \quad (45)$$

77%

Other nitrogen containing functional groups may be reduced to give amines. Nitriles may be hydrogenated over a platinum or palladium catalyst at 25°C. This process can give some secondary amine via addition of primary amine to the intermediate imine but that can be suppressed by providing an acid to make the salt of the primary amine. Oximes may also be hydrogenated to primary amines over rhodium on alumina. Lithium aluminum hydride or $NaAlH_2(OCH_2CH_2OCH_3)_2$ will reduce amides to secondary or tertiary amines and will also reduce nitriles, oximes, and nitroalkanes to primary amines.

4.7 ISOCYANATES

Isocyanates are commonly prepared by treating primary amine hydrochlorides with phosgene in a hot solvent. Gaseous hydrogen chloride is eliminated from the intermediate carbamoyl chlorides:[75]

$$NH_3Cl\ \ +\ COCl_2 \xrightarrow[reflux]{toluene} NCO\ \ +\ HCl \quad (46)$$

89%

Carboxylic acid chlorides may be converted to isocyanates via the Curtius rearrangement of the acyl azide. A solution of tetrabutylammonium azide (prepared by CH_2Cl_2 extraction from aqueous sodium azide and tetrabutylammonium hydroxide) in toluene or benzene reacts readily with acid chlorides to give acyl azides. These are heated in solution to afford the isocyanates and nitrogen:[76]

$$(CH_3)_2CHCH_2COCl + n\text{-}Bu_4N^+N_3^- \xrightarrow[25°C]{benzene} (CH_3)_2CHCH_2CON_3 \xrightarrow[50\text{–}90°C]{benzene} N_2 + (CH_3)_2CHCH_2N{=}C{=}O \quad (86\%) \tag{47}$$

4.8 ALKENES

Alkenes may be made from saturated compounds by various β-elimination reactions. A vicinal dihalide may be dehalogenated by sodium iodide or activated zinc to give a double bond specifically between the carbons that bore the halogens. One practical use of such a process is the inversion of configuration of alkenes. The anti addition of chlorine followed by syn elimination by using sodium iodide does so as in Eq. 48.[77] About 90% anti elimination may be obtained with activated zinc and acetic acid in DMF.

$$(CH_3)_2CH(CH_2)_4CH{=}CH(CH_2)_9CH_3\ (trans) + Cl_2 \xrightarrow[-78°C]{CH_2Cl_2} (CH_3)_2CH(CH_2)_4CHClCHCl(CH_2)_9CH_3 \xrightarrow[DMF,\ 153°C]{NaI} (CH_3)_2CH(CH_2)_4CH{=}CH(CH_2)_9CH_3\ (cis) + I_2 + NaCl \tag{48}$$

100% stereospecificity
95% yield

The more common circumstance is the elimination of a leaving group and a hydrogen. Here regiospecificity is the problem because there is usually a choice of β hydrogens. β-Dehydrohalogenation is usually done with strong bases. If substitution competes, sterically hindered, less nucleophilic bases such as potassium *tert*-butoxide are chosen. Anti elimination is usual where possible. Among anti possibilities, conjugated products predominate over nonconjugated, and more substituted alkenes predominate over less substituted. If the alkene will be conjugated, less potent bases may be used, especially when there are other base-sensitive functional groups in the molecule. The amidine 1,5-diazabicyclo[4.3.0]non-5-ene "DBN"[78] or LiBr and Li_2CO_3 in DMF (Eq. 49)[79] give good selectivity.

$$+ \text{LiBr} + \text{Li}_2\text{CO}_3 \xrightarrow[125°\text{C}]{\text{DMF}}$$

78% (49)

Alcohols may be eliminated by acid treatment. Carbocations are intermediates; therefore, rearrangements are likely:[80]

$$\xrightarrow[\text{heat}]{\text{KHSO}_4} \quad + \text{H}_2\text{O} \qquad (50)$$

Selenoxides eliminate rapidly without a base. They may be prepared from enolates and lead to α,β-unsaturated ketones and esters:[81,82]

$$\xrightarrow[\text{2. Se, HMPA, }-78\text{ to }-15\text{°C}]{\text{1. LDA, THF, }-78\text{°C}} \left[\ \text{Se}^-\ \right] \xrightarrow{\text{CH}_3\text{I}} \text{SeCH}_3 \xrightarrow[\text{rt}]{\text{H}_2\text{O}_2}$$

80%

$$\left[\ \text{SeCH}_3\ \right] \longrightarrow \qquad (51)$$

Ketones may be reductively eliminated via their tosylhydrazones[83] as illustrated in Eq. 52.[84] If there is a choice, the least substituted alkene will predominate.

$$\text{(4-}t\text{-butylcyclohexanone)} \xrightarrow[\text{TsNHNH}_2\text{, reflux}]{\text{EtOH}} \text{TsNHN=(4-}t\text{-butylcyclohexylidene)} \xrightarrow{2n\text{-BuLi}} \text{4-}t\text{-butylcyclohexene} \quad (52)$$

Trisubstituted alkenes require LDA in place of the alkyllithium.[85]

cis-Alkenes may be prepared by partial hydrogenation of appropriate alkynes. Various catalysts have been used, including colloidal nickel containing some boron from $NaBH_4$ reduction of nickel acetate. With ethylene diamine, this catalyst gives a 200:1 selectivity toward cis isomers:[86]

$$\text{alkyne} + H_2 \xrightarrow[25°\text{C}]{\text{Ni, } H_2NCH_2CH_2NH_2} \text{cis-alkene} \quad (53)$$

99.5% cis
96% yield

The corresponding trans isomers are available from reduction of alkynes with sodium in liquid ammonia. Carbanions will also add to alkynes to give alkenes (Chapter 5, Eq. 9).

4.9 OTHER FUNCTIONAL GROUPS

There are many other functional groups besides those detailed in this chapter. A further sampling of structures and names is as follows:

RNHOH	hydroxylamines
RONO	nitrites
RNO_2	nitroalkanes
$RONO_2$	nitrates
$R{-}N(\to O){=}C(R){-}R$	nitrones
RSH	mercaptans, thiols
$RN{\equiv}C$	isonitriles

RNHNH	hydrazines
R—N=N—R	azo compounds

The chemistry of these may be found in the general references of Chapter 1.

This chapter is again a selection of methods; you can find many more, especially methods that produce combinations of functional groups such as α-diketones or vinyl halides.

PROBLEMS

Show how you would prepare each of the following products from the given starting materials. Where more than one step is required, show each step distinctly.

1. 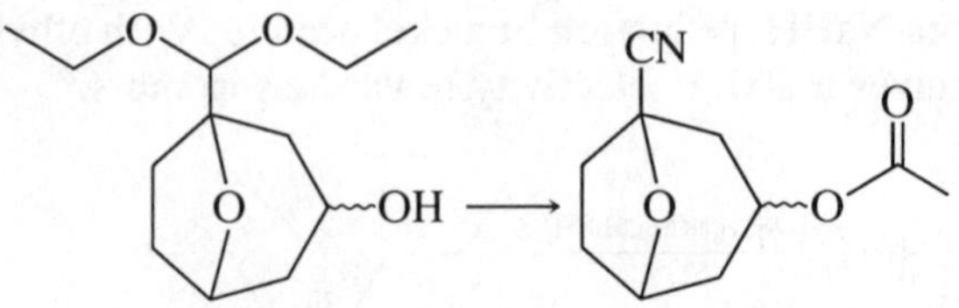 Ref. 87

2. 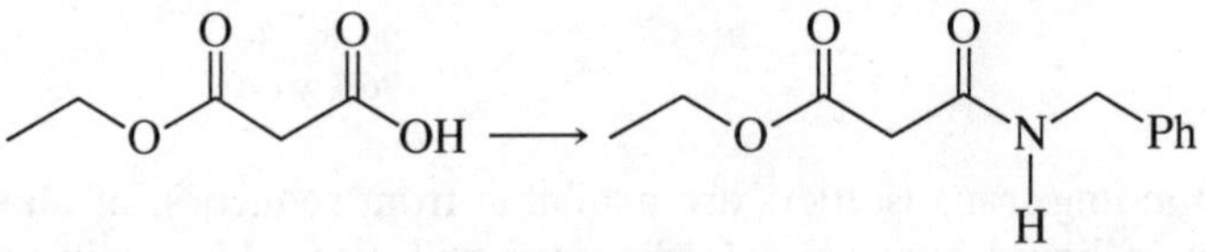 Ref. 88

3. 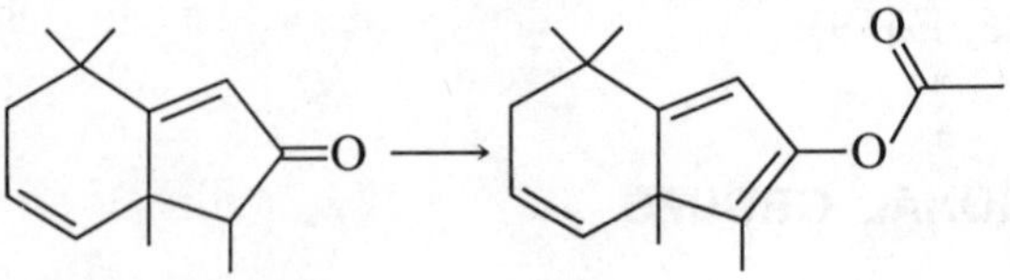 Ref. 89

4. 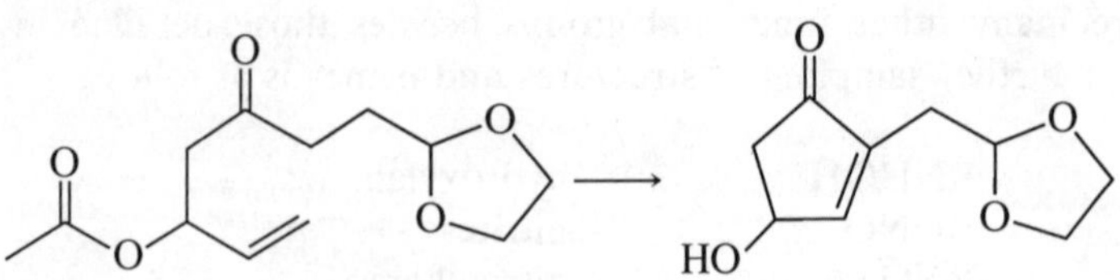 Ref. 90

5. 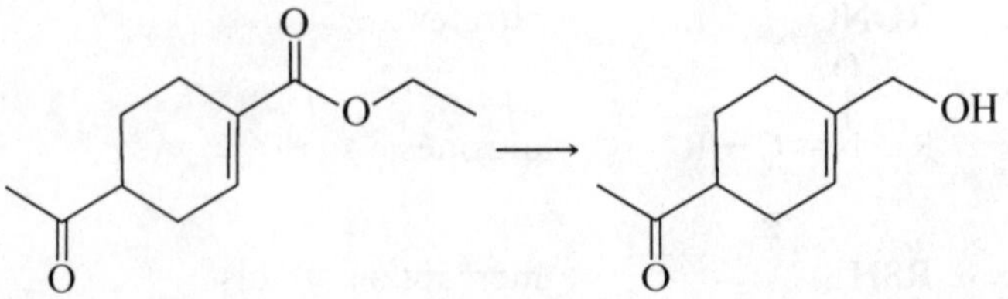 Ref. 91

6. H H → H H — Ref. 92

7. CH_3O O, OCH_3 → CH_3O O, OCH_3 — Ref. 93

8. O O → Br O O — Ref. 94

9. → HO O O OH →

$(CH_3)_2N$ $N(CH_3)_2$ — Ref. 95

10. O O OH → O O O — Ref. 96

11. O= O OH OH O → O= I I — Ref. 97

12. O → NH_2 — Ref. 98

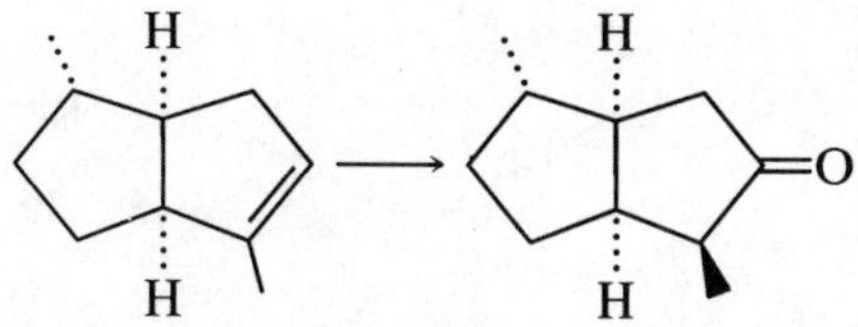

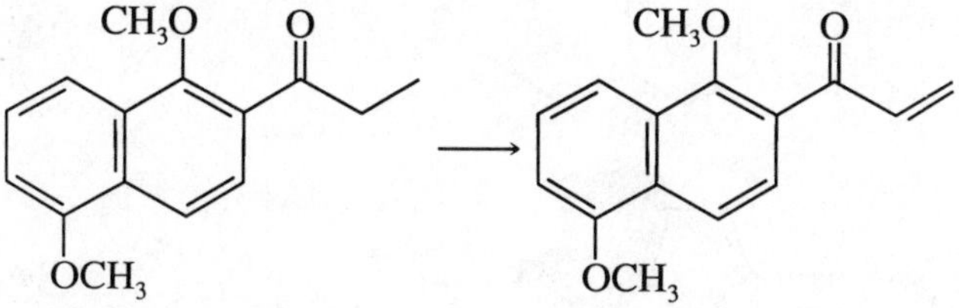

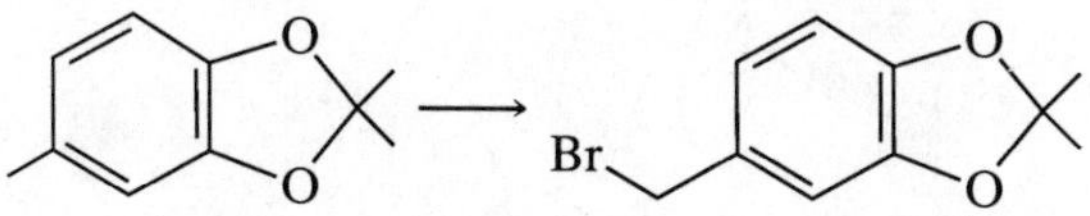

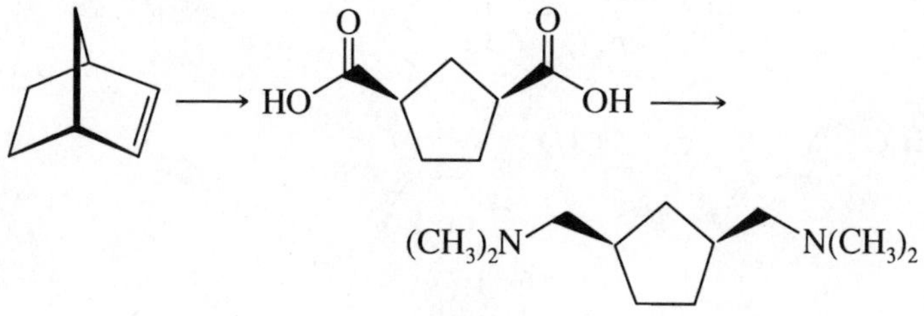

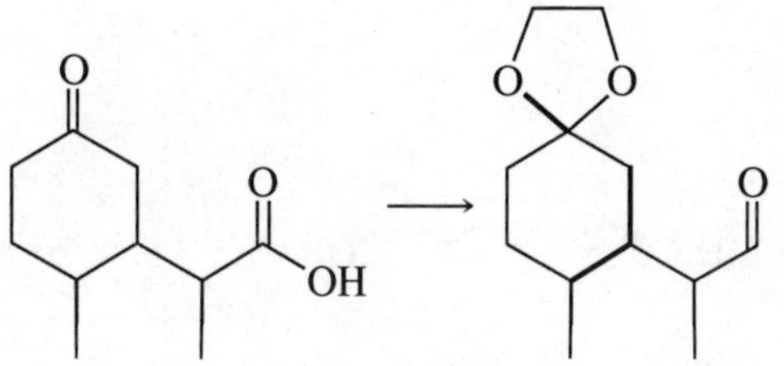

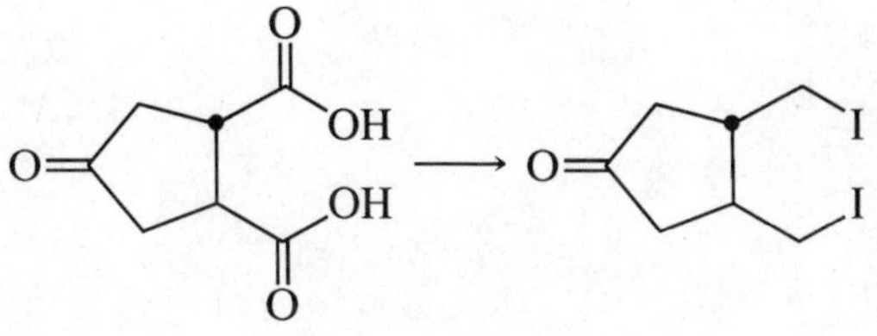

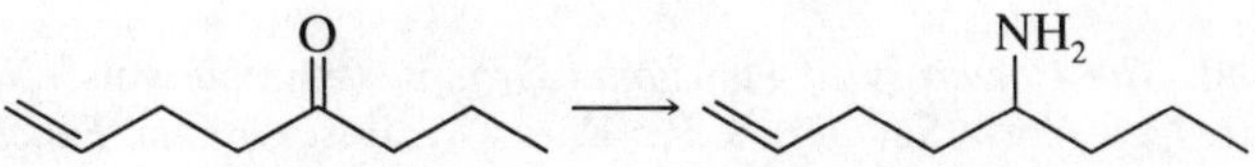

13. 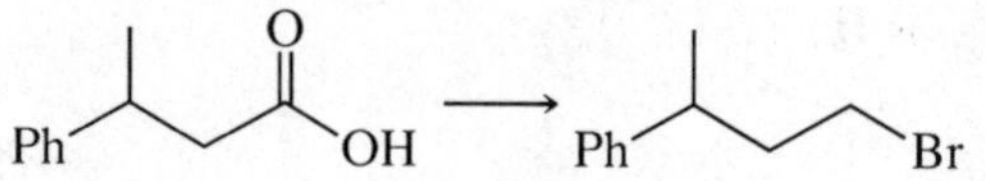 Ph–CH(CH$_3$)–CH$_2$–COOH ⟶ Ph–CH(CH$_3$)–CH$_2$–CH$_2$Br — Ref. 99

14. 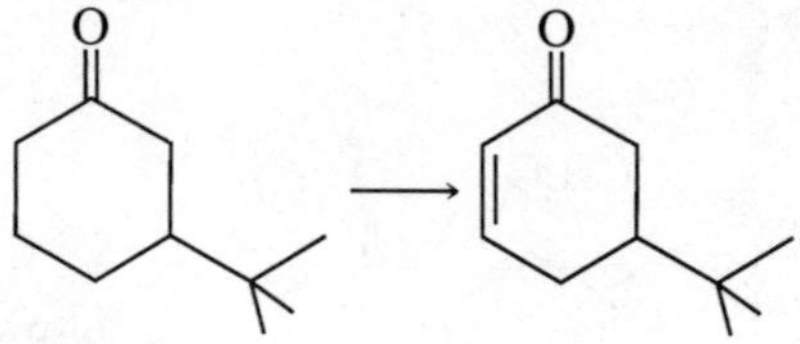 ⟶ — Ref. 100

15. 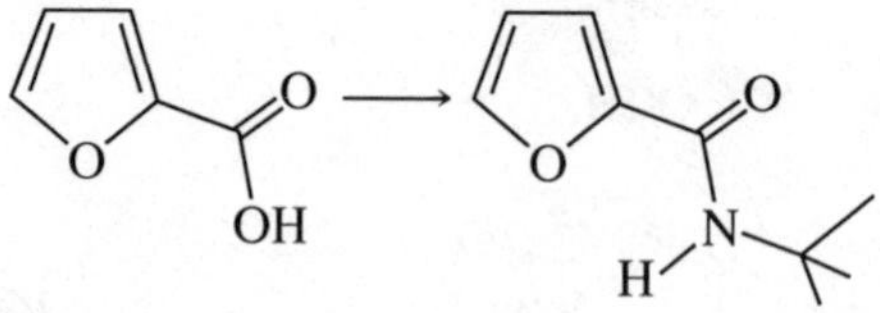 ⟶ — Ref. 101

16. 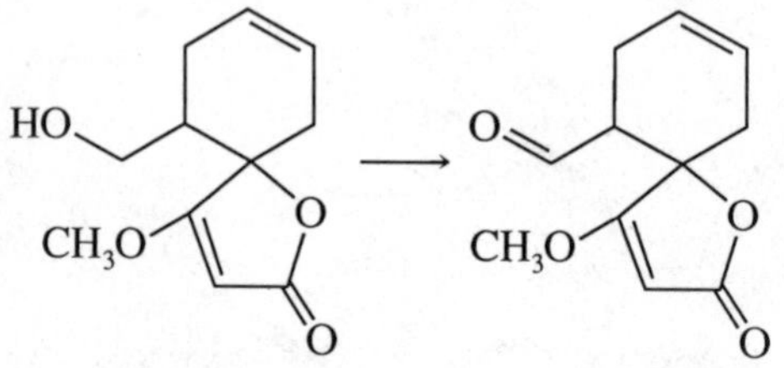 ⟶ — Ref. 102

17. 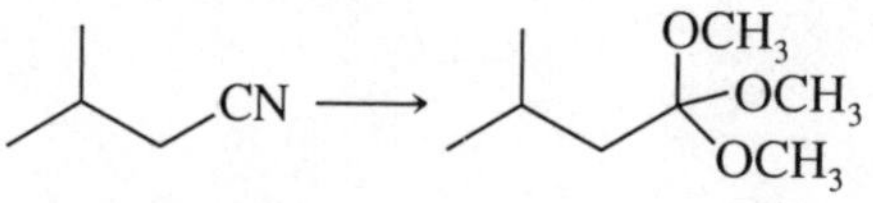 CN ⟶ $C(OCH_3)_3$ — Ref. 103

18. 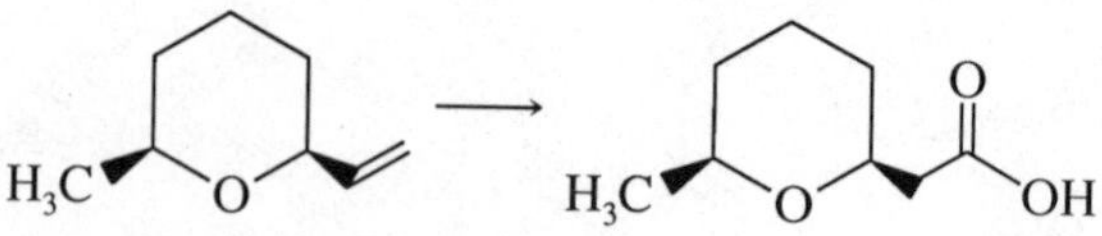 ⟶ — Ref. 104

19. 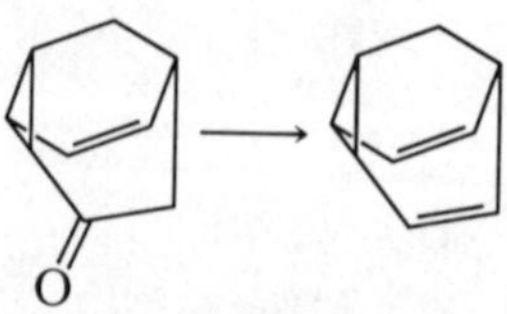 ⟶ — Ref. 105

REFERENCES

1. Patai, S., ed. *The Chemistry of Functional Groups,* (multiple vols.) Wiley-Interscience, New York; Sandler, S. R.; Karo, W., Eds. *Organic Functional Group Preparations,* Vols. 1–3, Academic, New York, 1971–1983; Barton,

D.; Ollis, W. D., Eds., *Comprehensive Organic Chemistry,* Pergamon, Oxford, 1979; Muller, E., Ed., *Methoden der Organischen Chemie* (*Houben-Weyl*) (multiple vols.) Georg Thieme Verlag, Stuttgart (general references on functional group synthesis).
2. Treley, W. F.; Marvel, C. S. *Org. Synth.* **1955,** *Coll. 3,* 822.
3. Hertz, W.; Mohanraj, S. *J. Org. Chem.* **1980,** *45,* 5417.
4. Starks, C. M. *J. Am. Chem. Soc.* **1971,** *93,* 195.
5. Matsuda, I.; Murata, S.; Izumi, Y. *J. Org. Chem* **1980,** *45,* 237.
6. Nicolaou, K. C.; Pavia, M. R.; Seitz, S. P. *Tetrahedron Lett.* **1979,** 2327.
7. Overman, L. E. *Tetrahedron Lett.* **1975,** 1149.
8. Hase, T. A.; Nylund, E.-L. *Tetrahedron Lett.* **1979,** 2633; Lindgren, B. O.; Nilsson, T. *Acta Chem. Scand.* **1973,** *27,* 888.
9. Welch, S. C.; Valdes, T. A. *J. Org. Chem.* **1977,** *42,* 2108.
10. Boesken, J. *Rec. Trav. Chim.* **1910,** *29,* 99.
11. Nugent, W. A.; McKinney, R. J. *J. Org. Chem.* **1985,** *50,* 5370.
12. Degenhardt, C. R. *J. Org. Chem.* **1980,** *45,* 2763.
13. Burgstahler, A. W.; Weigel, L. O.; Bell, W. J.; Rust, M. K. *J. Org. Chem.* **1975,** *40,* 3456.
14. Jones, G.; Raphael, R. A.; Wright, S. *J. Chem. Soc. Perkin I* **1974,** 1676.
15. Mehta, G.; Pandey, P. N. *Synthesis* **1975,** 404.
16. House, H. O.; Gall, M.; Olmstead, H. D. *J. Org. Chem.* **1971,** *36,* 2361.
17. Guthrie, J. L.; Rabjohn, N. *Org. Synth.* **1963,** *Coll 4,* 513.
18. Inami, K.; Shiba, T. *Bull. Chem. Soc. Jpn.* **1985,** *58,* 352.
19. Cacchi, S.; Misiti, D. *Synthesis* **1980,** 243.
20. Ravindranathan, M.; Kalyanam, N.; Sivaram, S. *J. Org. Chem.* **1982,** *47,* 4812.
21. Adams, R.; Ulich, L. H. *J. Am. Chem. Soc.* **1920,** *42,* 599.
22. Burgstahler, A. W.; Weigel, L. O.; Shaefer, C. G. *Synthesis* **1976,** 767.
23. Hurd, C. D.; Christ, R.; Thomas, C. L. *J. Am. Chem. Soc.* **1933,** *55,* 2589.
24. Krimen, L. I. *Org. Synth.* **1970,** *50,* 1.
25. Martin, S. F.; Benage, B. *Tetrahedron Lett.* **1984,** *25,* 4863.
26. Bargar, T. M.; Riley, C. M. *Synth. Commun.* **1980,** *10,* 479.
27. Olah, G. A.; Narang, S. C.; Fung, A. P.; Gupta, B. G. B. *Synthesis* **1980,** 657.
28. Olah, G. A.; Vankar, Y. D.; Garcia-Luna, A. *Synthesis* **1979,** 227.
29. Ho, T. L.; Wong, C. M. *Synth. Commun.* **1975,** *5,* 299.
30. DeWolfe, R. H. *Carboxylic Ortho Acid Derivatives,* Academic, New York, **1970;** DeWolfe, R. H. *Synthesis* **1974,** 153.
31. McElvain, S. M.; Nelson, J. W. *J. Am. Chem. Soc.* **1942,** *64,* 1825.
32. Corey, E. J.; Raju, N. *Tetrahedron Lett.* **1983,** *24,* 5571.

33. Johnstone, R. A. W.; Telford, R. P. *J. Chem. Soc. Chem. Commun.* **1978,** 354.
34. Kabat, M. M.; Kurek, A.; Wicha, J. *J. Org. Chem.* **1983,** *48,* 4248.
35. Ireland, R. E.; Courtney, L.; Fitzsimmons, B. J. *J. Org. Chem.* **1983,** *48,* 5186.
36. Malek, J.; Cerny, M. *Synthesis* **1972,** 217.
37. Corey, E. J.; Suggs, J. W. *Tetrahedron Lett.* **1975,** 2647.
38. Hecker, S. J.; Heathcock, C. H. *J. Org. Chem.* **1985,** *50,* 5159.
39. Guziec, T. S. Jr.; Luzzio, F. A. *J. Org. Chem.* **1982,** *47,* 1787.
40. Pappas, J. J.; Keaveney, W. P.; Gancher, E.; Berger, M. *Tetrahedron Lett.* **1966,** 4273.
41. Maas, D. D.; Blagg, M.; Wiemer, D. F. *J. Org. Chem.* **1984,** *49,* 853.
42. Wiberg, K. B. In *Oxidation in Organic Chemistry, Part A,* Wiberg, K. B., Ed., Academic, New York, 1965, Chapter 2.
43. Stevens, R. V.; Chapman, K. T.; Weller, H. N. *J. Org. Chem.* **1980,** *45,* 2030.
44. Mohrig, J. R.; Nienhuis, D. M.; Linch, C. F.; Van Zoeren, C.; Fox, B. G.; Mahaffy, P. G. *J. Chem Ed.* **1985,** *62,* 519.
45. Layer, R. W. *Chem. Rev.* **1963,** *63,* 489.
46. Pearce, G. T.; Gore, W. E.; Silverstein, R. M. *J. Org. Chem.* **1976,** *41,* 2797.
47. Meskens, F. A. J. *Synthesis* **1981,** 501.
48. Lansbury, P. T.; Mazur, D. J. *J. Org. Chem.* **1985,** *50,* 1632.
49. Bauduin, G.; Pietrasanta, Y. *Tetrahedron* **1973,** *29,* 4225.
50. Millar, J. G.; Oehlschlager, A. C.; Wong, J. W. *J. Org. Chem.* **1983,** *48,* 4404.
51. Corey, E. J.; Gras, J. L.; Ulrich, P. *Tetrahedron Lett.* **1976,** 809.
52. Frye, S. V.; Eliel, E. L. *J. Org. Chem.* **1985,** *50,* 3402.
53. Snowden, R.; Sonnay, P. *J. Org. Chem.* **1984,** *49,* 1464.
54. Benedict, D. R.; Bianchi, T. A.; Cate, L. A. *Synthesis* **1979,** 428.
55. Burgstahler, A. W.; Weigel, L. O.; Sanders, M. E.; Shaefer, C. G.; Bell, W. J.; Vuturo, S. B. *J. Org. Chem.* **1977,** *42,* 566.
56. Brown, H. C.; Kurek, J. T.; Kei, M.-H.; Thompson, K. L. *J. Org. Chem.* **1985,** *50,* 1171.
57. Landini, D.; Rolla, F. *J. Org. Chem.* **1980,** *45,* 3527.
58. Hepburn, D. R.; Hudson, H. R. *J. Chem. Soc. Perkin I* **1976,** 754.
59. Yoshihara, M.; Eda, T.; Sakaki, K.; Maeshima, F. *Synthesis* **1980,** 746.
60. Castro, B. R. *Org. React.* **1983,** *29,* 5; Lee, J. B.; Nolan, T. J. *Can. J. Chem.* **1966,** *44,* 1331.
61. Dolbier, W. R., Jr.; Dulcere, J.-P.; Sellers, S. F.; Korniak, H.; Shatkin, B. T.; Clark, T. L. *J. Org. Chem.* **1982,** *47,* 2298.
62. Kocienski, P. J.; Cernigliaro, G.; Feldstein, G. *J. Org. Chem.* **1977,** *42,* 353.
63. Jung, M. E.; Hatfield, G. L. *Tetrahedron Lett.* **1978,** 4483.

64. Ouertani, M.; Girard, P.; Kagan, H. B. *Bull. Soc. Chim. Fr.* **1982,** II–327.
65. Lauwers, M.; Regnier, B.; Van Enoo, M.; Denis, J. N.; Krief, A. *Tetrahedron Lett.* **1979,** 1801.
66. Olah, G. A.; Husain, A.; Singh, B. P.; Mehrota, A. K. *J. Org. Chem.* **1983,** *48,* 3667.
67. Carvalho, J. F.; Prestwich, G. D. *J. Org. Chem.* **1984,** *49,* 1251.
68. Middleton, J. *J. Org. Chem.* **1975,** *40,* 574.
69. Cox, D. P.; Terpinski, J.; Lawrynowicz, W. *J. Org. Chem.* **1984,** *49,* 3216.
70. Wolfe, S.; Hasan, S. K. *Can. J. Chem.* **1970,** *48,* 3572.
71. Lane, C. F. *Synthesis* **1975,** 135.
72. Jones, T. H.; Blum, M. S.; Fales, H. M.; Thompson, C. R. *J. Org. Chem.* **1980,** *45,* 4778.
73. Stowell, J. C.; Padegimas, S. J. *J. Org. Chem.* **1974,** *39,* 2448.
74. Timberlake, J. W.; Alender, J.; Garner, A. W.; Hodges, M. L.; Ozmeral, C.; Szilagyi, S.; Jacobus, J. O. *J. Org. Chem.* **1981,** *46,* 2082.
75. Ray, F. E.; Rieveschl, G., Jr.; *J. Am. Chem. Soc.* **1938,** *60,* 2676.
76. Brändström, A.; Lamm, B.; Palmertz, I. *Acta Chem. Scand B* **1974,** *28,* 699.
77. Sonnet, P. E.; Oliver, J. E. *J. Org. Chem.* **1976,** *41,* 3284, 3279.
78. Oediger, H.; Moller, F.; Eiter, K. *Synthesis* **1972,** 591.
79. Kametani, T.; Suzuki, K.; Nemoto, H. *J. Org. Chem.* **1980,** *45,* 2204.
80. Christol, H.; Jacquier, R.; Mousseron, M. *Bull. Soc. Chim. Fr.* **1958,** 248.
81. Liotta, D.; Zima, G.; Barnum, C.; Saivdane, M. *Tetrahedron Lett.* **1980,** *21,* 3643.
82. Liotta, D. *Acc. Chem. Res.* **1984,** *17,* 28.
83. Shapiro, R. H. *Org. React.* **1976,** *23,* 405.
84. Bellucci, G.; Ingrosso, G.; Marsili, A.; Mastrorilli, E.; Morelli, I. *J. Org. Chem.* **1977,** *42,* 1079.
85. Kolonko, K. J.; Shapiro, R. H. *J. Org. Chem.* **1978,** *43,* 1404.
86. Brown, C. A.; Ahuja, V. K. *J. Chem. Soc. Commun.* **1973,** 553.
87. La Belle, B. E.; Knudsen, M. J.; Olmstead, M. M.; Hope, H.; Yanuck, M. D.; Schore, N. E. *J. Org. Chem.* **1985,** *50,* 5215.
88. Moyer, M. P.; Feldman, P. L.; Rapoport, H. *J. Org. Chem.* **1985,** *50,* 5226.
89. Ishihara, M.; Tsuneya, T.; Shiota, H.; Shiga, M.; Nakatasu, K. *J. Org. Chem.* **1986,** *51,* 491.
90. Ellison, R. A.; Lukenbach, E. R.; Chiu, C. *Tetrahedron Lett.* **1975,** 499.
91. Mandai, T.; Yokoyama, H.; Miki, T.; Fukuda, H.; Kolata, H.; Kawada, M.; Otera, J. *Chem. Lett.* **1980,** 1057.
92. Matthews, R. S.; Whitesell, J. K. *J. Org. Chem.* **1975,** *40,* 3312.
93. Sih, C. J.; Massuda, D.; Corey, P.; Gleim, R. D.; Suzuki, F. *Tetrahedron Lett.* **1979,** 1285.
94. Ogura, K.; Tsuchihashi, G. *Tetrahedron Lett.* **1971,** 3151.

95. Wiberg, K. B.; Bailey, W. F.; Jason, M. E. *J. Org. Chem.* **1976,** *41,* 2711.
96. Ficini, J.; Eman, A.; Touzin, A. M. *Tetrahedron Lett.* **1976,** 679.
97. Baraldi, P. G.; Pollini, G. P.; Simoni, D.; Barco, A.; Benetti, S. *Tetrahedron* **1984,** *40,* 761.
98. Harding, K. E.; Burks, S. R. *J. Org. Chem.* **1984,** *49,* 40.
99. Kogura, T.; Eliel, E. L. *J. Org. Chem.* **1984,** *49,* 577.
100. Gorthey, L. A.; Vairamani, M.; Djerassi, C. *J. Org. Chem.* **1985,** *50,* 4173.
101. Carpenter, A. J.; Chadwick, D. J. *J. Org. Chem.* **1985,** *50,* 4362.
102. Takeda, K.; Shibata, Y.; Sagawa, Y.; Urahata, M.; Funaki, K.; Hori, K.; Sasahara, H.; Yoshii, E. *J. Org. Chem.* **1985,** *50,* 4673.
103. Wiberg, K.; Martin, E. J.; Squires, R. R. *J. Org. Chem.* **1985,** *50,* 4717.
104. Kim, Y.; Mundy, B. P. *J. Org. Chem.* **1982,** *47,* 3556.
105. Henkel, J. G.; Hane, J. T. *J. Org. Chem.* **1983,** *48,* 3858.

5

Carbon–Carbon Bond Formation

For two carbons to be mutually attractive and join together,[1] they usually begin with opposite charge polarizations. One has available electrons and is termed the *nucleophile* (seeking a plus charge or nucleus), and the other carries a partial or full positive charge and is called the *electrophile*. In basic solution the nucleophile carries a negative charge by virtue of being bonded to, or associated as ions with, a more electropositive element, that is, a metal. This connection may be direct or may include intervening π-conjugated atoms. The electrophile has a partially positive carbon because of a dipolar bond with a more electronegative atom. In acid solution the nucleophile is an alkene or an arene with its projecting electrons, either polarized or unpolarized (acid solutions simply protonate carbanions). In this case the electrophile is a complexed or free carbocation. Examples of each are shown in Table I.

Some carbon–carbon bond-forming reactions can occur with little or no charge polarization. Examples include the Diels–Alder, Cope, and Claisen reactions where concerted bond reorganization occurs (Chapter 8). Others involve the transient neutral but electrophilic intermediates, carbenes, and arynes.

Example applications of these in the formation of single and double bonds are shown in the following sections and in Chapter 6.

TABLE 1. Examples of Nucleophilic and Electrophilic Carbon–Carbon Bond Forming Agents

Nucleophiles	Electrophiles
CH_3MgBr	$\overset{\delta+}{C}H_3\overset{\delta-}{Br}$
$R-\bar{C}HCN\ \ Li^+$	$CH_3-\overset{\delta+}{C}(=O^{\delta-})-CH_3$
$R-C(-O-Li^{\delta+})=CH_2^{\delta-}$	$>\overset{\delta+}{C}=C<-\overset{\delta+}{C}(=O^{\delta-})-$
Cyclopentadienide$^-$ Na^+	$R-\overset{\delta+}{C}(=O^{\delta-})-Cl^{\delta-}$
$(CH_3)_3SiO-C(CH_3)=CH_2$	$C_6H_5-\overset{+}{C}H-CH_3\ \ AlCl_4^-$

5.1 CARBON–CARBON SINGLE BOND FORMATION

5.1.1 Reactions in Basic Solution

Basic solutions are generally prepared by introduction of a reducing agent.[2] The more electropositive metals, in the metallic state, are strong reducing agents, that is, electron donors. They will react with most organohalides, reducing them to halide ion and negatively charged, strongly basic carbon. These carbons have a complete octet of electrons at the expense of the metals. The new carbon–metal bond is substantially ionic, and the reagents are referred to as *carbanions,* with the metals essentially cations. Magnesium, lithium, zinc, and sometimes sodium and potassium are used in this way. These carbanions are powerful nucleophiles toward most electrophiles. In some cases the reactivity is lowered by exchanging less electropositive metal cations (e.g., copper, mercury, or cadmium) for those initially used in order to obtain selectivity on polyfunctional electrophiles. This direct use of reducing metals is commonly the source of simple alkyl, aryl, and vinyl carbanions.

The second routine method of preparing carbanions begins, not with organohalides, but with reagents where carbon carries a hydrogen of sufficient acidity to be removed by a stronger base. This is used where the resulting carbanionic charge is delocalized by resonance, especially to oxygen or nitrogen atoms, or by inductive effects from adjacent neutral or positively charged sulfur or phosphorus atoms. Strong bases are also used to remove protons from carbons that are sp- or sp^2-hybridized[3] as exemplified in Eq. 1. Here the stronger base is itself a carbanion.

$$HC{\equiv}CH + n\text{-BuLi} \xrightarrow[-78^\circ]{THF} HC{\equiv}C^-Li^+ + n\text{-}C_4H_{10} \quad (1)$$

Protons can be removed similarly from aromatic rings, specifically ortho-to-cation chelating substituents.[4] A selection of bases for this purpose is given in Fig. I. The weakest bases are used to generate small equilibrium concentrations of carbanions that are often sufficient for high-yield overall carbon–carbon bond formation. Sodium ethoxide will give nearly complete formation of carbanions that are resonance-delocalized to two oxygens as in diethyl malonate anion. Ketone and ester enolates are usually prepared by using lithium diisopropylamide (LDA) This stronger base has little nucleophilicity because of the high steric hinderance. LDA will remove two protons from carboxylic acids to give *O*,α-dianions, which then react at the α position with electrophiles. The strongest bases, the alkyllithiums are also strongly nucleophilic and limited to use with carbanion precursors that have no electrophilic carbonyl groups.

The stronger bases in Fig. I are again formed by a redox reaction of a metal or by electrochemical reduction. Thus in this second method, carbanions are again prepared in basic solution with a reducing agent, but a proton transfer step is included.

A third method is metal–halogen exchange. An aryl or vinyl halide may be treated with *tert*-butyllithium, which gives the aryl or vinyllithium and the *tert*-butyl halide:[3]

$$CH_2{=}CH{-}Br + t\text{-BuLi} \xrightarrow[-78C^\circ]{Et_2O} CH_2{=}CH{-}Li + t\text{-BuBr} \quad (2)$$

The carbanions from any of these methods are usually combined with electrophiles immediately after they are prepared or even during their preparation. The four common kinds of electrophiles give alkylation, acylation, addition, and conjugate addition.

Simple primary or secondary alkyl chlorides, bromides, iodides, or tosylates will give alkylation. For the reaction in Eq. 3 a weak base was used

Et_3N	K_2CO_3	NaOH	$NaOC_2H_5$	$KOC(CH_3)_3$	LiN*i*-Pr_2	*n*-BuLi	*t*-BuLi
Weakest							Strongest

Figure I. Some bases suitable for preparing carbanions from their conjugate acids.

to generate a small concentration of the resonance delocalized enolate anion.[5] In this case the electrophile, methyl iodide, is present from the start. It approached the carbanion from the least hindered side to give the diastereomer shown. The alkylation of simple Grignard and organolithium compounds requires copper catalysis.

$$+ K_2CO_3 + CH_3I \xrightarrow[\text{reflux}]{\text{acetone}} \qquad (3)$$

97%

The cyanide ion is only weakly basic and is thus available as salts which can be used in water solution. It can be alkylated most readily under phase-transfer conditions by using a quaternary ammonium catalyst (Eq. 4).[6] The alkylations also proceed well under homogeneous conditions in DMSO.

$$+ \text{KCN} \xrightarrow[\text{benzene–water reflux}]{R_4N^+Cl^-} \quad + \text{KOTs} \qquad (4)$$

Trimethylsilyl ethers may serve as electrophiles and can be prepared *in situ* from alcohols. Heating an alcohol at 65°C with sodium cyanide, trimethylchlorosilane, and a catalytic amount of sodium iodide in acetonitrile–DMF gives the nitriles in a single operation.[7] Good yields are obtained with primary, secondary, and tertiary alcohols, and inversion of configuration has been demonstrated in a secondary case.

Carbanions will add to ketones and aldehydes to give alcohols (Eq. 5).[3] Some enolate anions will add reversibly and give only partial conversions.

$$\text{HC}{\equiv}\text{CLi} + \xrightarrow{-78°\text{C to rt}} \xrightarrow{NH_4Cl} \qquad (5)$$

70%

However, the addition of a chelating Zn^{2+} salt will favor the product and will also suppress competing polycondensation and dehydration:[8]

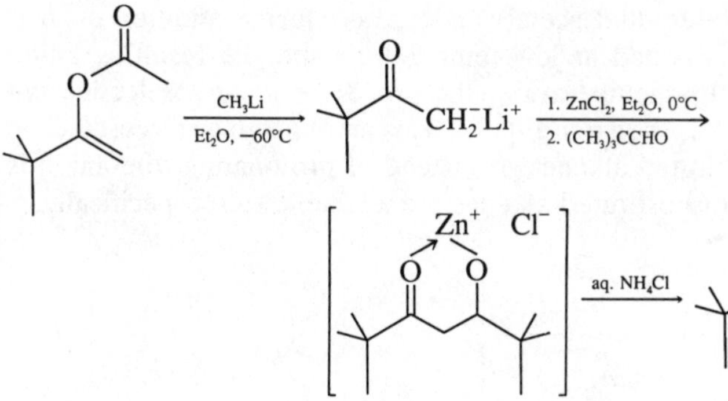

$\xrightarrow[\text{Et}_2\text{O}, -60°\text{C}]{\text{CH}_3\text{Li}}$ $CH_2^-Li^+$ $\xrightarrow[\text{2. (CH}_3)_3\text{CCHO}]{\text{1. ZnCl}_2, \text{Et}_2\text{O, 0°C}}$

Zn^+ Cl^- O O $\xrightarrow{\text{aq. NH}_4\text{Cl}}$ O OH (6)

82%

Organocuprates and highly resonance delocalized carbanions will attack the β position of α,β-unsaturated ketones or esters to give an enolate anion (Eqs. 7 and 8).[9,10] This may become protonated to give a saturated ketone or tricarbonyl compound and is called *conjugate addition* or *Michael addition.*

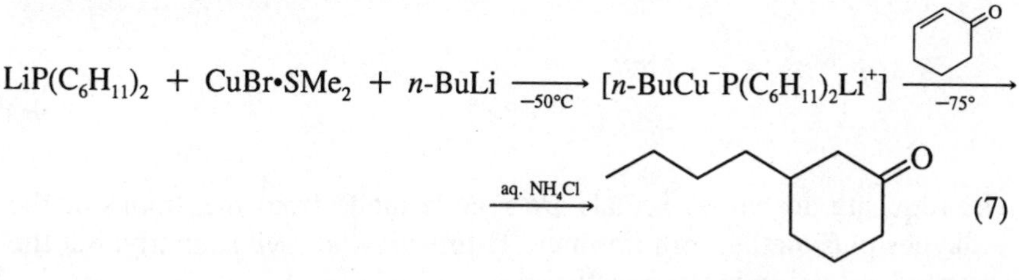

$$LiP(C_6H_{11})_2 + CuBr{\bullet}SMe_2 + n\text{-BuLi} \xrightarrow[-50°C]{} [n\text{-BuCu}^-P(C_6H_{11})_2Li^+] \xrightarrow[-75°]{}$$

$\xrightarrow{\text{aq. NH}_4\text{Cl}}$ (7)

83%

Lithium dibutyl cuprate gives conjugate addition, but only one of the butyl groups is transferred and also a large excess is required. The reagent in Eq. 7 is more efficient since an excess is not necessary and only one butyl group is needed in the reagent. This is particularly significant for more valuable organolithium starting materials.

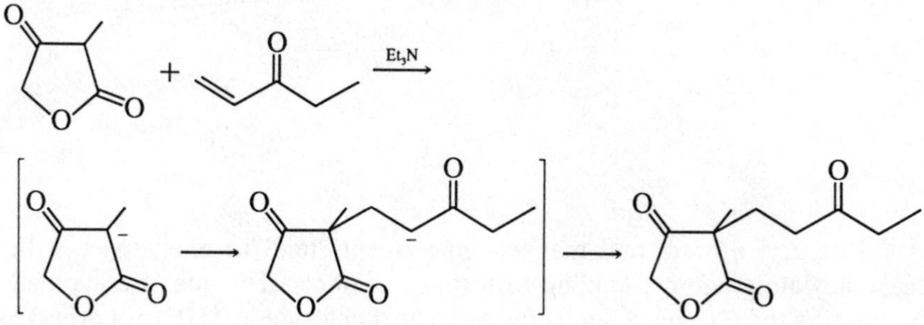

$\xrightarrow{\text{Et}_3\text{N}}$

98%

(8)

α,β-Acetylenic esters and acetals undergo conjugate addition also. If an organocuprate is added at low temperature and the resulting anion protonated also at low temperature, the process is a stereoselective syn addition of R and H as shown in Eq. 9.[11] This amounts to a stereoselective synthesis of trisubstituted alkenes. If instead of protonating, the anion is alkylated, even tetrasubstituted alkenes are available stereospecifically.

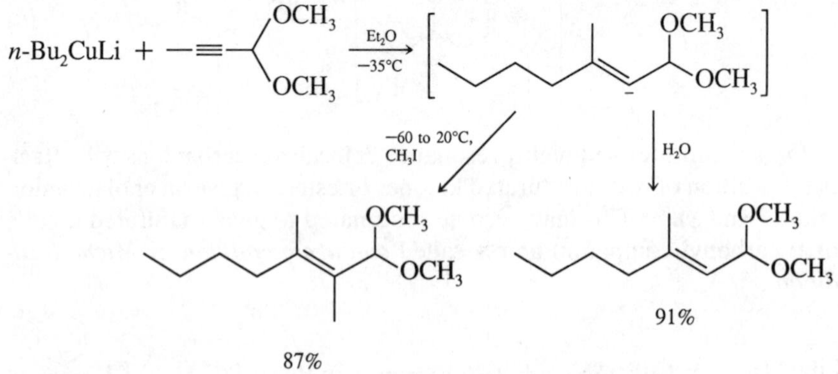

n-Bu_2CuLi + [acetylenic acetal with OCH_3, OCH_3] $\xrightarrow[-35°C]{Et_2O}$ [vinyl anion intermediate with OCH_3, OCH_3]

−60 to 20°C, CH_3I → tetrasubstituted alkene acetal (OCH_3, OCH_3), 87%

H_2O → trisubstituted alkene acetal (OCH_3, OCH_3), 91%

(9)

The requisite acetylenic acetals are readily made from the anions of the 1-alkynes plus methyl orthoformate. Esters may be used similarly, but the intermediate anion is less reactive.[12]

Acid chlorides, anhydrides, and esters react with carbanions to acylate them, affording ketones. An intramolecular example is shown in Eq. 10.[13] The base gave the lactone enolate which was then acylated by the methyl ester.

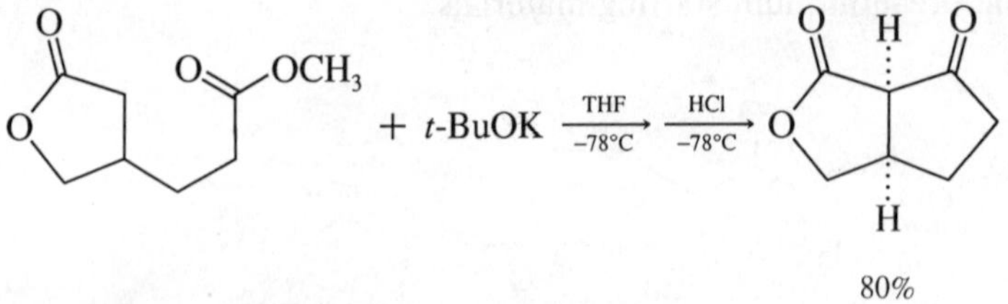

[lactone with $CH_2CH_2CO_2CH_3$ side chain] + t-BuOK $\xrightarrow[-78°C]{THF}$ $\xrightarrow[-78°C]{HCl}$ [bicyclic keto lactone, cis H, H] (10)

80%

Alkyl or aryl lithium and magnesium reagents tend to react twice with these acylating agents, leading to tertiary alcohols. This may be avoided by treating the Grignard reagent with an acid chloride in THF (not ether)[14] at −78°C or by converting the organolithium reagent to the cuprate and then treating with the acid chloride at −78°C. The less reactive nitriles or tertiary amides will also acylate the Grignard reagents:[15]

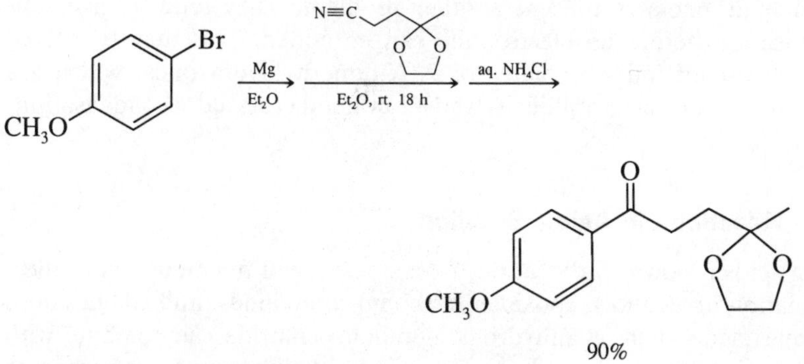

(11)

90%

In each of the above carbon–carbon bond-forming reactions, part of the driving force is the formation of a less basic anion in place of the carbanion, that is, a Cl^- or Br^- from alkylation and acylation, an alkoxide from addition or acylation, or an enolate from conjugate addition.

Ketones can present a problem in specificity. Under basic conditions they may react with two or more molecules of the electrophile to give a mixture of products. Furthermore, unsymmetrical ketones may present a choice of two enolate sites so that control is necessary to direct to the desired one. Many alternatives have been developed for this problem. One solution is to incorporate a temporary group on one enolate site to render that site more acidic so that the electrophile will react there. The familiar β-ketoester reactions (acetoacetic ester synthesis) are widely used. For another alternative, the ketone is first converted to an imine (Section 4.2.3) or a dimethyl hydrazone and the enolate of that derivative is used with electrophiles. The enolate forms selectively on the least substituted α carbon and also gives selective monoalkylation or addition. In Eq. 12 we see an example of a directed aldol condensation.[16]

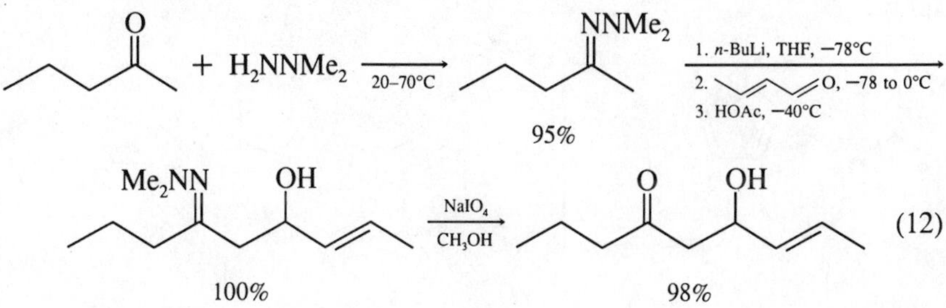

(12)

Alkylation of unsymmetrical ketones without derivatizations will give mostly reaction at the more substituted enolate site under reversible deprotonating conditions.

Aldehyde enolates present another problem. They tend to give self-condensation before an electrophile can be added. This may be solved again by use of imine enolates or *N,N*-dimethylhydrazones, which are themselves of low electrophilicity and allow good cross aldol condensations and alkylations.

5.1.2 Reactions in Acidic Solution

Strong acids produce carbocations from a variety of functional molecules. Protonation of alcohols, epoxides, carbonyl compounds, and alkenes does so. Lewis acids such as anhydrous aluminum chloride can combine with the foregoing substrates and can also remove halide ions from carbon to give carbocations. Diazotization of primary amines in acid solution is another source.

The carbocations are transient intermediates, generated in the presence of alkene or arene nucleophiles to give carbon–carbon bond formation. This gives a new carbocation requiring a second step that may be deprotonation, bonding to an oxygen or a halide ion, loss of a silyl group, or abstraction of a hydride. Carbocations may react with alkenes and dienes to give new carbocations that may do likewise repetitively to give polymers.[17] Some varieties of synthetic rubber are produced in this way. Although σ bonding electrons are more tightly held than π electrons, they will react with carbocations particularly when they are arranged close by for intramolecular (rearrangement) processes. Migration of an atom or group with the σ bonding pair from an adjacent carbon to the initial carbocationic site will occur if the new carbocation has greater stability (delocalization) from electron-donating alkyl groups, adjacent nonbonding electron pairs, or resonance to allylic sites, or if strain energy is released.

The example in Eq. 3 shows the formation of a carbocation from an epoxide, reaction with an alkene, and finally aromatic substitution on a furan.[18]

$ZnI_2 \cdot Et_2O$, CH_2Cl_2 → [$I_2\bar{Z}nO$ … $\oplus$] →

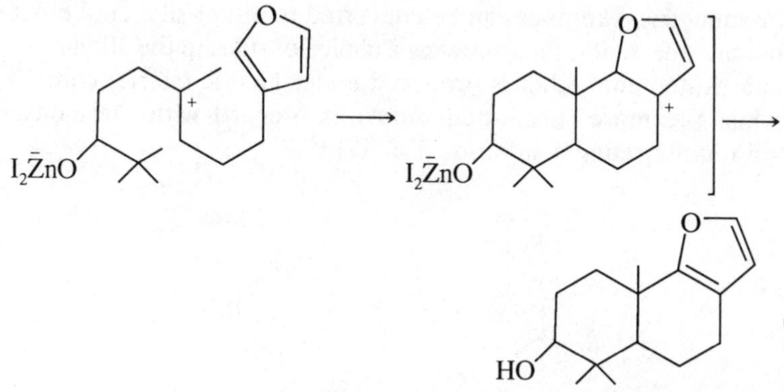

$I_2\bar{Z}nO$ $\longrightarrow$ $I_2\bar{Z}nO$ $\longrightarrow$

HO

62%

(13)

Trimethylsilyl enol ethers are especially valuable nucleophiles toward carbocations.[19] After attachment of the carbocation the trimethylsilyl group is readily removed by a halide ion to afford a ketone (Eqs. 14 and 15)[20,21]:

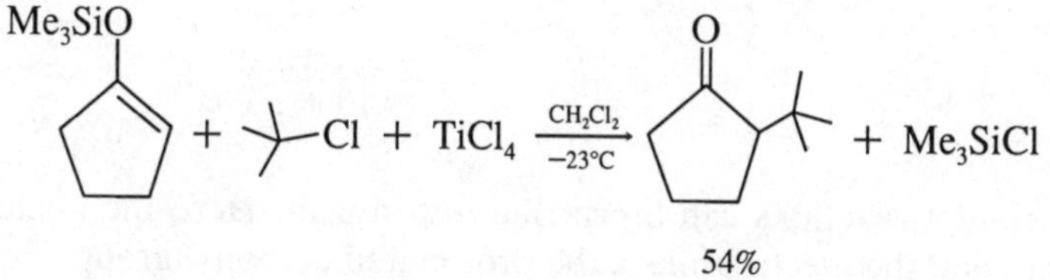

Me_3SiO + Cl + $TiCl_4$ $\xrightarrow[-23°C]{CH_2Cl_2}$ O + Me_3SiCl (14)

54%

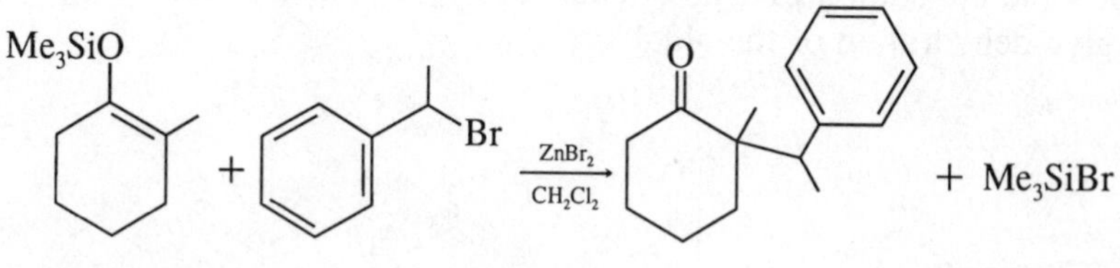

Me_3SiO + Br $\xrightarrow[CH_2Cl_2]{ZnBr_2}$ O + Me_3SiBr (15)

71%

This is analogous to the alkylation of ketone enolate anions but differs in several ways. Here a specific enol ether can be used, restricting the alkylation to one site and giving no dialkylation (which sometimes competes in enolate anion alkylation). Most significantly, it allows attachment of tertiary alkyl groups and others that would have given mostly elimination in basic solutions. The reaction in Eq. 14 required 1 equiv of Lewis acid while the benzylic case in Eq. 15 required only 0.02 equiv of catalyst. Even 2-methyl-2-*tert*-butylcyclohexanone can be prepared in this way in 48% yield.

Unsymmetrical ketones can be converted to either silyl enol ether with reasonable selectivity, thus allowing a choice of sides in the alkylation. The less substituted enol ether is produced under kinetic (steric) control (Eq. 16) while the more substituted isomer is prepard with thermodynamic control (equilibrating conditions, Eq. 17).[22]

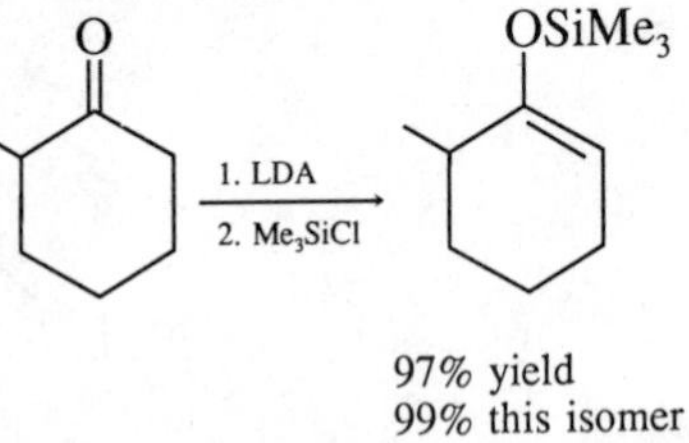

$$\text{2-methylcyclohexanone} \xrightarrow[\text{2. } Me_3SiCl]{\text{1. LDA}} \text{6-methyl-1-}(OSiMe_3)\text{cyclohexene} \quad (16)$$

97% yield
99% this isomer

$$\text{2-methylcyclohexanone} \xrightarrow[Me_3SiCl,\ \text{DMF},\ 130°C]{Et_3N} \text{2-methyl-1-}(OSiMe_3)\text{cyclohexene} \quad (17)$$

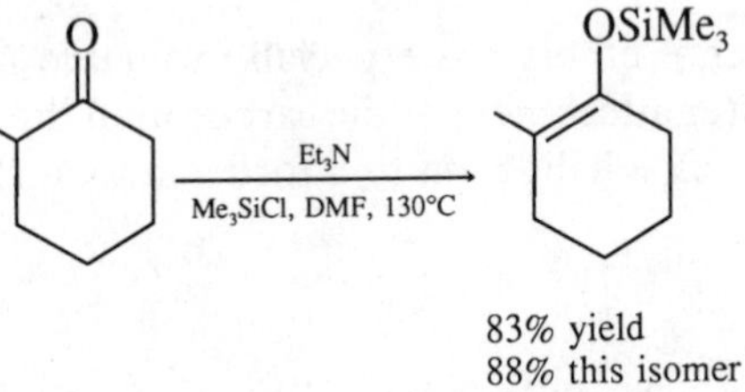

83% yield
88% this isomer

Some aldol reactions can be carried out in acid. Here the nucleophile is an enol and the electrophile is the protonated carbonyl group. Equation 18 shows the cyclization of a keto aldehyde.[23] The acidic conditions generally give dehydration of the aldol.

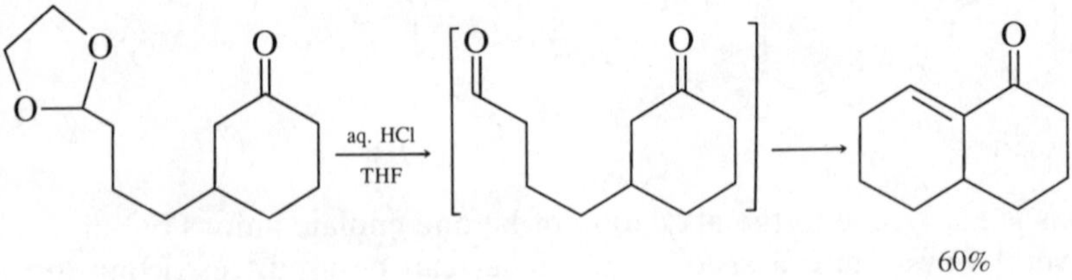

$$\text{3-[2-(1,3-dioxolan-2-yl)ethyl]cyclohexanone} \xrightarrow[\text{THF}]{\text{aq. HCl}} [\text{keto aldehyde}] \longrightarrow \text{bicyclic enone} \quad (18)$$

60%

Cross aldol condensations[24] between dissimilar ketones may be carried out under Lewis acid conditions using the silyl enol ether of that ketone intended as the nucleophile. This affords the aldols without dehydration or polycondensation:[25]

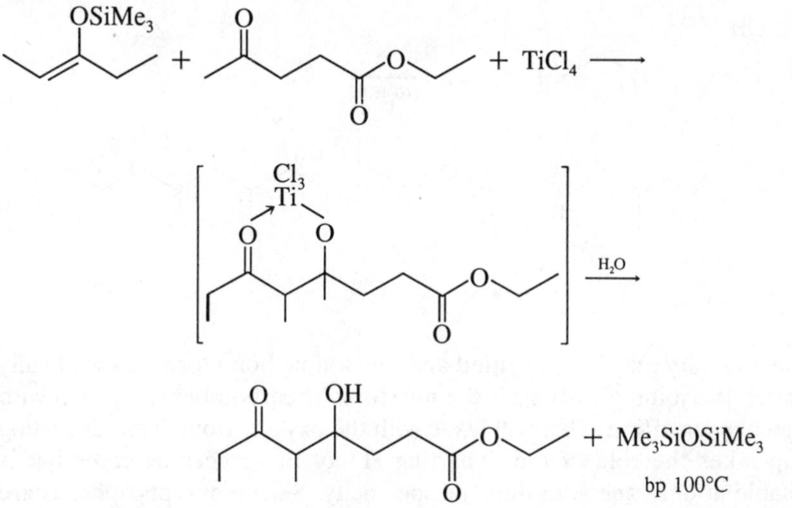

$OSiMe_3$ + O + $TiCl_4$ ⟶

Cl_3 Ti ... $\xrightarrow{H_2O}$

O OH + $Me_3SiOSiMe_3$ (19)

bp 100°C

86%

Simple ketones and 1,3-diketones give conjugate addition in acidic solution as shown in Eq. 20.[26] Here, too, the silyl enol ethers and $TiCl_4$ may be used.[27]

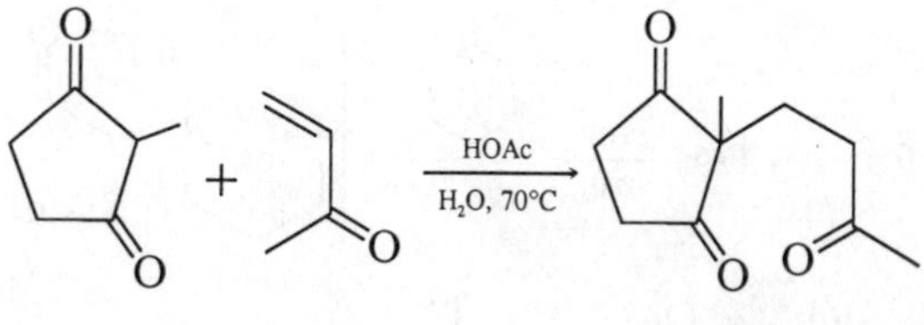

$\xrightarrow[H_2O,\ 70°C]{HOAc}$ (20)

100%

5.2 CARBON–CARBON DOUBLE BOND FORMATION

Two carbons may be brought together and joined by a double bond. Typically the electrophilic side will be a ketone or an aldehyde. If the nucleophilic side is simply an alkylmagnesium halide or an alkyllithium, the secondary or tertiary alcohol intermediate may be dehydrated, but in many cases there is a choice of sites for the double bond and it may arise elsewhere than between the newly joined carbons. Even the β-hydroxyesters from the Reformatsky reaction often give mixtures of alkenes:[28]

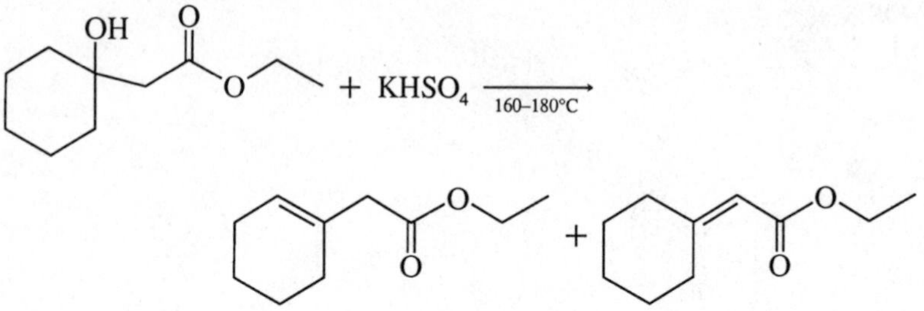

$$+ KHSO_4 \xrightarrow[160\text{–}180°C]{}$$

(21)

33% 27%

This variability may be prevented and the double bond formed specifically between the joining carbons if the nucleophilic carbon bears a group with a high oxygen affinity that will leave with the oxygen atom. That departing group takes the role of the departing H^+ of the preceding cases but is available at only one site; thus the specificity. Silicon and phosphorus are excellent in that role.

Trimethylsilylacetate esters may be converted to the enolate by treatment with lithium dialkylamide bases in THF at −78°C. These will add to ketones or aldehydes quickly at −78°C, followed by elimination of Me_3SiOLi and formation of α,β-unsaturated esters in high yields, uncontaminated by β,γ-unsaturated isomers:[29,30]

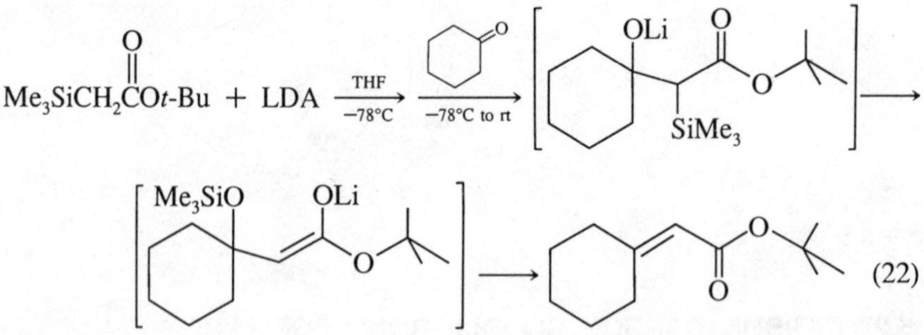

$$Me_3SiCH_2\overset{O}{\overset{\|}{C}}Ot\text{-}Bu + LDA \xrightarrow[-78°C]{THF} \xrightarrow[-78°C \text{ to rt}]{}$$

OLi, SiMe$_3$, Me$_3$SiO, OLi

(22)

90%

The reaction mixture is quenched with aqueous HCl, extracted, and distilled. The by-product hexamethyldisiloxane (bp 100°C) is easily removed. This is known as the *Peterson reaction.*

The requisite ethyl trimethylsilylacetate was made by the reaction of chlorotrimethylsilane, ethyl bromoacetate, and zinc.[31] It and the *tert*-butyl ester can also be made by treating ethyl or *tert*-butyl acetate with LDA followed by chlorotrimethylsilane. Esters of longer chain acids give mostly

O-silylation under these conditions, but diphenylmethylchlorosilane gives *C*-silylation selectively. The diphenylmethylsilylated esters give the Peterson reaction as well:[32]

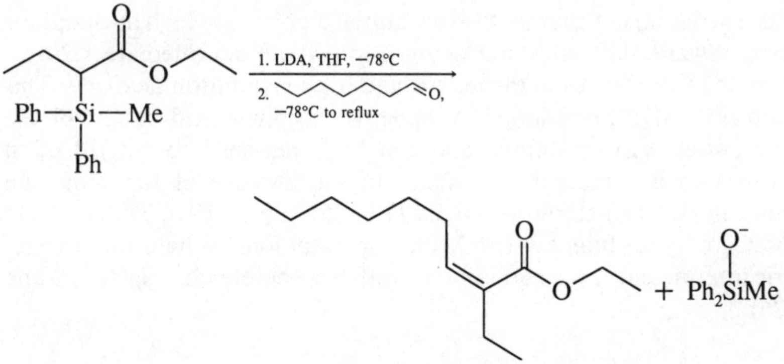

(23)

67% yield
78% *Z*, 22% *E*

Nonconjugated alkenes may be assembled by using a siloxide elimination, but the nucleophile is usually made in a different way since bases are unable to remove a proton α to a silicon without conjugative stabilization (unless it is a $SiCH_3$ site). Organolithium reagents will add to triphenylvinylsilane and may then be used with an aldehyde or ketone as exemplified by the synthesis of the alkene precursor of the sex pheromone of the gypsy moth:[33]

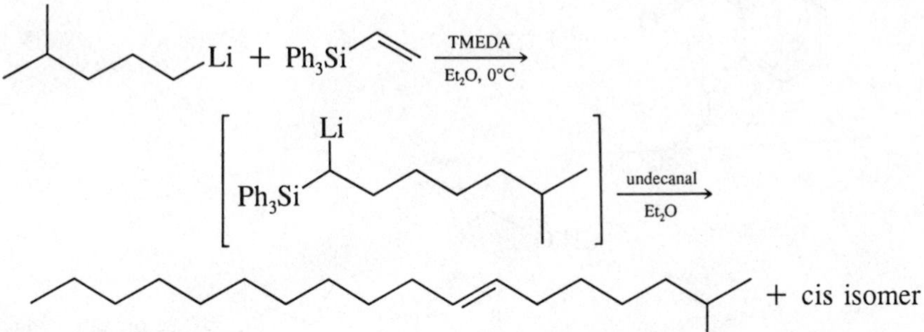

1 : 1
50% yield

(24)

Phosphorus has been used to a far greater extent for specific olefin synthesis.[34,35] Alkyl chlorides and bromides may be treated with triphenylphosphine to give quaternized salts. A base will remove a proton from a carbon α to the phosphorus to generate an ylide. The plus-charged phosphorus allows that proton removal in contrast to neutral silicon. Although the ylide carries no net charge, the substantial dipole gives high nucleophilic reactivity toward aldehydes and ketones to give an intermediate 1,2-oxaphosphetane that cleaves to the alkene and triphenylphosphine oxide. This is known as the *Wittig reaction*. The triphenyl phosphine oxide is nonvolatile and somewhat organic soluble and can be a nuisance to get rid of in comparison to hexamethyl disiloxane. In the absence of Li^+ ions, the reaction can give (*Z*)-stereoselectivity (Eq. 25)[36] up to 95%. The opposite stereoselectivity is obtained with Schlosser conditions[37] where the diastereomeric intermediates are equilibrated with base before cleavage to alkene (Eq. 26).[36]

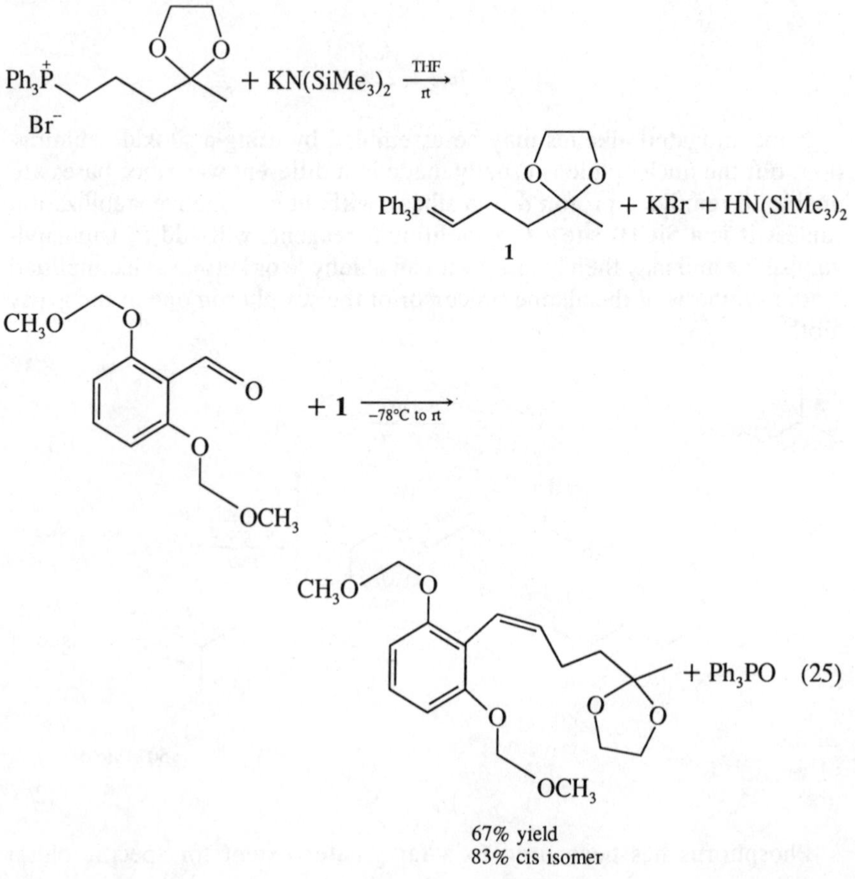

$Ph_3\overset{+}{P}$ Br^- + $KN(SiMe_3)_2$ $\xrightarrow[\text{rt}]{\text{THF}}$ Ph_3P (**1**) + KBr + $HN(SiMe_3)_2$

+ **1** $\xrightarrow{-78°C \text{ to rt}}$ + Ph_3PO (25)

67% yield
83% cis isomer

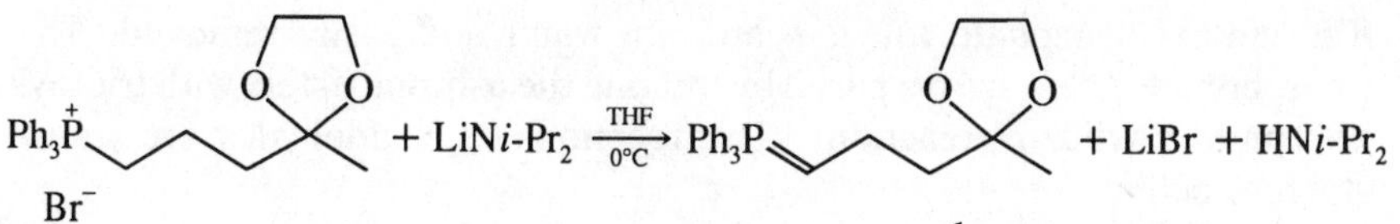

$Ph_3\overset{+}{P}$ Br^- + $LiNi\text{-}Pr_2$ $\xrightarrow[0°C]{THF}$ Ph_3P **1** + LiBr + $HNi\text{-}Pr_2$

CH_3O ... O ... OCH_3

1. 1, –78°C
2. LDA, –30°C
3. MeOH, –30°C to rt

CH_3O ... O ... OCH_3 (26)

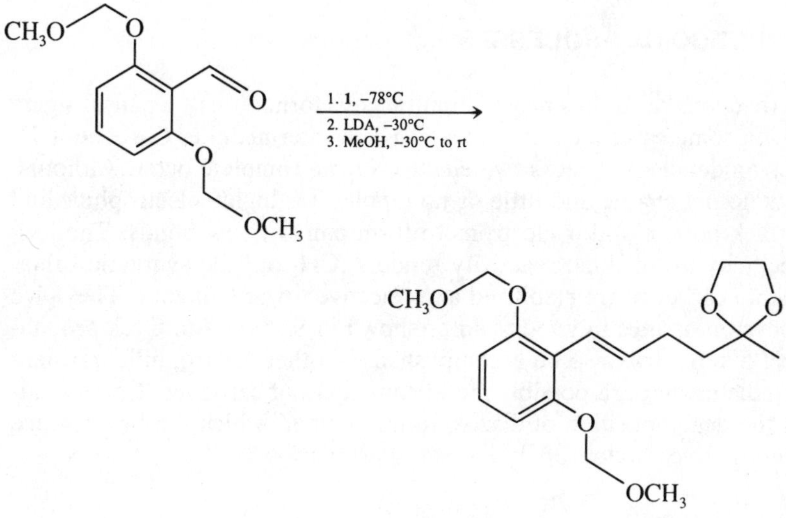

57% yield
95% trans isomer

α,β-Unsaturated esters can be prepared by means of the Horner–Wadsworth–Emmons modification of the Wittig reaction.[38] In this case the leaving group with oxygen affinity is a phosphonate, and the nucleophile is a net anion with resonance stabilization by the ester carbonyl group:[10]

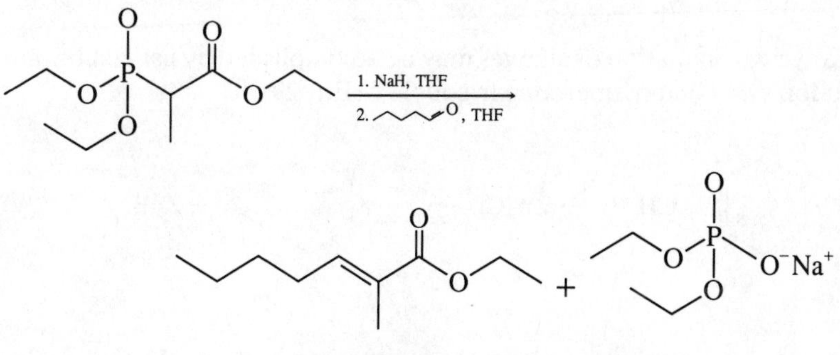

1. NaH, THF
2. O, THF

O⁻Na⁺ (27)

91% yield
all *E* isomer

The diethyl phosphate salt is soluble in water and easily removed. The phosphonate esters are prepared by treating the α-bromoesters with triethyl phosphite (Arbuzov reaction). The reactions with aldehydes are trans-stereoselective.

5.3 MULTIBOND PROCESSES

Many reactions result in a nearly simultaneous formation of a pair of sigma bonds. In some cases a carbene is a transient intermediate. Carbene $:CH_2$ is electron deficient; it lacks two electrons for a complete octet. Although there is no net charge and little or no dipole, it is highly electrophilic and will attack both π and σ electrons to form pairs of new bonds. The lack of specificity in this high reactivity renders $:CH_2$ of little synthetic value, but dihalo carbenes are stabilized and selective toward alkenes. They give dihalocyclopropanes in good yield as shown in Section 9.6. Cyclopropanation of alkenes can also be accomplished via other electrophilic transient intermediates that are possibly metal complexes of carbenes. Copper catalyzes the decomposition of diazoketones or esters which, in the presence of alkenes, gives cyclopropyl ketones or esters:[39]

CH_3O $+ N_2CHCOC_2H_5$ (with C=O) $\xrightarrow[140°C]{Cu}$ CH_3O ... $+ N_2$ (28)

57%

Simple cyclopropanation of alkenes may be accomplished by using dibromo or diiodomethane and zinc–copper couple[40] (Eq. 29):[41]

$+ CH_2Br_2 + Zn–Cu \xrightarrow[\text{sonication}]{\text{ether}} + ZnBr_2$ (29)

60%

Concerted reactions are commonly used to join carbons. For example, the Diels–Alder reaction is the formation of a cyclohexene from a diene and an alkene. Usually the alkene is rendered electrophilic by conjugation with a carbonyl group, and the diene may be rendered nucleophilic by electron-donating substituents. In the case shown in Eq. 30 the alkene is further electron depleted by association with a Lewis acid,[42] a common technique for accelerating Diels–Alder reactions.

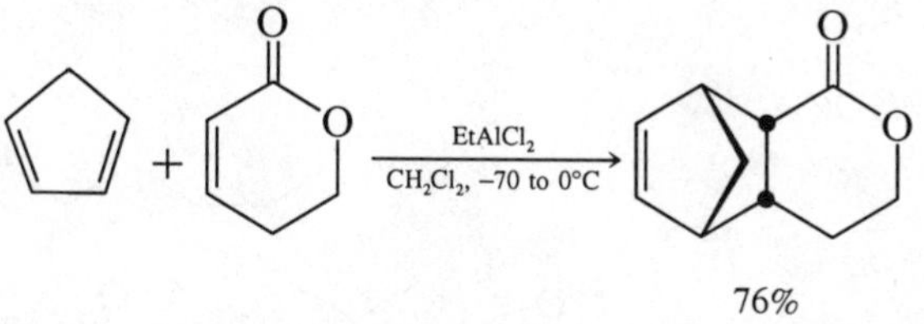

$EtAlCl_2$, CH_2Cl_2, −70 to 0°C

76%

(30)

It is important to point out here that the concerted reactions differ from the foregoing in that no carbanion or cation intermediate is involved, and in many cases the electrophilic and nucleophilic factors are not present, as in the very favorable dimerization of cyclopentadiene. These reactions are covered in more detail in Chapter 8.

PROBLEMS

Show how you would prepare each of the following products from the given starting materials. Where more than one step is required, show each step distinctly.

1.

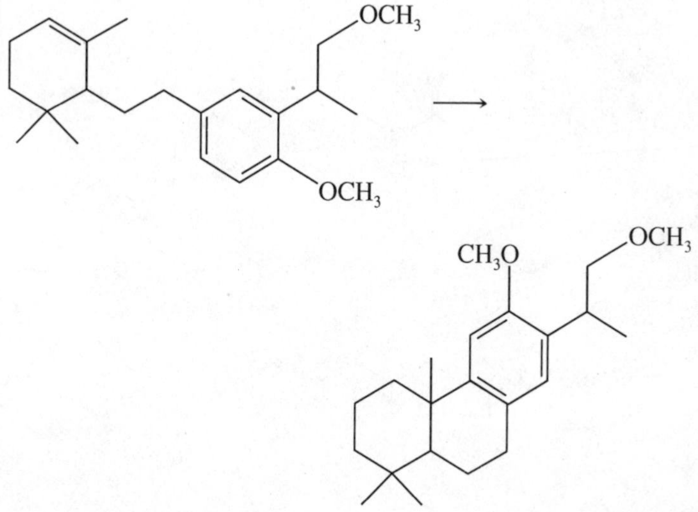

OCH_3

OCH_3

→

CH_3O

OCH_3

Ref. 43

2. 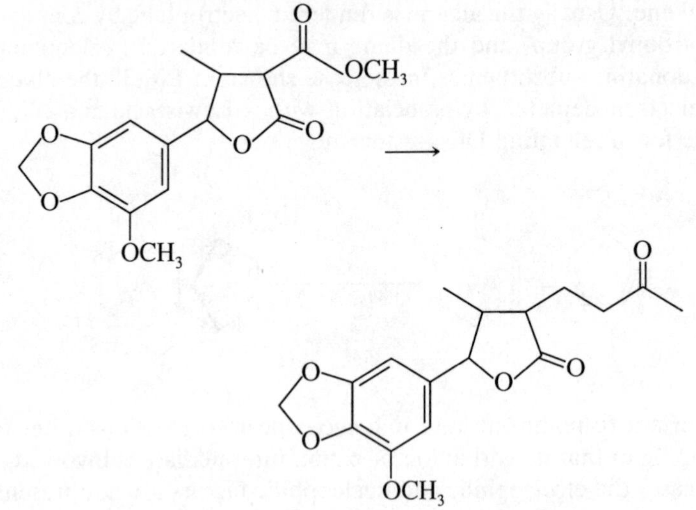

Ref. 44

3. 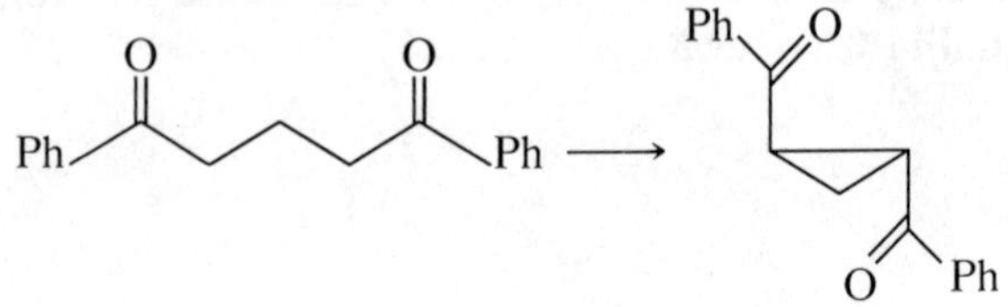 Ref. 45

4. 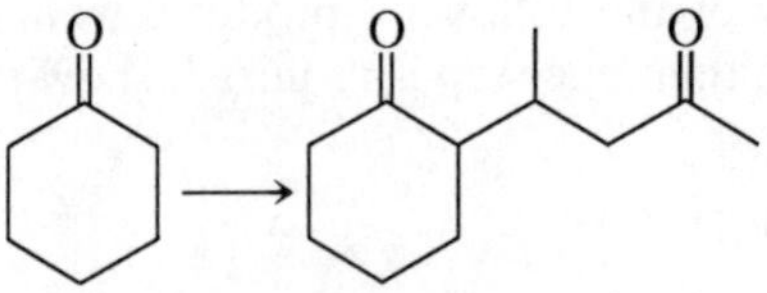 Ref. 46

5. 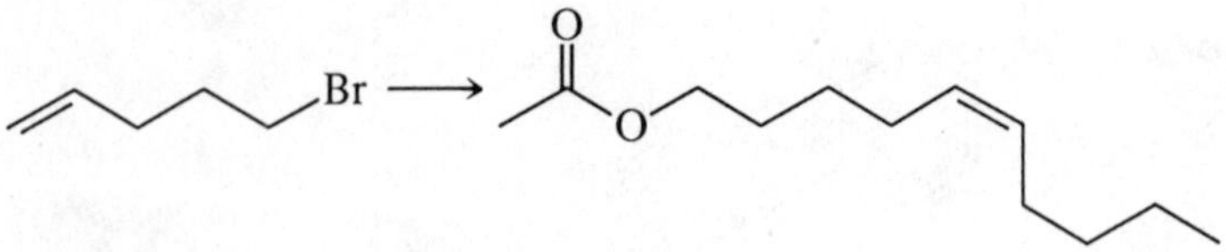 Ref. 47

6. 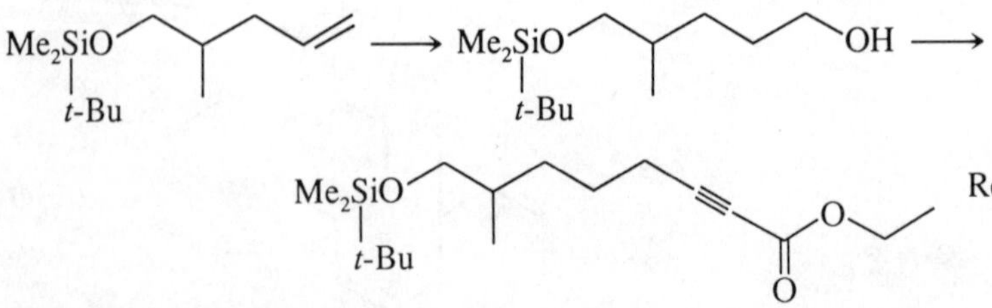

Ref. 48

7. CH_3O ⟶ CH_3O ⟶ CH_3O Ref. 49

8. n-$C_{11}H_{23}$ Br O O ⟶ O O $C_{11}H_{23}$-n Ref. 50

9. OCH_3 O OCH_3 ⟶ OCH_3 OCH_3 O O Ref. 51

10. O O ⟶ O O Br Br ⟶ O Ref. 52

11. → Ref. 53

OCH_3 → CH_3, OCH_3

12. CH_3O … CN → HO … O

Ref. 54

13. $CH_3C{\equiv}CCO_2C_2H_5$ → OH →

Ref. 55

14. OH →

O

Ref. 56

15. O → OH → O, OH

Ref. 57

16. $PhCCH_3$ (O) → Ph … OCH_3 (O, O) Ref. 27

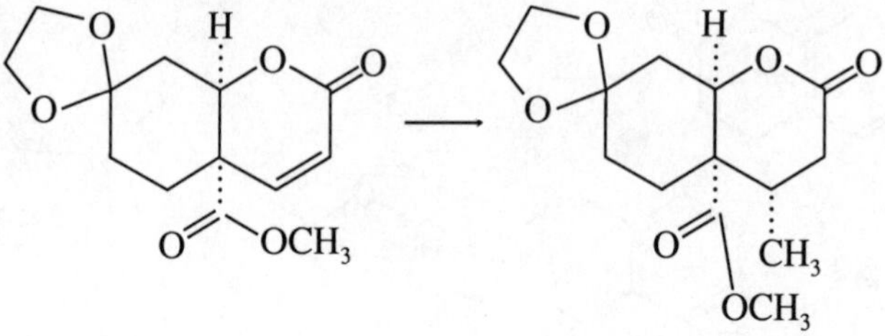

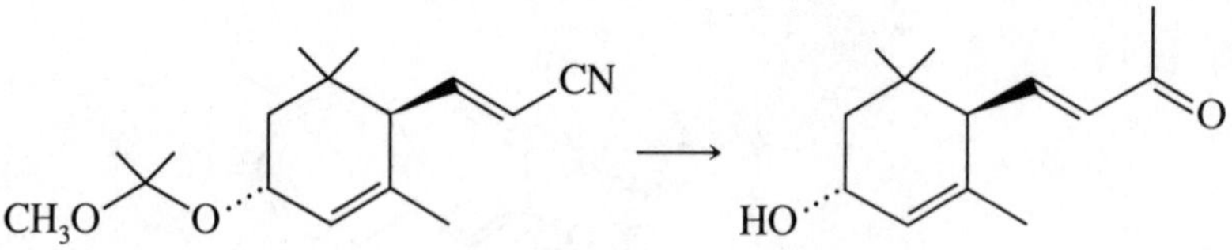

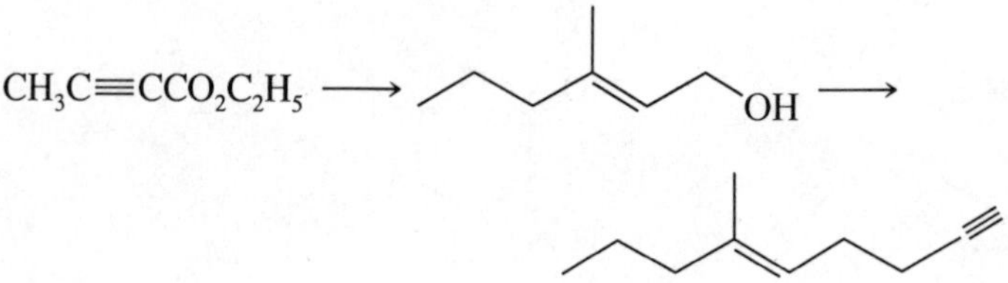

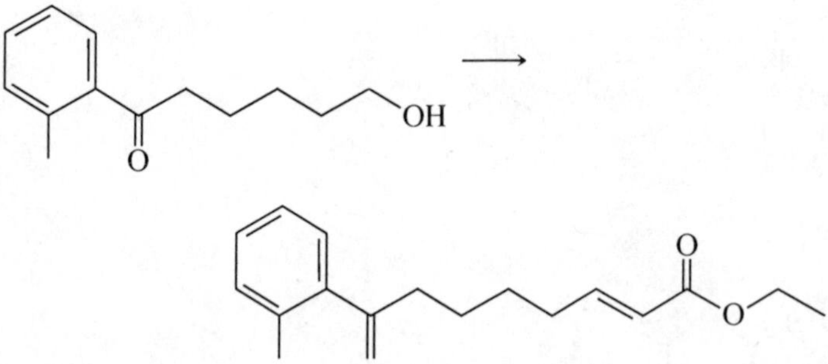

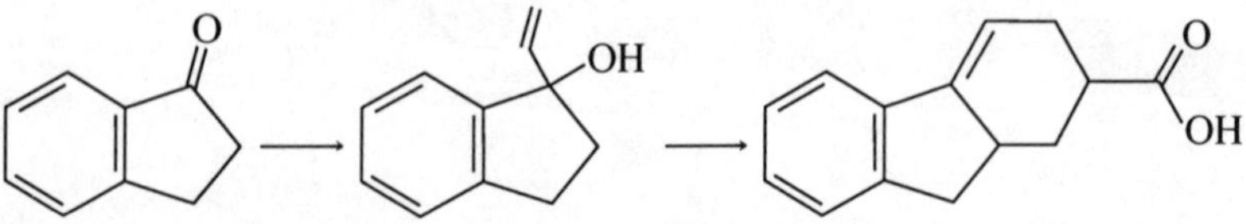

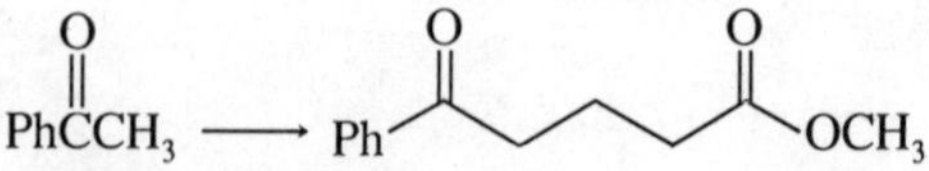

17. [structure: 4-heptanone] ⟶ [structure: 3-tert-butyl-4-heptanone] Ref. 20

REFERENCES

1. Augustine, R. L., Ed. *Carbon–Carbon Bond Formation,* Vol. 1, Marcel Dekker, New York, 1979.
2. Stowell, J. C. *Carbanions in Organic Synthesis,* Wiley-Interscience, New York, 1979.
3. Hecker, S. J.; Heathcock, C. H. *J. Org. Chem.* **1985,** *50,* 5159.
4. Beak, P.; Snieckus, V. *Acc. Chem. Res.,* **1982,** *15,* 306. Gschwend, H. W.; Rodriguez, H. R. *Org. React.* **1979,** *26,* 1.
5. Inokuchi, T.; Asanuma, G.; Torii, S. *J. Org. Chem.* **1982,** *47,* 4622.
6. Foos, J.; Steel, F.; Rizvi, S. Q. A.; Fraenkel, G. *J. Org. Chem.* **1979,** *44,* 2522.
7. Davis, R.; Untch, K. G. *J. Org. Chem.* **1981,** *46,* 2985.
8. House, H. O.; Crumrine, D. S.; Teranishi, A. Y.; Olmstead, H. D. *J. Am. Chem. Soc.* **1973,** *95,* 3310.
9. Bertz, S. H.; Dabbagh, G. *J. Org. Chem.* **1984,** *49,* 1119.
10. White, J. D.; Takabe, K.; Prisbylla, M. P. *J. Org. Chem.* **1985,** *50,* 5233.
11. Alexakis, A.; Commercon, A.; Coulentianos, C.; Normant, J. F. *Tetrahedron* **1984,** *40,* 715.
12. Corey, E. J.; Katzenellenbogen, J. A. *J. Am. Chem. Soc.* **1969,** *91,* 1851.
13. Boeckman, R. K., Jr.; Naegley, P. C.; Arthur, S. D. *J. Org. Chem.* **1980,** *45,* 752.
14. Sato, F.; Inoue, M.; Oguro, K.; Sato, M. *Tetrahedron Lett.* **1979,** 4303.
15. Martin, S. F.; Puckette, T. A.; Colapret, J. A. *J. Org. Chem.* **1979,** *44,* 3391.
16. Corey, E. J.; Enders, D. *Chem. Ber.* **1978,** *111,* 1337, 1362.
17. Kennedy, J. P. *Cationic Polymerization of Olefins: A Critical Inventory,* Wiley-Interscience, New York, 1975.
18. Tanis, S. P.; Herrinton, P. M. *J. Org. Chem.* **1983,** *48,* 4572.
19. Brownbridge, P. *Synthesis* **1983,** 1.
20. Chan, T. H.; Paterson, I.; Pinsonnault, J. *Tetrahedron Lett.* **1977,** 4183.
21. Paterson, I. *Tetrahedron Lett.* **1979,** 1519.
22. House, H. O.; Czuba, L. J.; Gall, M.; Olmstead, H. D. *J. Org. Chem.* **1969,** *34,* 2324. Paterson, I.; Fleming, I. *Tetrahedron Lett.* **1979,** 995.
23. Abbott, R. E.; Spencer, T. A. *J. Org. Chem.* **1980,** *45,* 5398.
24. Mukaiyama, T. *Org. React.* **1982,** *28,* 203–335.

25. Banno, K. *Bull. Chem. Soc. Jpn.* **1976,** *49,* 2284.
26. Hajos, Z. G.; Parrish, D. R. *Org. Synth.* **1985,** *63,* 26.
27. Narasaka, K.; Soai, K.; Aikawa, Y.; Mukaiyama, T. *Bull. Chem. Soc. Jpn.* **1976,** *49,* 779.
28. Kon, G. A.-R.; Nargund, K. S. *J. Chem. Soc.* **1932,** 2461.
29. Hartzell, S. L.; Sullivan, D. F.; Rathke, M. W. *Tetrahedron Lett.* **1974,** 1403.
30. Taguchi, H.; Katsuchi, S.; Yamamoto, H.; Nozaki, H. *Bull. Chem. Soc. Jpn.* **1974,** *47,* 2529.
31. Fessenden, R. J.; Fessenden, J. S. *J. Org. Chem.* **1967,** *32,* 3535.
32. Larson, G. L.; Fernandez de Keifer, C.; Seda, R.; Torres, L. E.; Ramirez, J. R. *J. Org. Chem.* **1984,** *49,* 3385.
33. Chan, T. H.; Chang, E. *J. Org. Chem.* **1974,** *39,* 3264.
34. Maercker, A. *Org. React.* **1965,** *14,* 270.
35. Bestmann, H. J.; Vostrowsky, O. *Topics Current Chem.* **1983,** *109,* 85.
36. Koreeda, M.; Hulin, B.; Yoshihara, M.; Townsend, C. A.; Christensen, S. B. *J. Org. Chem.* **1985,** *50,* 5426.
37. Schlosser, M.; Christman, K. F.; Piska, A. *Chem. Ber.* **1970,** *103,* 2814.
38. Wadsworth, W. S., Jr. *Org. React.* **1977,** *25,* 73.
39. Schiehser, G. A.; White, J. D. *J. Org. Chem.* **1980,** *45,* 1864.
40. Simmons, H. E.; Cairns, T. L., Vladuchick, S. A.; Hoiness, C. M. *Org. React.* **1973,** *20,* 1.
41. Friedrich, E. C.; Domek, J. M.; Pong, R. Y. *J. Org. Chem.* **1985,** *50,* 4640.
42. Ikeda, T.; Yue, S.; Hutchinson, C. R. *J. Org. Chem.* **1985,** *50,* 5193.
43. Matsumoto, T.; Imai, S.; Miuchi, S.; Sugibayashi, H. *Bull. Chem. Soc. Jpn.* **1985,** *58,* 340.
44. Matsumoto, T.; Imai, S.; Yamaguchi, T.; Morihira, M.; Murakami, M. *Bull. Chem. Soc. Jpn.* **1985,** *58,* 346.
45. Colon, I.; Griffin, G. W.; O'Connell, E. J., Jr. *Org. Synth.* **1972,** *52,* 33.
46. Huffman, J. W.; Potnis, S. M.; Satish, A. V. *J. Org. Chem.* **1985,** *50,* 4266.
47. Bestmann, H. J.; Vostrowski, O.; Koschatsky, K. H.; Platz, H.; Brosche, T.; Kantardjiew, I.; Rhinewald, M.; Knauf, W. *Angew. Chem. Internatl. Ed.* **1978,** *17,* 768.
48. Walba, D. M.; Stoudt, G. S. *J. Org. Chem.* **1983,** *48,* 5404.
49. McChesney, J. D.; Swanson, R. A. *J. Org. Chem.* **1982,** *47,* 5201.
50. Kikukawa, T.; Tai, A. *Chem. Lett.* **1984,** 1935.
51. Coburn, C. E.; Anderson, D. K.; Swenton, J. S. *J. Org. Chem.* **1983,** *48,* 1455.
52. Taylor, M. D.; Minaskanian, G.; Winzenberg, K. N.; Santone, P.; Smith, A. B., III. *J. Org. Chem.* **1982,** *47,* 3960.

53. White, J. D.; Matsui, T.; Thomas, J. A. *J. Org. Chem.* **1981,** *46,* 3376.
54. Mayer, H.; Ruttimann, A. *Helv. Chim. Acta* **1980,** *63,* 1451.
55. Bowlus, S. B.; Katzenellenbogen, J. A. *Tetrahedron Lett.* **1973,** 1277.
56. Hornback, J. M.; Barrows, R. D. *J. Org. Chem.* **1983,** *48,* 90.
57. Ranu, B. C.; Sarkar, M.; Chakraborti, P. C.; Ghatak, U. R. *J. Chem. Soc. Perkin Trans. I* **1982,** 865.

6

Planning Multistep Syntheses

The challenge in synthesis is to devise a set of reactions that will convert inexpensive, readily available materials into complex, valuable products. Ordinarily this is not an obvious following of a roadmap, but rather a complex puzzle requiring much strategy. This chapter gives samples of the planning process with actual syntheses of relatively simple cases. Enough of this planning procedure is provided to enable you to analyze and devise syntheses for many molecules. A more thorough treatment is given in an excellent book by Warren.[1]

6.1 RETROSYNTHETIC ANALYSIS

You should familiarize yourself with what sorts of compounds are readily available by perusing commercial catalogs, but the actual process begins at the end of the synthesis; that is, you must study the desired structure and work *backward.* What penultimate intermediate would be readily convertible to that product, and then what before that? This process is called *retrosynthetic analysis,* and each backward step is indicated by a double-shafted arrow ($\Rightarrow$). With this a backward scheme is drawn, and then a forward process is developed with actual reagents, indicated with ordinary arrows. In more complicated syntheses you will need to look ahead toward steps in the middle of the process, but still a backward approach is most practical.

The steps include functional group interconversions as given in Chapter

4 and carbon backbone construction as illustrated in Chapter 5. Viewed as the disassembly of the product, you should first disconnect the parts that are joined by functional groups; for example, esters should be separated to acid and alcohol parts. The carbon–carbon bonds should be disconnected at or near functional groups and at branch points in the backbone. There are often a great many choices of dividing points and starting materials. For example, jasmone and dihydrojasmone have been made by hundreds of routes.[2] In selecting among choices, the number of steps should be minimized, cheaper starting materials selected, high-yield reactions favored, and the scheme should converge instead of following a long linear sequence of steps. Sometimes a closely related molecule will be available, requiring a minimum of construction effort, as in making other steroids from cholesterol.

6.2 DISCONNECTION AT A FUNCTIONAL GROUP OR BRANCH POINT

Carbon–carbon bonds are frequently built by using carbonyl compounds. A carbonyl group normally confers a pattern of alternating potential electrophilic or nucleophilic reactivity along a carbon chain:[3]

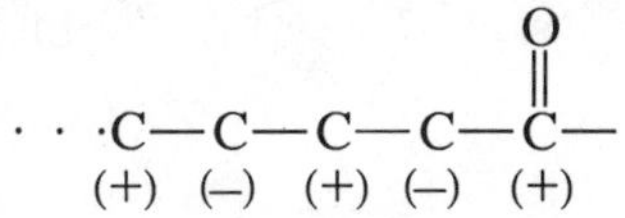

$$\cdots \underset{(+)}{C}-\underset{(-)}{C}-\underset{(+)}{C}-\underset{(-)}{C}-\underset{(+)}{\overset{O}{\overset{\|}{C}}}-$$

1

The electrophilic character exists in the carbonyl compounds themselves, continuing along the chain as far as conjugating *p*-orbitals are present to transmit it:

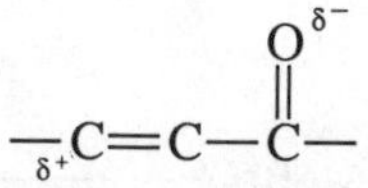

$$-\underset{\delta^+}{C}=C-\overset{O^{\delta-}}{\overset{\|}{C}}- \qquad -\overset{O^{\delta-}}{\overset{\|}{\underset{\delta^+}{C}}}-$$

2 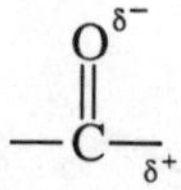 **3**

The nucleophilic character exists in the derived enolate (**4, 5**) or enol form.

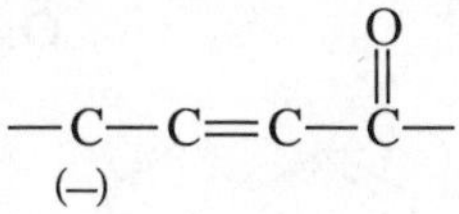

$$-\underset{(-)}{C}-C=C-\overset{O}{\overset{\|}{C}}- \qquad -\underset{(-)}{C}-\overset{O}{\overset{\|}{C}}-$$

4 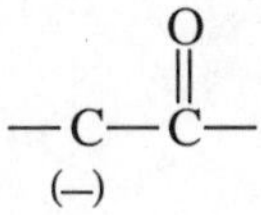 **5**

The charges at these sites serve to attract another carbon reagent of opposite charge and give a new bond.

If we consider a disconnection somewhere along the chain, we can decide whether the reactive site backed by the carbonyl group will be nucleophilic or electrophilic. Three different choices are taken in the following examples to illustrate the rationale.

Compound **6** is useful for the synthesis of hemlock alkaloids. The carbamate functional group is made from a chlorocarbonate and the amine; therefore, that disconnection is the first retro step (Scheme I).

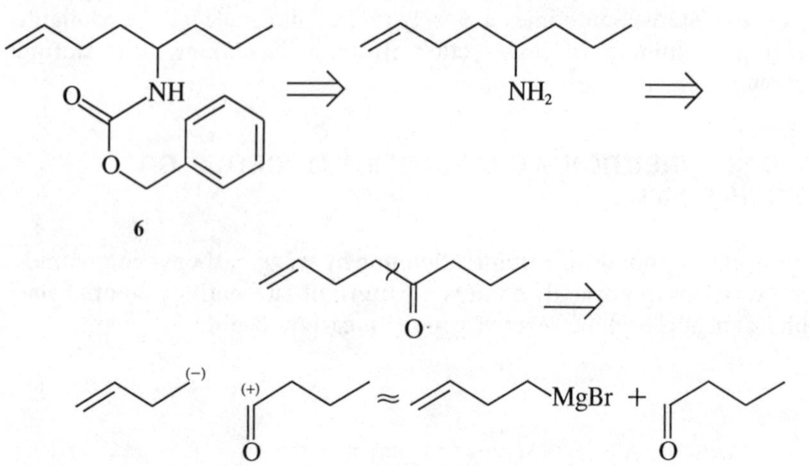

O NH ⟹ NH_2 ⟹

6

O ⟹

(−) (+) O ≈ MgBr + O

Scheme I

Since amines are often made from ketones, we go on to that key intermediate. Disconnection of the α carbon gives fragments with a (+) on the carbonyl (in accord with **1**) and requiring a (−) on the other reagent. We can now write a forward scheme with commercially available starting materials:[4]

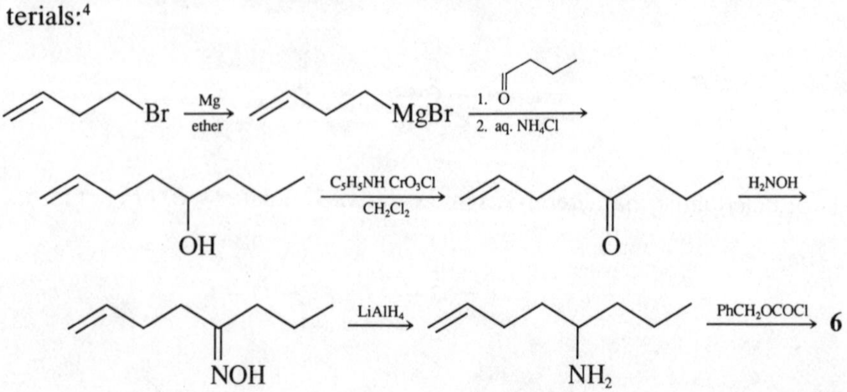

Br $\xrightarrow[\text{ether}]{\text{Mg}}$ MgBr $\xrightarrow[\text{2. aq. } NH_4Cl]{\text{1. O}}$

OH $\xrightarrow[CH_2Cl_2]{C_5H_5NH\ CrO_3Cl}$ O $\xrightarrow{H_2NOH}$

NOH $\xrightarrow{LiAlH_4}$ NH_2 $\xrightarrow{PhCH_2OCOCl}$ **6**

(1)

The 4-bromo-1-butene is available or can be prepared from the alcohol, which may, in turn, be prepared from vinylmagnesium chloride plus ethylene oxide. The Grignard reagent may be treated with an acylating agent, or as these authors chose, an aldehyde followed by an oxidizing agent. In this example the amine functional group suggested a carbonyl intermediate. This same retro step should be suggested by many other functional groups, including alcohols and halides.

A very similar ketone (**7**) was made with a disconnection between the α and β carbons, in fact on both sides (Scheme II).

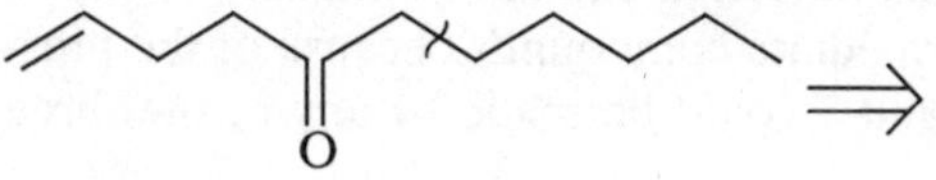

O

7

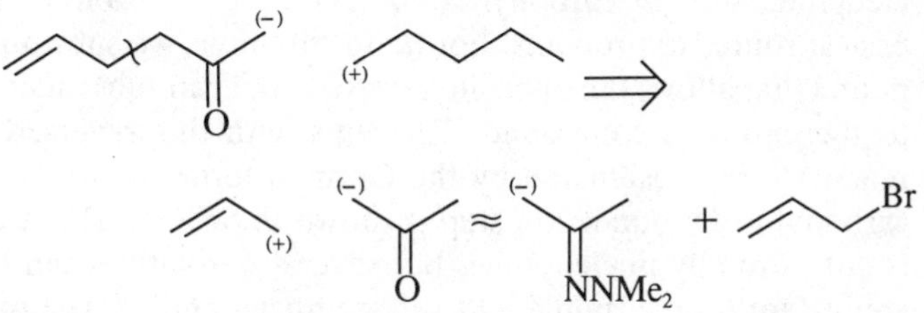

(–) (+) O

(–) (+) O ≈ (–) NNMe$_2$ + Br

Scheme II

In this disconnection the carbonyl fragment is the nucleophile, specifically an enolate of a hydrazone derivative (Section 5.1.1). Ths synthesis is shown in Eq. 2.[5]

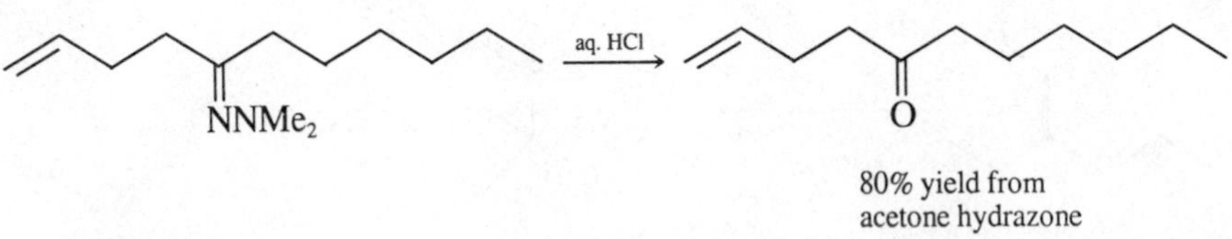

O + Me_2NNH_2 $\xrightarrow[\text{benzene}]{CF_3CO_2H}$ NNMe$_2$ 1. *n*-BuLi, THF, -5°C 2. I

NNMe$_2$ 1. *n*-BuLi, -5°C 2. Br

NNMe$_2$ aq. HCl O

80% yield from acetone hydrazone

(2)

Disconnection one atom further, that is, between the β and γ carbons, requires a conjugate addition as in the synthesis of **8** (Scheme III). The actual reaction is illustrated in Chapter 5, Eq. 7.

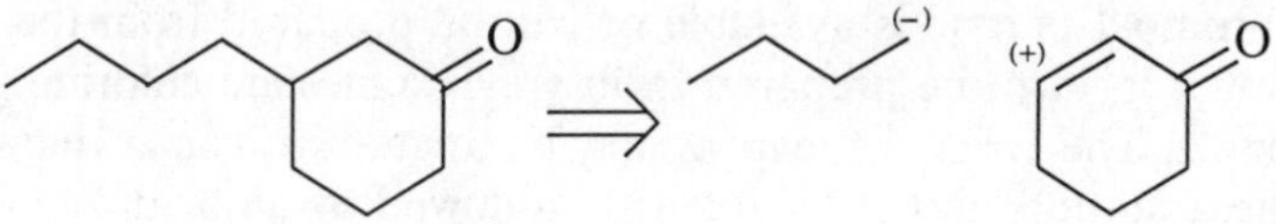

8

Scheme III

Which one of the three disconnections we select depends on other structural features of the particular product and on availability of materials. In the synthesis of **8** the other two disconnection choices would have required more steps and difunctional intermediate compounds because of the presence of the ring. Compounds **6** and **7** could be made by any of the three methods.

There are some circumstances where the normal electrophilic or nucleophilic sites in carbonyl compounds are unusable or do not give the easiest routes to products. For these situations we substitute another compound that allows the opposite reactivity but can subsequently be converted to the carbonyl compound. Reagents with this reversed or "abnormal" reactivity are designated by the German term *umpolung*.[6] The abnormal carbonyl α disconnection step is shown in Scheme IV. A carbonyl carbon is not normally nucleophilic, but several derivatives can be used that are equivalent to a carbonyl and yet are nucleophilic.[7] The dithioacetal monoxide anion **9** is an example. A preparation of 3-heptanone (Eq. 3) demonstrates the use of this reagent.[8]

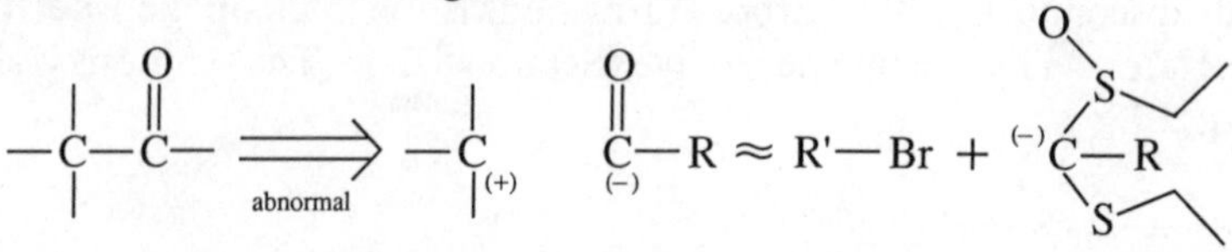

abnormal

C—R ≈ R'—Br + $^{(-)}$C—R

9

Scheme IV

Disconnecting one bond farther from the carbonyl, the α,β abnormal case is provided indirectly by the reaction of Grignard reagents with epoxides.

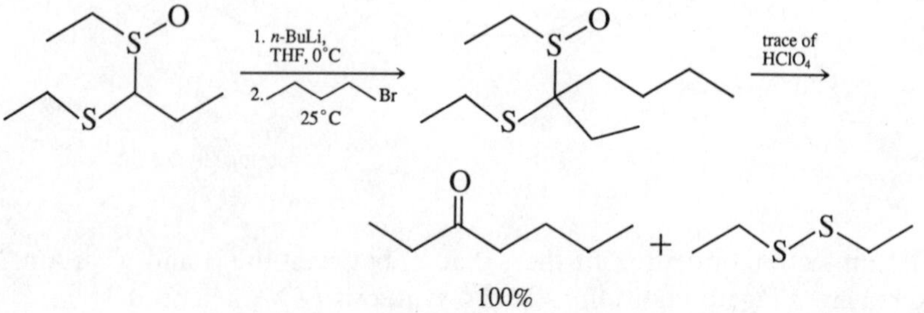

1. *n*-BuLi, THF, 0°C
2. Br, 25°C

trace of $HClO_4$

100%

(3)

(This is followed by oxidation if the carbonyl group is the desired functionality.) The β,γ disconnection with abnormal polarity is readily arranged if the α carbon is sp^3-hybridized and insulates the carbonyl group from the β carbon. In this case the carbonyl group must be protected from the nucleophilic β carbon:[9]

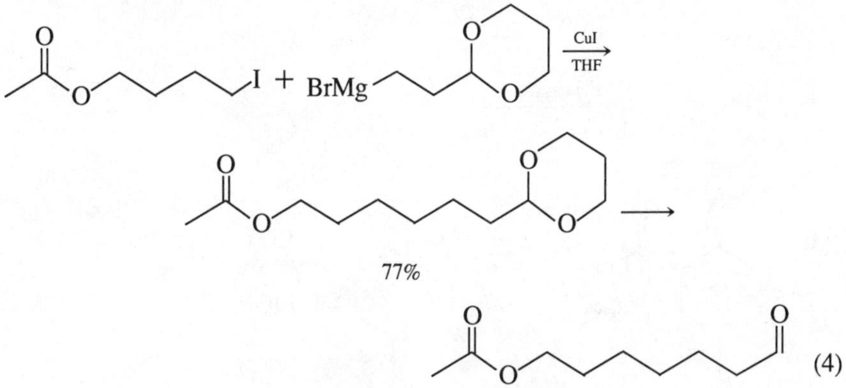

CuI THF

77%

(4)

Next we use these choices in some longer sequences. *exo*-Brevicomin (**10**) is a pheromone from the western pine beetle. Examining the functionality, we see a carbon attached to two oxygens, that is, an acetal derivative of a ketone. Dividing this functionality to the components, we find a ketodiol. 1,2-Diols are usually made from alkenes. Bellas and coworkers[10] chose to make the double bond by the Wittig reaction and to use an α,β disconnection on the ketone (Scheme V).

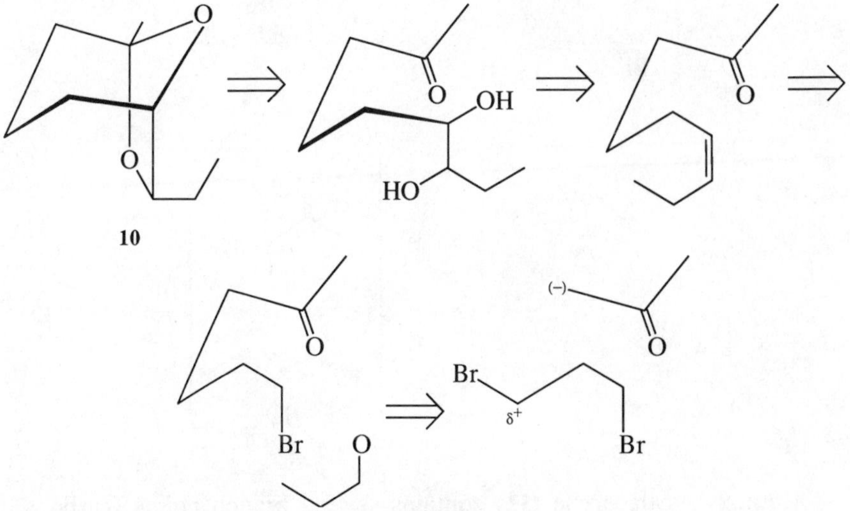

10

Scheme V

Of the various ways available for controlling monoalkylation of acetone, they chose to use the acetoacetic ester synthesis (Eq. 5). Notice also that the exo stereochemistry of the pheromone requires the threo diol. This could be made by syn glycolization of the trans alkene or by anti glycolization of the cis alkene. The latter was used here.

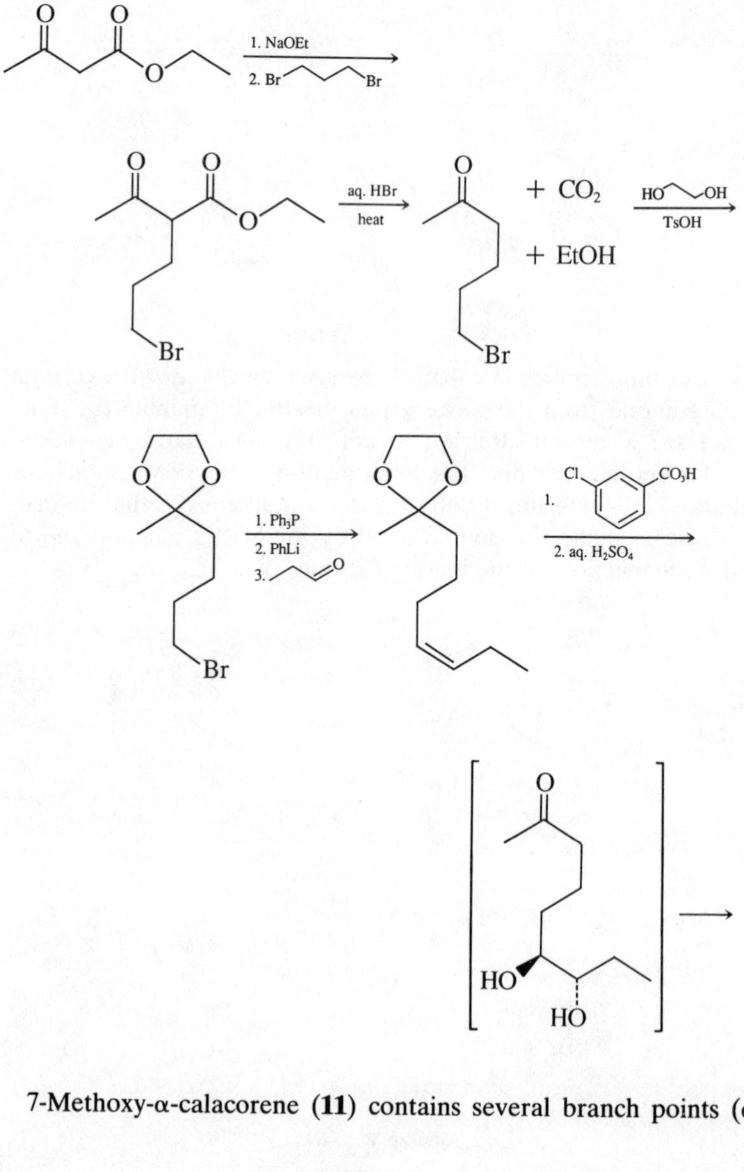

1. NaOEt
2. Br Br

aq. HBr, heat → + CO_2 + EtOH

HO OH, TsOH →

Br

1. Ph_3P
2. PhLi
3. O

1. Cl, CO_3H
2. aq. H_2SO_4

HO HO

→ **10** (5)

7-Methoxy-α-calacorene (**11**) contains several branch points (carbons with three other carbons directly attached) where we may consider disconnecting with the help of temporary functional groups. To aid in choosing a retro starting point, we should look over the whole structure and consider steps that may be required. This structure includes a benzene ring that has other alkyl carbons attached forming another ring. The Friedel–Crafts acylation reaction would be a way to assemble this, using the carbonyl groups to incorporate the methyl and isopropyl groups. The methoxy group is a strong ortho, para director; therefore, the attchment at the para position should be developed first. This idea is elaborated in Scheme VI.

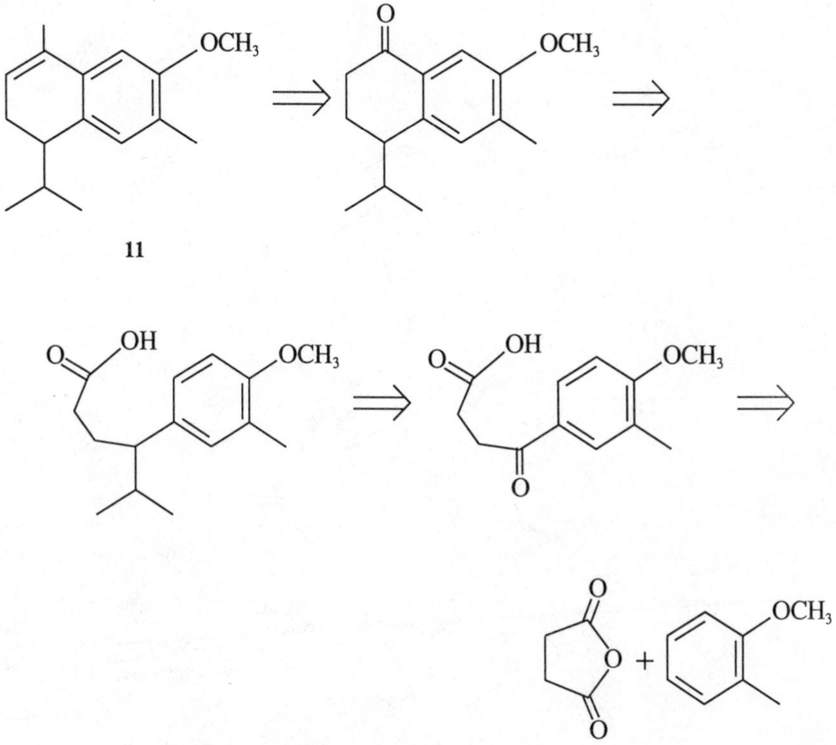

OCH$_3$ ⇒ O OCH$_3$ ⇒

11

O OH OCH$_3$ ⇒ O OH OCH$_3$ O ⇒

O O O + OCH$_3$

Scheme VI

Beginning with the readily available succinic anhydride and 2-methylanisole, the synthesis was carried out as in Eq. 6.[11] The ketoester intermediate is sufficiently more reactive at the keto group to give the desired product from the Grignard treatment.

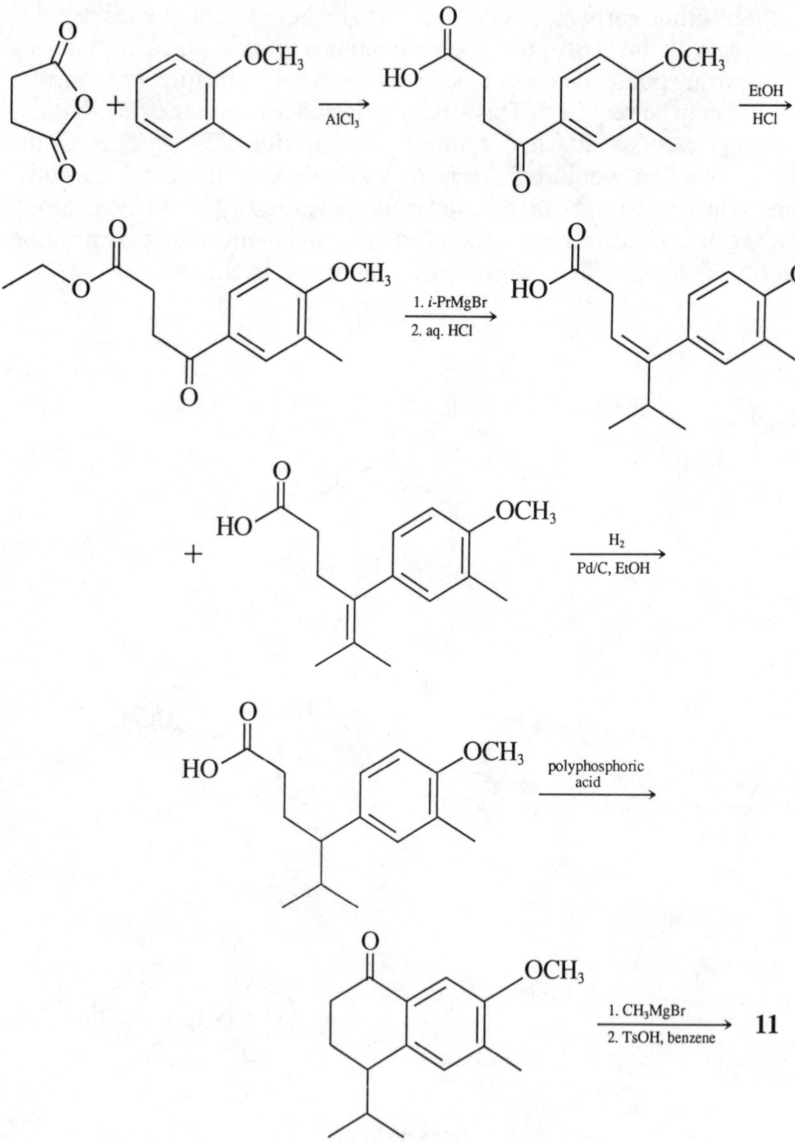

$$\xrightarrow{AlCl_3} \quad \xrightarrow[HCl]{EtOH}$$

$$\xrightarrow[2.\ aq.\ HCl]{1.\ i\text{-}PrMgBr}$$

$$+ \quad \xrightarrow[Pd/C,\ EtOH]{H_2}$$

$$\xrightarrow{\text{polyphosphoric acid}}$$

$$\xrightarrow[2.\ TsOH,\ benzene]{1.\ CH_3MgBr} \mathbf{11} \quad (6)$$

6.3 COOPERATION FOR DIFUNCTIONALITY

Molecules that contain two functional groups at particular distances apart are assembled considering the electronic influence of both groups together. As with the monofunctional compounds, consider carbonyl groups to be primary intermediates and examine their influence on the fragments from disconnection between the groups. Beginning with a 1,3-difunctional chain, we find fragments **12** and **13.**

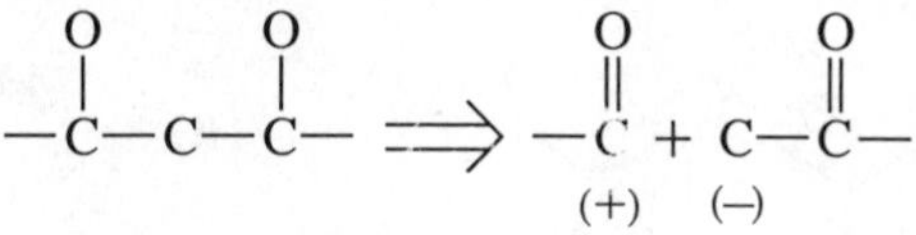

$$-\overset{\overset{\displaystyle O}{|}}{C}-C-\overset{\overset{\displaystyle O}{|}}{C}- \Longrightarrow \underset{(+)}{-\overset{\overset{\displaystyle O}{\|}}{C}} + \underset{(-)}{C}-\overset{\overset{\displaystyle O}{\|}}{C}-$$

12 **13**

Writing the normal charges expected in each fragment under the influence of the carbonyl group, we find opposite charges, which means that they would attract each other and this would be a favorable approach to the synthesis of 1,3-difunctional compounds. To be specific, consider the synthesis of **14.** This hydroxyester may be disconnected between the α and β carbons to give appropriate fragments (Scheme VII).

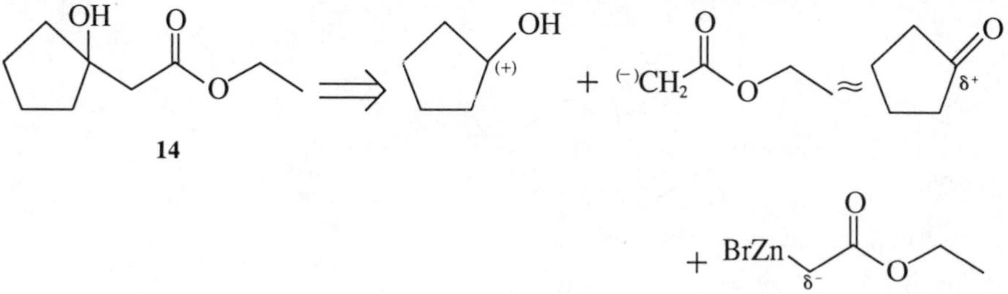

OH O OH (+) + $^{(-)}CH_2$ O O ≈ O δ+

14

+ BrZn O O δ−

Scheme VII

The actual reagents could be the ester enolate in the form of a Reformatsky reagent and the ketone. β-Hydroxycarbonyl compounds are often dehydrated; thus the α, β-, or β,γ-unsaturated compounds should be approached in planning by first rehydrating. The initial adduct could also be reduced or oxidized or further converted to other functional groups; therefore, compounds **15** and **16** would also be approached with the same intermediates and disconnection in mind.

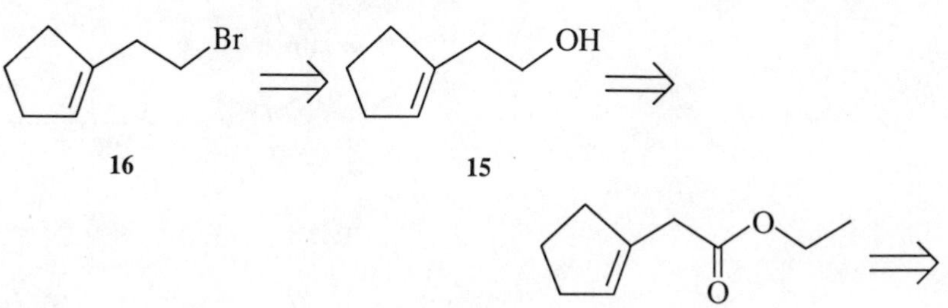

Br ⟹ OH ⟹

16 **15**

O O ⟹ **14**

The actual synthesis of **16** was carried out as[12]

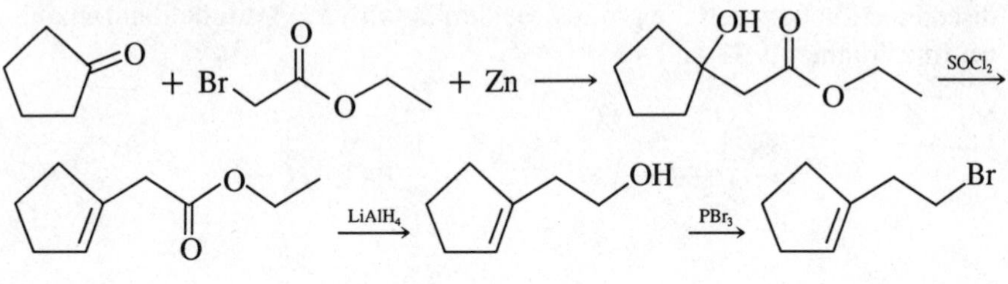

+ Br + Zn ⟶ OH O SOCl$_2$

O LiAlH$_4$ OH PBr$_3$ Br

O

16

(7)

Bromoalcohol **17** is 1,3-difunctional also. Maximum simplifications would result from disconnecting the C_3 alcohol fragment, but first a carbonyl function should probably be considered. This gives a β-hydroxyester, so again the Reformatsky reaction is suggested (Scheme VIII). Equation 8 shows the sequence used[13] to prepare **17.**

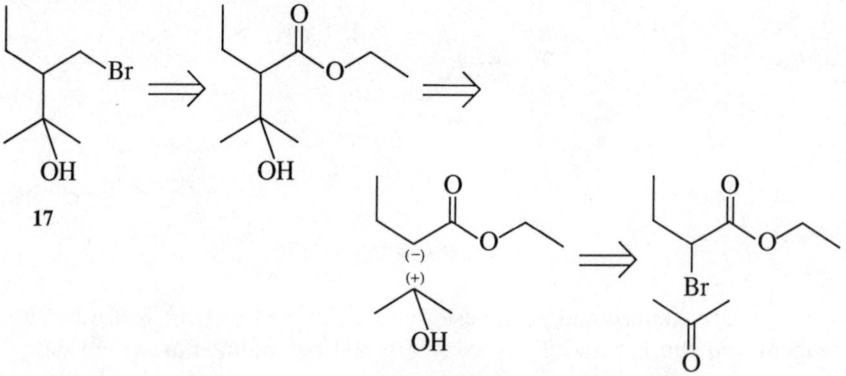

Br ⟹ O O ⟹

OH OH

17

O O

O ⟹ O

(−)

(+) Br

OH O

Scheme VIII

O O

O + + Zn $\xrightarrow[\text{benzene reflux}]{I_2}$ $\xrightarrow{\text{aq. } H_2SO_4}$

Br

O

O $\xrightarrow[\text{2. } H_2O]{\text{1. LiAlH}_4\text{, ether}}$

OH

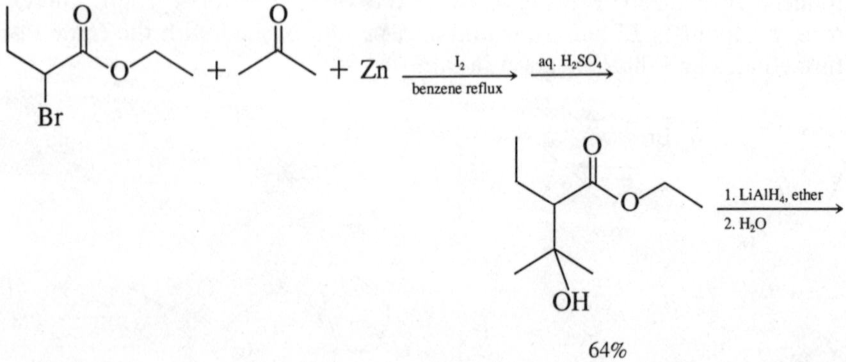

64%

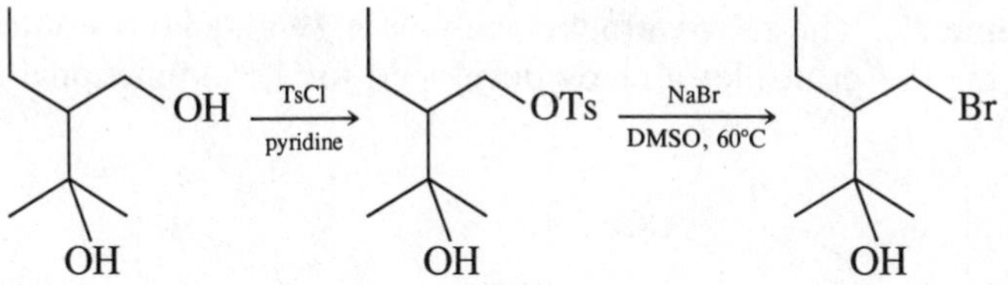

(8)

100% 86% **17**

Turning next to a 1,4-difunctional chain, we find that the disconnections do not give mutually attracting fragments (Scheme IX).

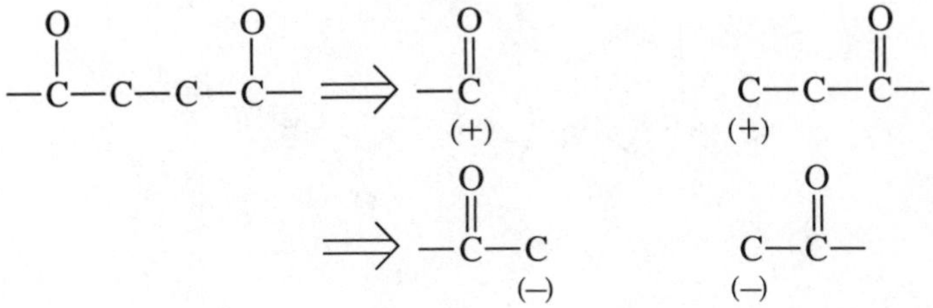

Scheme IX

It is thus necessary to use an *umpolung* (reverse polarity) reagent in place of one of the usual components. A nitro group on a carbon facilitates the removal of a proton from that carbon, giving a nucleophilic nitronate anion. Later the nitro group and carbon may be converted to a carbonyl group (which itself would have been electrophilic). 2,5-Heptanedione (**18**) was prepared in this way:[14]

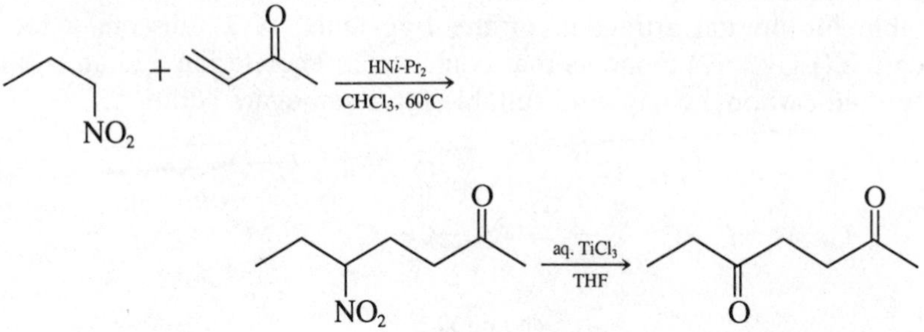

(9)

85%

18

The 2,3-disconnection of 1,4-difunctional molecules is also of value. γ-Ketoester **19** disconnects to a pair of mutually unattractive enolate ions, but one may be rendered electrophilic by placing a bromine on the α carbon

(Scheme X). The retrosynthetic analysis of **19** may be continued to **20** by applying the principles already developed for 1,3-difunctional cases.

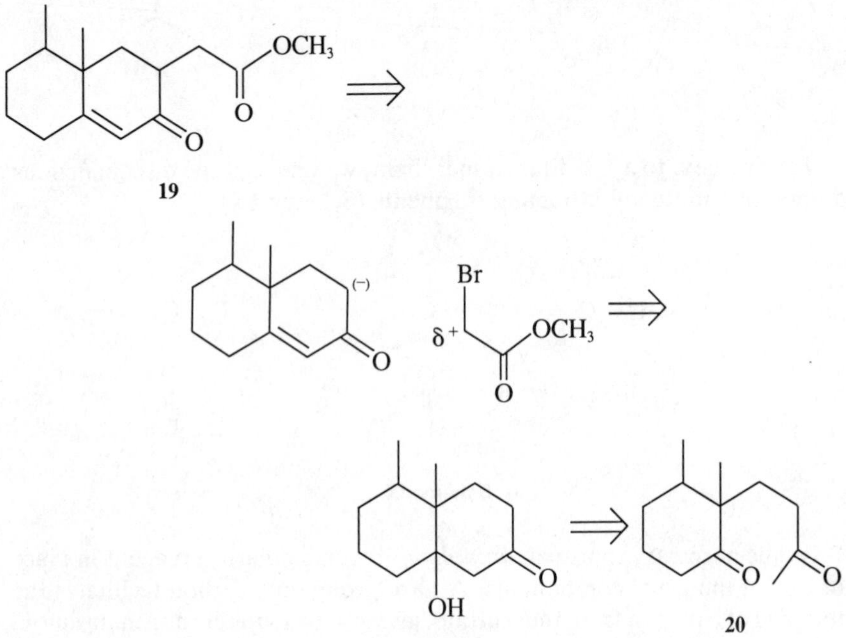

OCH_3 O O $\Longrightarrow$

19

(−) Br δ^+ OCH_3 O O $\Longrightarrow$

OH O $\Longrightarrow$ O O

20

Scheme X

Disconnections of 1,5-difunctional chains give normal charges that are favorable for mutual attraction of the fragments. A 2,3-disconnection (Scheme XI) gives components that could be an enolate ion and an α,β-unsaturated carbonyl compound suitable for a conjugate addition.

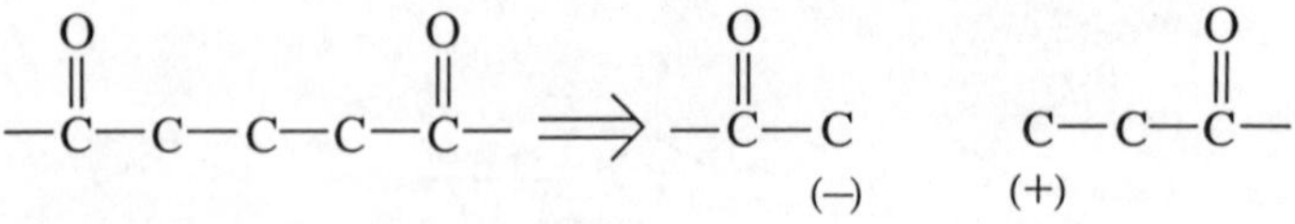

$$-\overset{O}{\overset{\|}{C}}-C-C-C-\overset{O}{\overset{\|}{C}}- \Longrightarrow -\overset{O}{\overset{\|}{C}}-\underset{(-)}{C} \qquad \underset{(+)}{C}-C-\overset{O}{\overset{\|}{C}}-$$

Scheme XI

The retrosynthetic analysis of **20** continues this way. We may expect 2,3-dimethylcyclohexanone to give enolate or enol preferentially at carbon 2. Equation 10 shows this step under acid conditions and also the continuation to compound 19.[15]

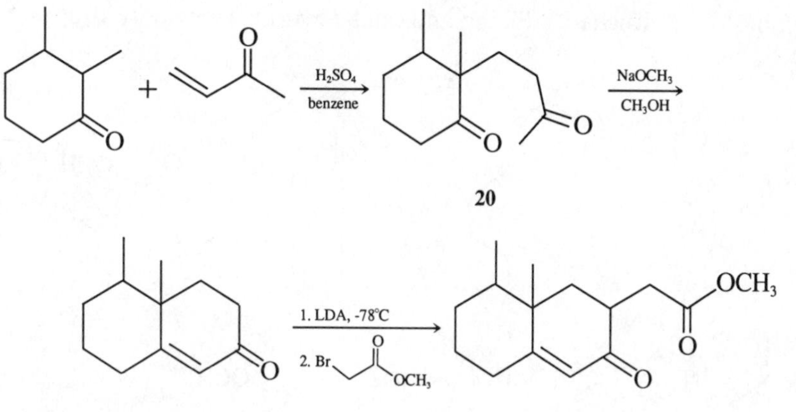

$\xrightarrow[\text{benzene}]{H_2SO_4}$ 20 $\xrightarrow[CH_3OH]{NaOCH_3}$

33% $\xrightarrow[\text{2. Br}\diagdown\!\!\diagup\!\!\overset{O}{\|}\text{OCH}_3]{\text{1. LDA, -78°C}}$ 19 (10)

Reviewing the reactive character of the fragments of disconnection of difunctional molecules, we can see a pattern. The 1,3 and 1,5 molecules give normally attractive fragments, while the 1,2 and 1,4 molecules require an umpolung reagent. Favorable and unfavorable normal charges alternate with increasing separation of the functional groups.

Lactone **21** is a 1,5-difunctional molecule, lower in oxidation state than **20,** and is accessible by the same disconnection rationale (Scheme XII). First disconnect the functionality and then the carbon chain at the 3,4-bond.

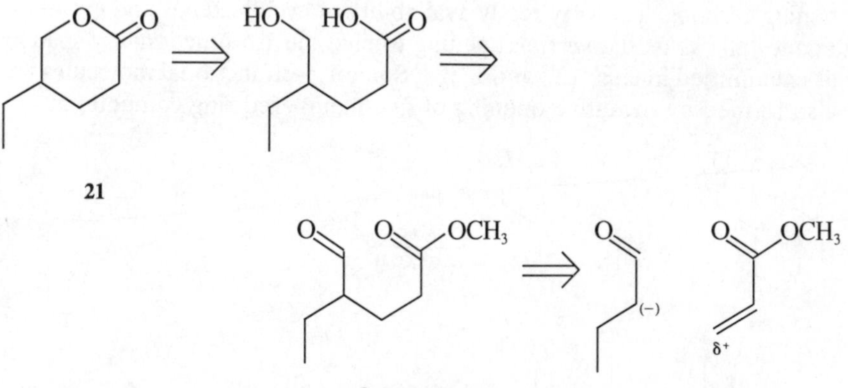

21 ⟹ HO HO O ⟹

O O OCH₃ ⟹ O (−) O OCH₃ δ+

Scheme XII

The butyraldehyde enolate called for here is not practical because of self-condensation (Section 5.1.1); therefore, a nonelectrophilic derivative of the aldehyde is used instead. The piperidine enamine is sufficiently nucleophilic for the Michael addition and can be hydrolyzed readily after that step:[16]

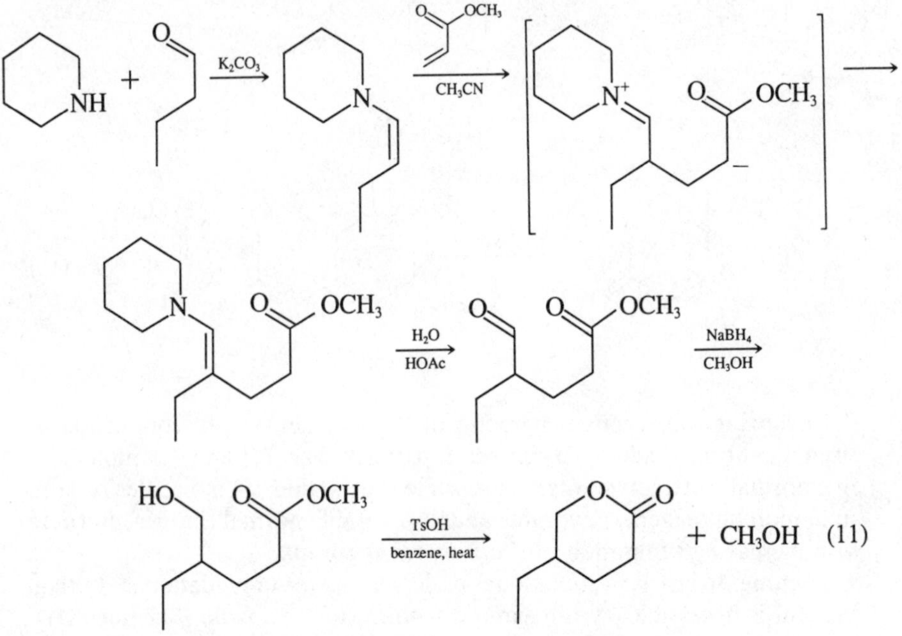

K_2CO_3 ; CH_3CN ; OCH_3 ; NH ; N ; N^+ ; H_2O, HOAc ; $NaBH_4$, CH_3OH ; HO ; TsOH, benzene, heat

$$+ \; CH_3OH \quad (11)$$

21

1,6-Difunctional molecules are less often formed by connecting intervening carbons. The very ready availability of cyclohexenes and cyclohexanones allows oxidative ring opening to give the 1,6-functionality spacing as exemplified in Eqs. 12[17] and 13.[18,19] Some 1,5-difunctional molecules are also formed by oxidative opening of five-membered ring compounds.

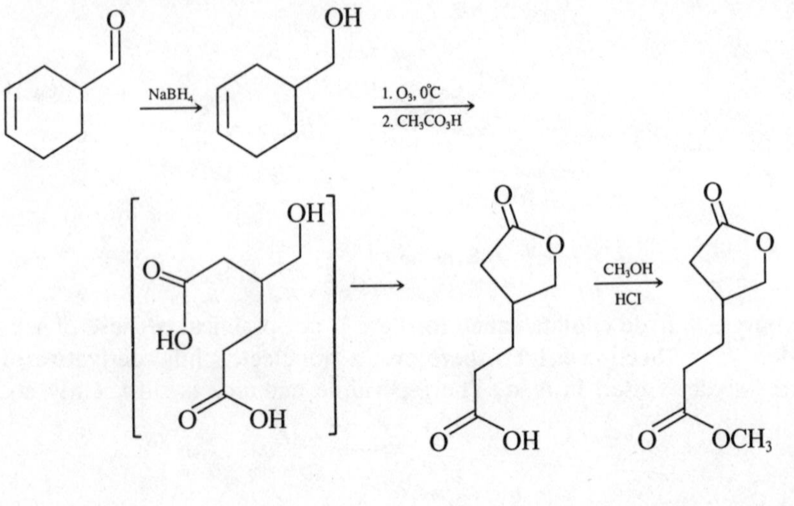

$NaBH_4$; 1. O_3, 0°C 2. CH_3CO_3H ; OH ; HO ; CH_3OH, HCl ; OCH_3 (12)

$$\text{2,2-dimethylcyclohexanone} + m\text{-}ClC_6H_4CO_3H \xrightarrow[CH_2Cl_2]{NaOAc} \text{lactone} + m\text{-}ClC_6H_4CO_2H \quad (13)$$

Difunctional molecules are sometimes assembled, ignoring the polarizing possibilities of one of the functional groups. Nucleophilic character can be developed at a site on a chain where it is insulated from a carbonyl group by intervening sp^3 carbons. The aldehydes or ketones then need protection as acetals. Equations 14 and 15 show preparation of 1,4 and 1,5 difunctionality this way.[20,21]

$$RCOCl + BrMg\text{-}CH_2CH_2\text{-(1,3-dioxan-2-yl)} \xrightarrow[-78°C]{THF}$$

$R = n\text{-}C_6H_{13}$

$$R\text{-}CO\text{-}CH_2CH_2\text{-(1,3-dioxan-2-yl)} \; (92\%) \xrightarrow[H^+]{H_2O} R\text{-}CO\text{-}CH_2CH_2\text{-}CHO \; (89\%) \quad (14)$$

$$\text{HO-C(CH}_3)_2\text{CH}_2\text{CH}_2\text{CH}_2\text{COCH}_3 \; (0.10\text{ mol}) + ClMg\text{-}CH_2CH_2CH_2\text{-C(CH}_3)(OCH_2CH_2O) \; (0.25\text{ mol}) \xrightarrow{THF} \xrightarrow{aq. NH_4Cl}$$

$$\text{HO-C(CH}_3)_2(CH_2)_3C(OH)(CH_3)(CH_2)_3C(CH_3)(OCH_2CH_2O) \; (51\%) \xrightarrow[TsOH]{CH_3COCH_3}$$

$$\text{HO-C(CH}_3)_2(CH_2)_3C(OH)(CH_3)(CH_2)_3COCH_3 \; (42\%) \quad (15)$$

6.4 RING CLOSURE

Many difunctional molecules in this chapter and Chapter 5 react to give rings. A competition may exist between intermolecular and intramolecular reactions. In most cases the formation of three-, five-, and six-membered rings is more favorable than polymerization because the intermolecular process requires a bimolecular collision while cyclization requires only conformational alignment. This becomes less probable for rings larger than six; therefore, high dilution is used to lessen the frequency of intermolecular collisions. The acyloin condensation[22] is efficient for such large-ring closures.

In some cases there is a competition within a molecule for closure to different size rings. If the closure is an irreversible reaction such as alkylation of an enolate, kinetic control may allow three-membered ring closure to predominate over five-membered as in Eq. 16.[23] The electrophilic carbon closer to the nucleophilic site is found first, in spite of the incorporation of ring strain.

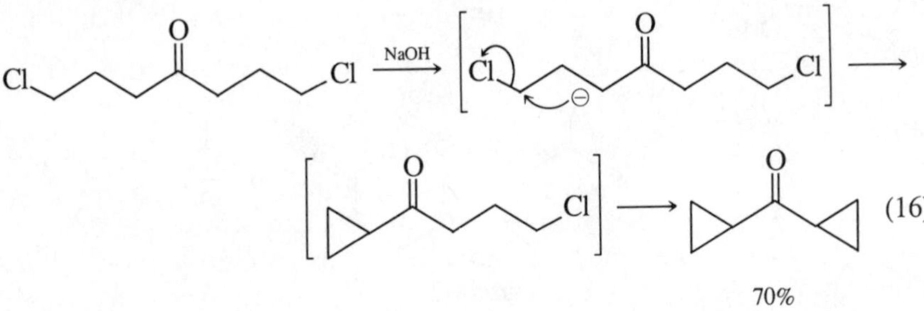

Cl ... Cl —NaOH→ [...] → [...] → ... (16)

70%

With reversible reactions, the thermodynamic product predominates. That is, the low strain five- or six-membered rings will form to the exclusion of three- or four-membered alternatives (Scheme XIII).

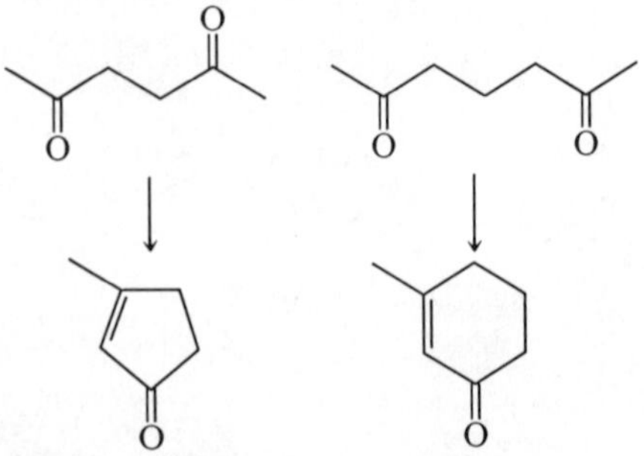

Scheme XIII. Products expected from aldol cyclizations of 1,4- to 1,7-diketones.

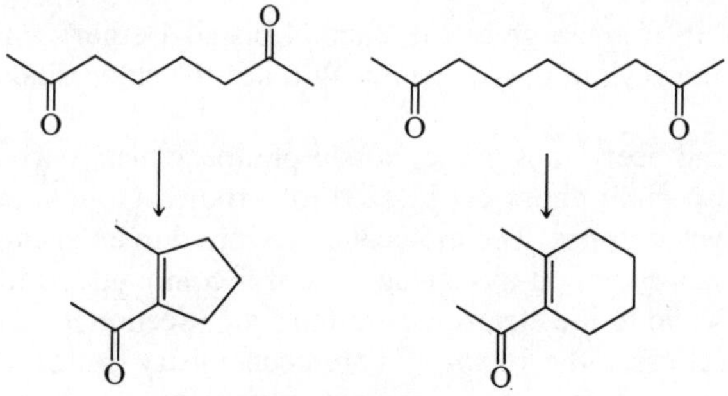

Scheme XIII. *(Continued)*

In Claisen and aldol cyclizations, five- and six-membered rings are not in competition with each other. Seven-membered rings may sometimes be closed readily without high dilution:[24]

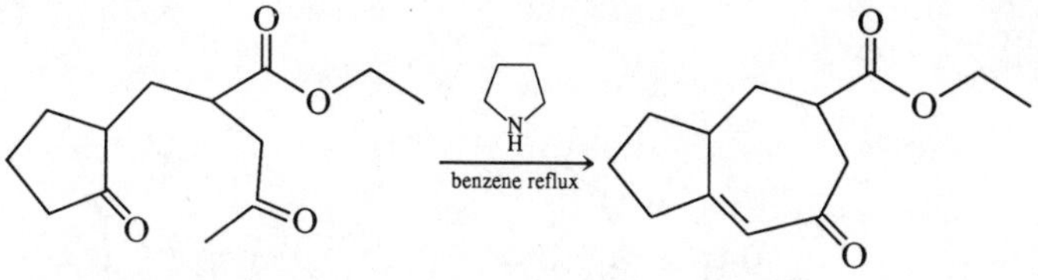

benzene reflux

$+ H_2O$ (17)

81%

Heterocyclic rings show the same size preferences. Halohydrins and base give epoxides under irreversible conditions. Hemiacetals, acetals, and lactones under reversible conditions favor five- or six-membered rings, where flexibility permits, over intermolecular polymerization.

The closure of four-membered rings requires special methods. Two reactions are frequently used: the acyloin[22] and photochemical cycloaddition. Treatment of a 1,4-diester (Eq. 18)[25] with sodium metal and chlorotri-

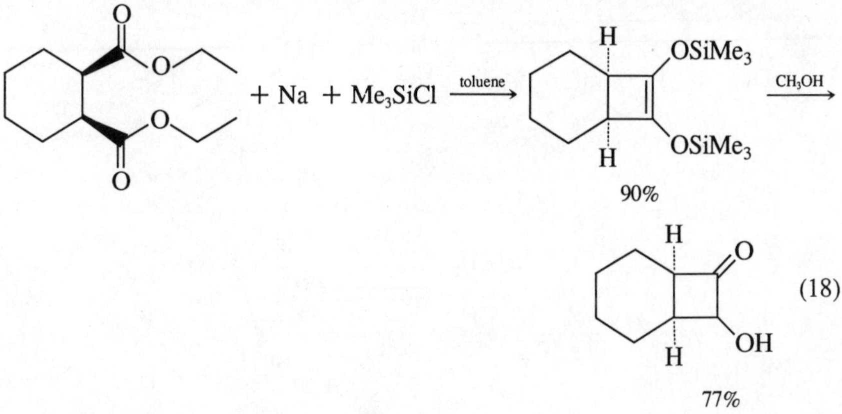

$+ Na + Me_3SiCl$ toluene → ($OSiMe_3$, $OSiMe_3$) 90% $\xrightarrow{CH_3OH}$

(OH) 77% (18)

methylsilane in toluene gives the enediol bis silyl ether, which is methanolyzed to the acyloin in good yield. Without the chlorosilane, yields are poor.

Alkenes and acetylenes will cycloadd photochemically to other alkene molecules, especially those conjugated to carbonyl groups, to give cyclobutanes or cyclobutenes. The molecules are raised to an excited electronic state often via a radiation-absorbing sensitizer compound, add to form the ring, and descend to the electronic ground state (Section 8.3.2.). In doubly unsymmetrical cases the regio- and stereochemistry can be complex and dependent on conditions. Nevertheless, many are synthetically useful. A few examples are:[26–28]

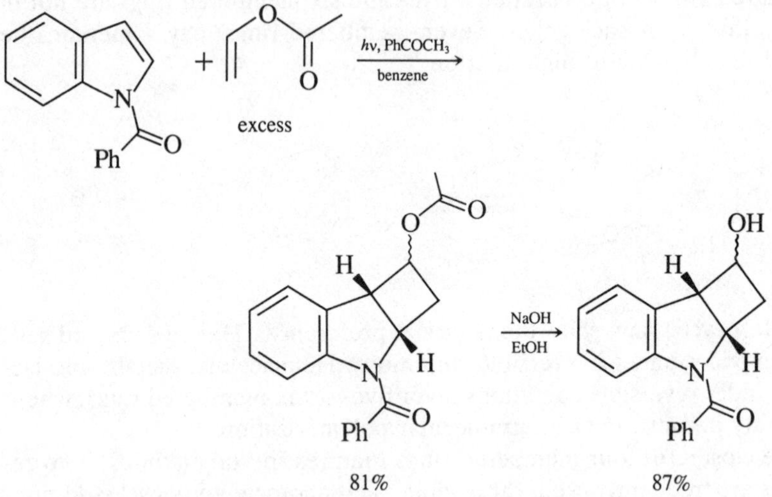

excess

$h\nu$, $PhCOCH_3$ / benzene

81%

NaOH / EtOH

87% (19)

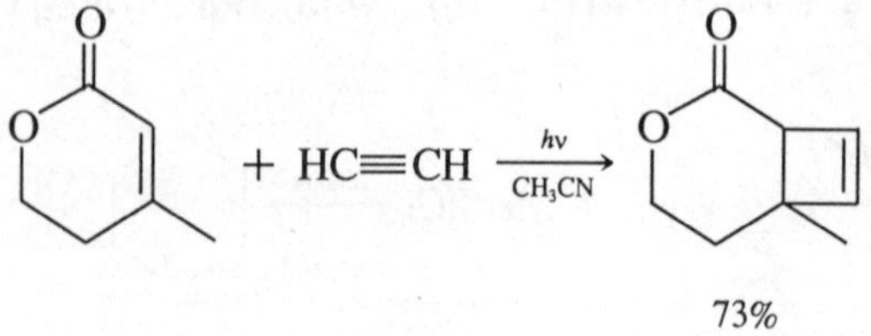

+ HC≡CH $\xrightarrow[CH_3CN]{h\nu}$ 73% (20)

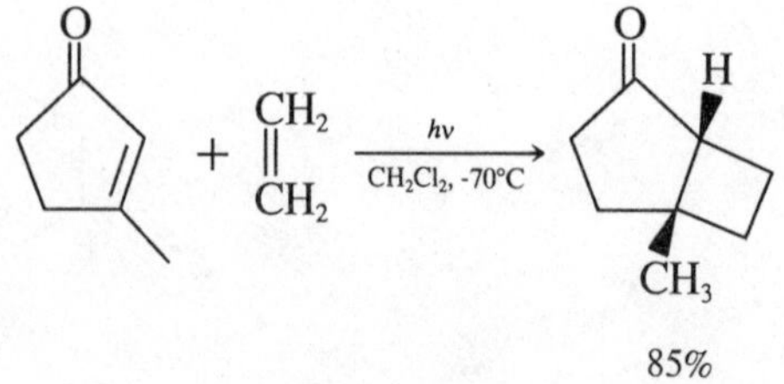

+ $CH_2=CH_2$ $\xrightarrow[CH_2Cl_2,\ -70°C]{h\nu}$ 85% (21)

6.5 ACETYLIDE ALKYLATION AND ADDITION

In the remainder of the chapter, particular reactions are selected for examination of their synthetic potential. Acetylide ions are useful for linking carbon chains, particularly where a double bond is desired with stereoselectivity. Acetylene and l-alkynes may be deprotonated with strong bases such as LDA and then treated with alkyl halides or carbonyl compounds. Preformed lithium acetylide complexed with ethylenediamine is available as a dry powder. Several alkynes derived from acetylide and carbon dioxide or formaldehyde are available, including propargyl alcohol ($HC{\equiv}CCH_2OH$), propargyl bromide ($HC{\equiv}CCH_2Br$) and methyl propiolate ($HC{\equiv}CCO_2CH_3$).

A disconnection between a double bond and an allylic carbon should suggest acetylide chemistry, while disconnection between the double-bonded carbons should suggest the Wittig and allied reactions (Section 5.2). Consider structure **22.** The cis double bond could come from an acetylene—in fact, propargyl alcohol. Following the retrosynthetic analysis, the amide should be disconnected first. The secondary amine could come from benzylamine and a halide, and that halide could be made via propargyl alcohol and ethylene oxide (Scheme XIV). This entails accessory steps to reduce the triple bond to a double bond and to distinguish the propargylic alcohol from the alcohol arising from the ethylene oxide opening.

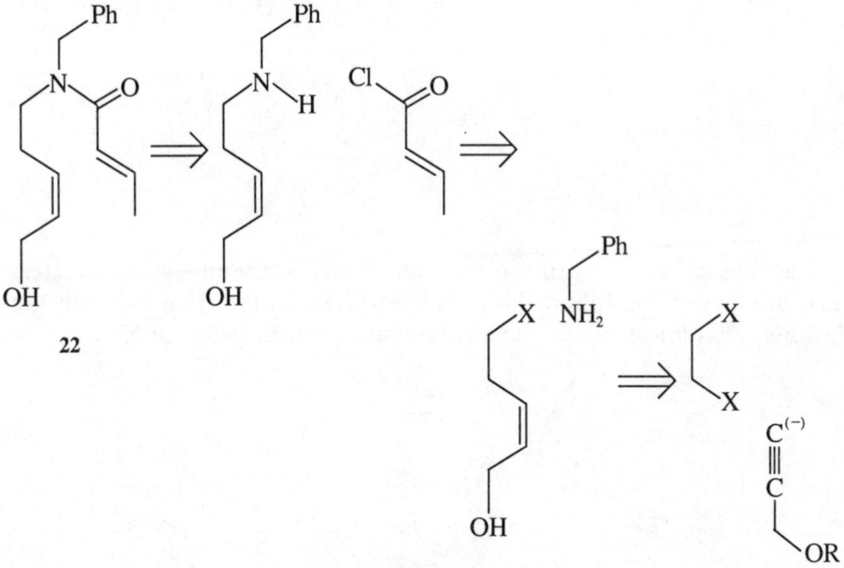

22

Scheme XIV

The actual synthesis is[29]

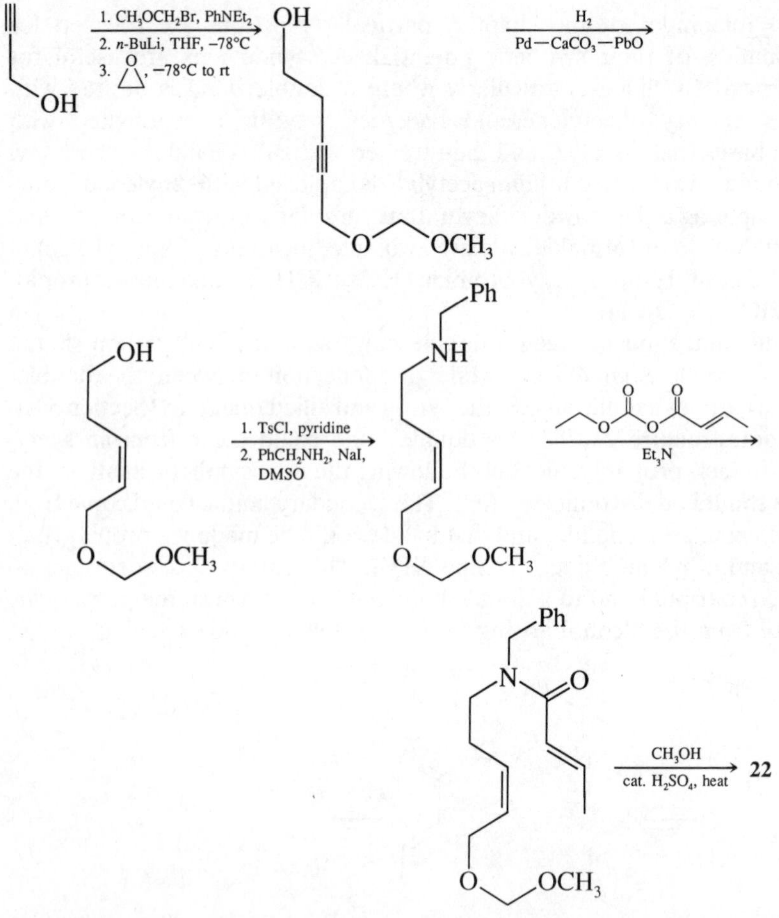

1. CH_3OCH_2Br, $PhNEt_2$
2. *n*-BuLi, THF, −78°C
3. ethylene oxide, −78°C to rt

H_2
Pd—$CaCO_3$—PbO

OH
OCH$_3$

1. TsCl, pyridine
2. $PhCH_2NH_2$, NaI, DMSO

Ph
NH

Et_3N

CH_3OH
cat. H_2SO_4, heat

→ **22** (22)

Lactone **23** was constructed by using acetylide chemistry at two sites. The disconnections follow the principles given earlier, that is, open the lactone, disconnect beside the alcohol, and so forth (Scheme XV).

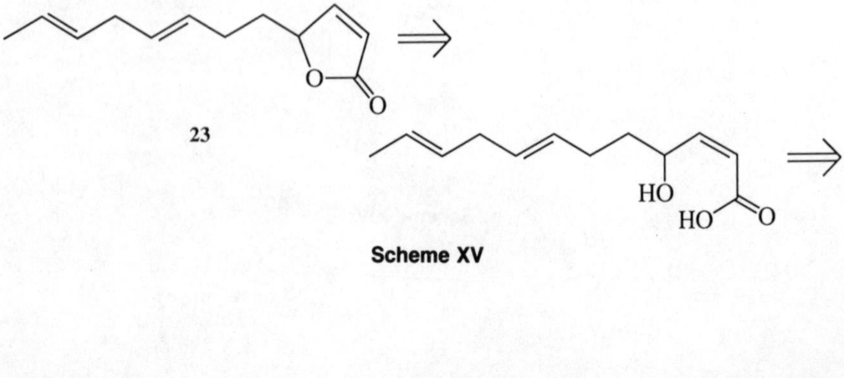

23

HO
HO O

Scheme XV

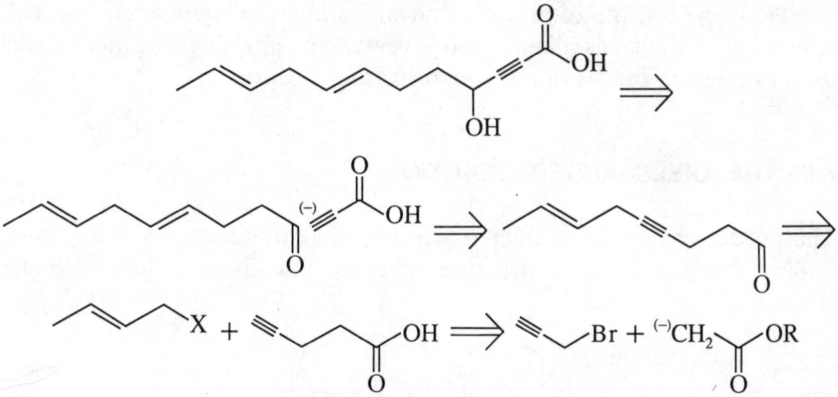

Scheme XV. (*Continued*)

The synthesis devised by Jakubowski and co-workers is shown in Eq. 23;[30] 2 equiv of ethylmagnesium bromide were used in order to deprotonate the carboxylic acid and the acetylide. The reactivity of dianions is generally

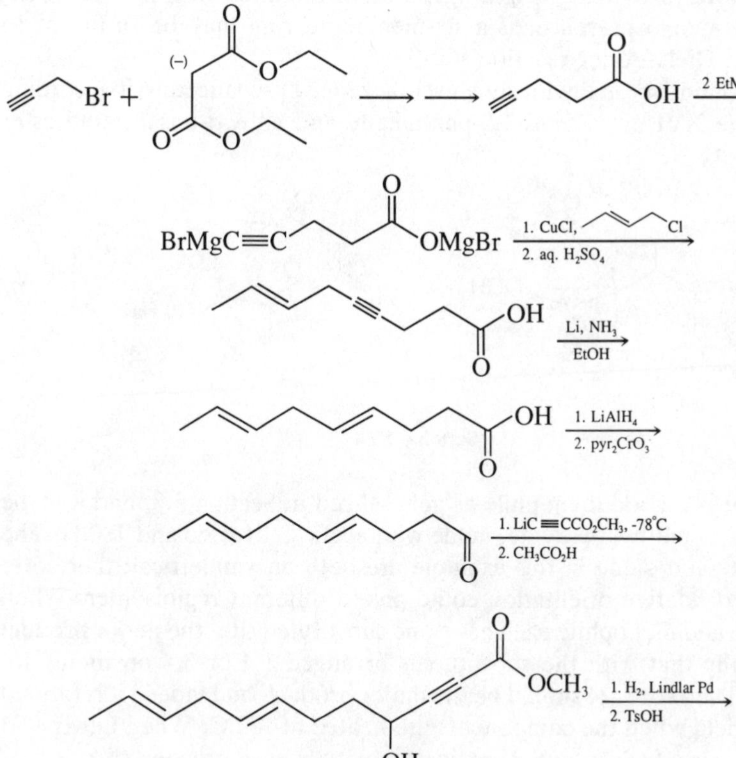

23 (23)

greater at the last site of proton removal. Lithium in liquid ammonia with ethanol gave the trans alkene, and several steps later, hydrogenation was used to prepare the cis double bond for ring closure.

6.6 THE DIELS–ALDER REACTION

The product of a Diels–Alder reaction is generally a cyclohexene; thus, finding that feature in a structure suggests that disconnection (Scheme XVI).

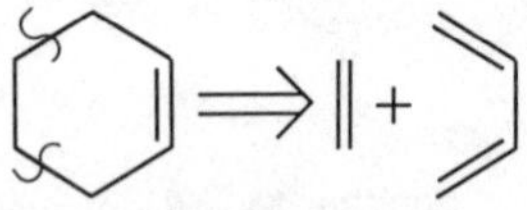

Scheme XVI

The cyclohexene may be part of a bicyclic or fused ring structure, or it may be a cyclohexadiene as when an activated acetylene is the dienophile. Furthermore, if the ring is hydrogenated or modified with new functionality, the simple presence of a six-membered ring may be sufficient to propose a Diels–Alder reaction step.

Compound **24**[31] is obviously a cyclohexene. Disconnecting it according to Scheme XVI gives *trans*-1,3-pentadiene and an α,β-unsaturated ester (Scheme XVII).

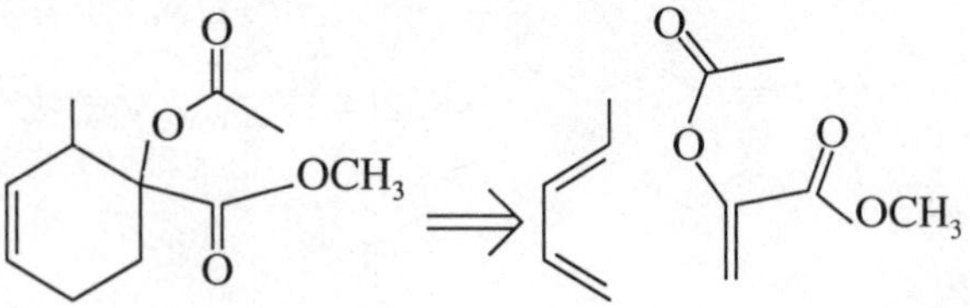

24

Scheme XVII

The latter is a good dienophile as generalized in Section 5.3, and it is the enol ester of methyl pyruvate, made with acetic anhydride and TsOH. The dienophile and diene in this example are both unsymmetrical; therefore, a reversed relative orientation could give a different regioisomer. When the diene and dienophile each have one substituted site, the major product is generally that with the substituents arranged 1,4 or 3,4 on the cyclohexene. Therefore, **24** should be the major product, and indeed it is formed in 97% yield when the components are heated at 160°C. When Lewis acid catalysis is used (Section 5.3) regioselectivity is even greater.

In compound **25** the cyclohexene is part of a bicyclic structure; therefore, disconnection gives a monocyclic diene (Scheme XVIII). With diethylaluminum chloride catalysis, this reaction gave **25** in 84% yield.[32]

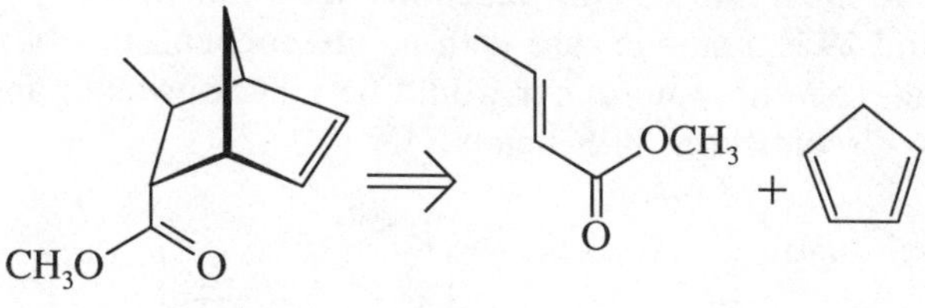

CH_3O O OCH_3 + O

25

Scheme XVIII

The Diels–Alder reaction is stereospecific. The diene and dienophile approach with one face of the π bonding of each, merging to form σ bonds, and the original geometry is minimally shifted (Section 8.3.2). This is a syn addition on both components. Because of this, groups that are cis in the dienophile remain cis in the cyclohexene, and the groups that are cis, cis (or trans, trans) in the diene become cis in the cyclohexene. Although four chiral centers are formed in many Diels–Alder reactions, often only one or two pairs of enantiomers are formed in appreciable amounts. Of the maximum of 32 regio- and stereoisomers imaginable with four chiral centers, the stereochemistry of the diene and dienophile limit it to a maximum of 8. Since both faces of cyclopentadiene are identical (Section 3.9), there is no regiochemistry in the formation of **25,** further reducing the possibilities to 4 stereoisomers (**25–28**). There is generally a favoring among diastereomers toward those with the final π bonds facing each other as in **25** and **26,** where the carbonyl and alkene are close (endo rule).

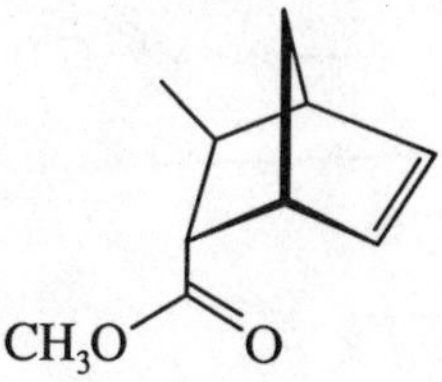

CH_3O O O OCH_3

25 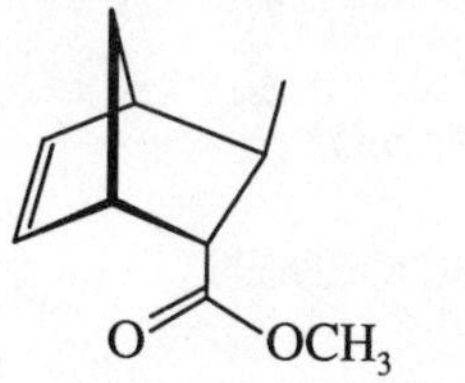 **26**

O O

CH_3O OCH_3

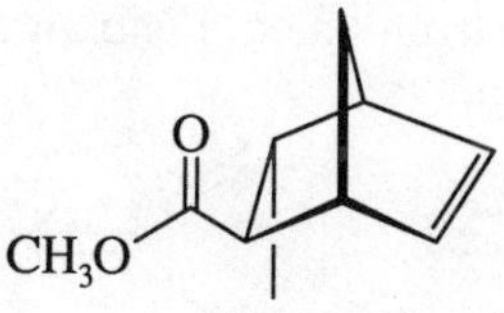

27 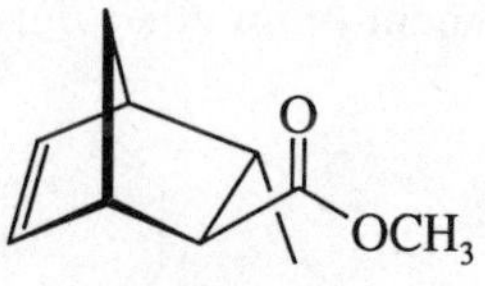 **28**

In this case the ratio of (**25** + **26**) to (**27** + **28**) was 15:1. In other cases steric hinderance or intramolecular restrictions will limit the number of isomers. Altogether, stereospecificity, regiospecificity, and the endo selectivity make most Diels–Alder reactions quite practical.

Compound **29** is a simple case with no stereochemistry, but it is not a cyclohexene. The enol tauntomer would be a cyclohexene, and with this idea we can disconnect as in Scheme XIX.

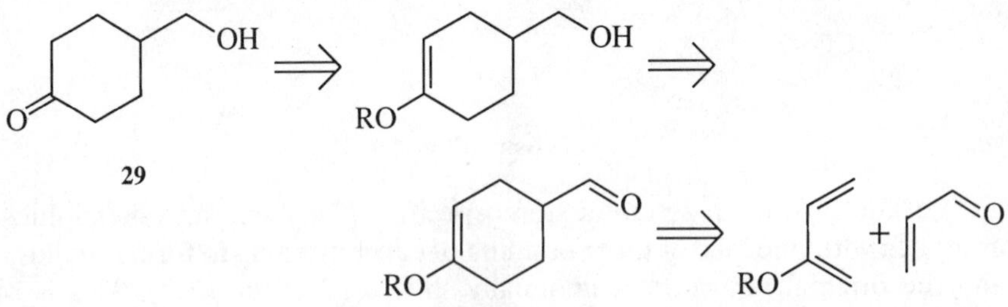

OH ⇒ OH ⇒

O RO

29

O ⇒ + O

RO RO

Scheme XIX

Allyl alcohol is not much of a dienophile, but acrolein has the needed carbonyl group that can be reduced later. The silyl enol ether of 1-buten-3-one is a good diene:[33]

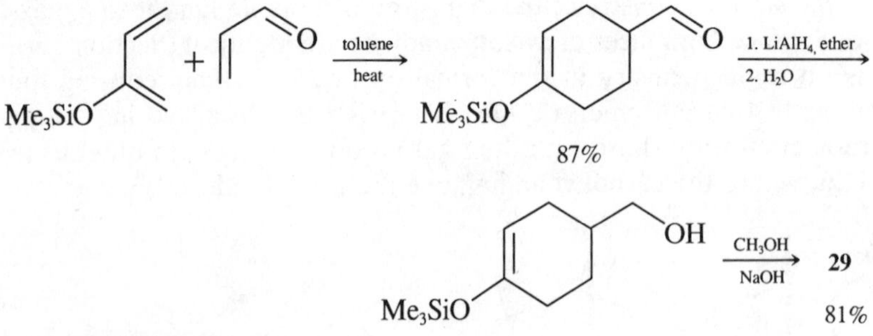

Me_3SiO + O —toluene, heat→ Me_3SiO O (87%) —1. $LiAlH_4$, ether; 2. H_2O→

Me_3SiO OH —CH_3OH, NaOH→ **29** (81%) (24)

6.7 THE CLAISEN REARRANGEMENT

The Claisen rearrangement is used for the preparation of γ,δ-unsaturated aldehydes, ketones, acids, esters, and amides.[34] It is a thermal rearrangement of an ether derived from an enol and an allyl alcohol (Scheme XX).

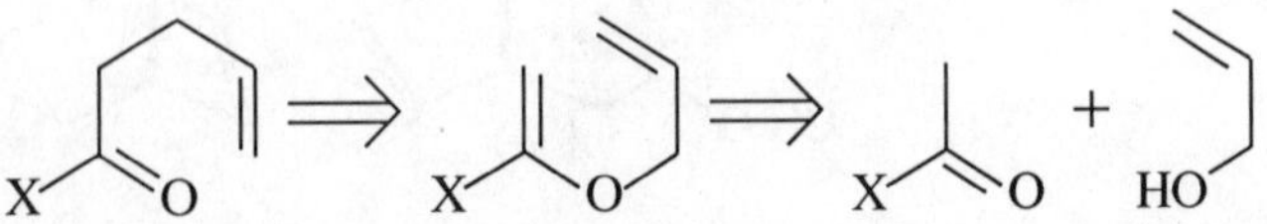

X O ⇒ X O ⇒ X O + HO

Scheme XX

In effect, a rearranged allyl group becomes attached to the carbon α to a carbonyl group. The formation of the enol ether requires a dehydrating reagent or a derivative of the carbonyl compound into which the allyl alcohol can be exchanged. These include another enol ether, an orthoester, or an amide acetal. Examples are[35–38]

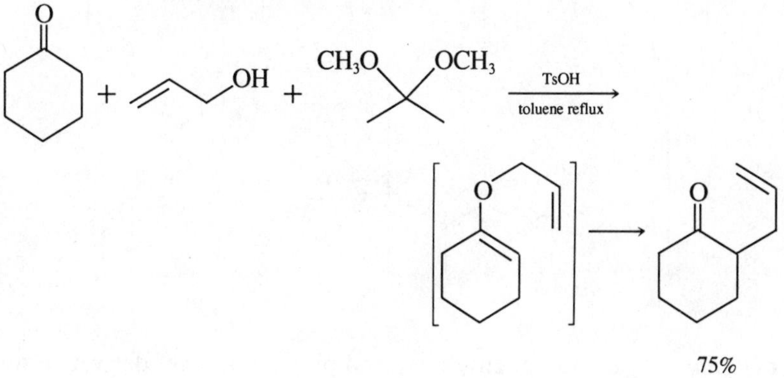

O + OH + CH_3O OCH_3 —TsOH, toluene reflux→ [O] → O (25)

75%

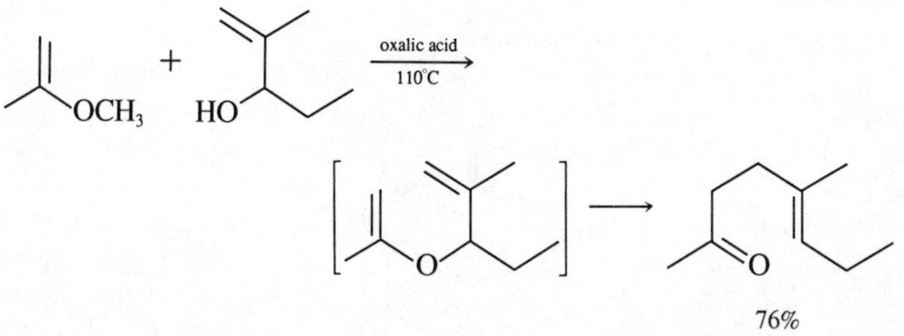

+ OCH_3 HO —oxalic acid, 110°C→ [O] → O (26)

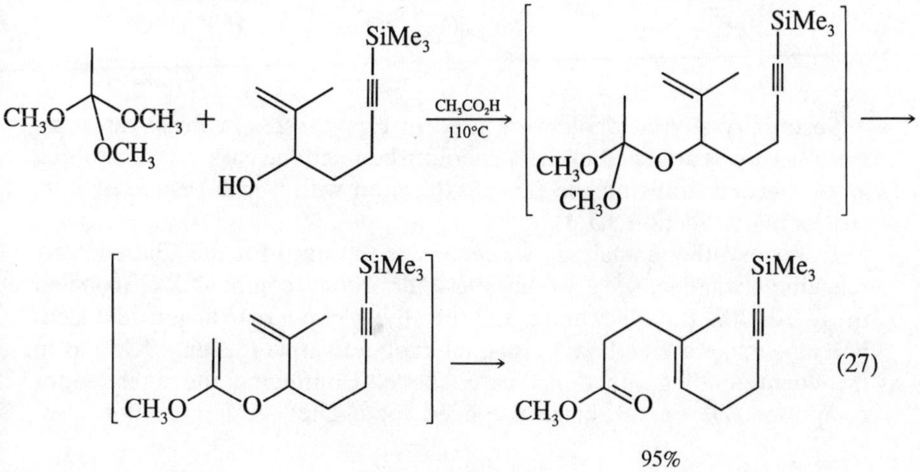

76%

CH_3O OCH_3 OCH_3 + $SiMe_3$ HO —CH_3CO_2H, 110°C→ [$SiMe_3$ CH_3O O CH_3O] →

[$SiMe_3$ CH_3O O] → $SiMe_3$ CH_3O O (27)

95%

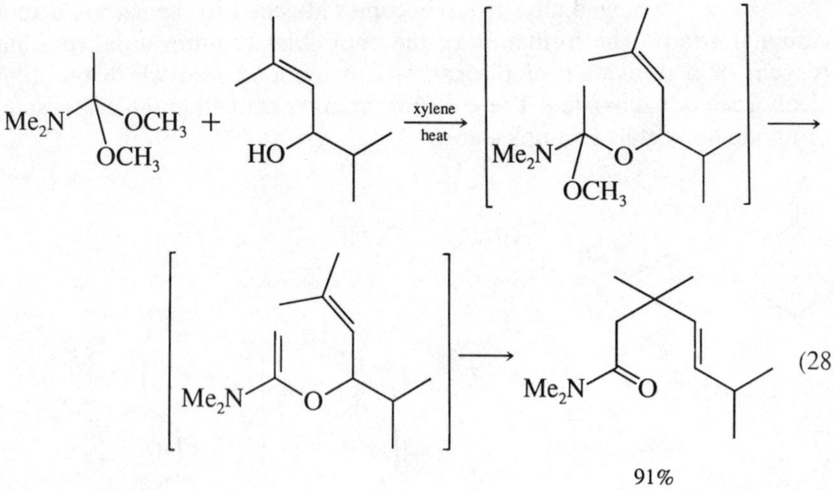

Me_2N, OCH_3, OCH_3 + HO — xylene, heat → [Me_2N, O, OCH_3] →

[Me_2N, O] → Me_2N, O (28)

91%

One may also begin with an allyl ester and prepare an enol derivative for rearrangement:[39]

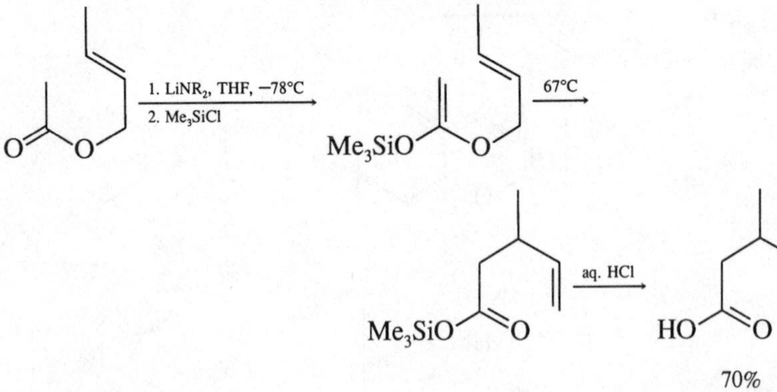

O, O — 1. $LiNR_2$, THF, −78°C; 2. Me_3SiCl → Me_3SiO, O — 67°C →

Me_3SiO, O — aq. HCl → HO, O (29)

70%

Secondary allyl alcohols were used in Eqs. 26–28. In these cases the group that was attached to the alcohol carbon and the carbonyl containing group become trans on the β,γ unsaturation with a high degree of stereoselectivity (Section 3.7.4).

In retrosynthetic analysis, we recognize the need for the Claisen rearrangement when we see a γ,δ-unsaturated carbonyl compound. Disconnect the α from the β carbon and make the allyl group a rearranged allyl alcohol; also make the carbonyl group an enol derivative (Scheme XXI) as in the corresponding functional case above. Continuing the analysis for compound **30,** we recognize the need for a Diels–Alder reaction also.

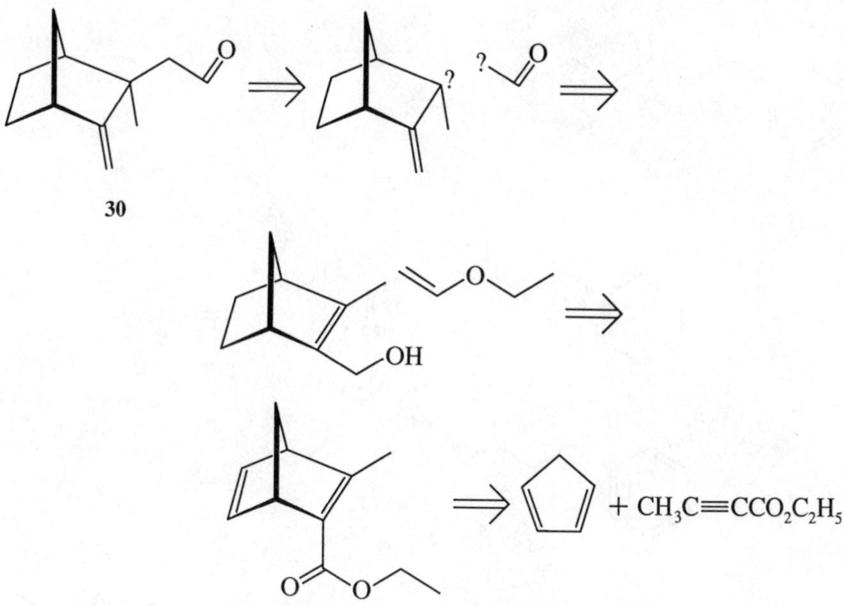

30

OH

+ $CH_3C{\equiv}CCO_2C_2H_5$

O O

Scheme XXI

The actual synthesis of **30** is[40]

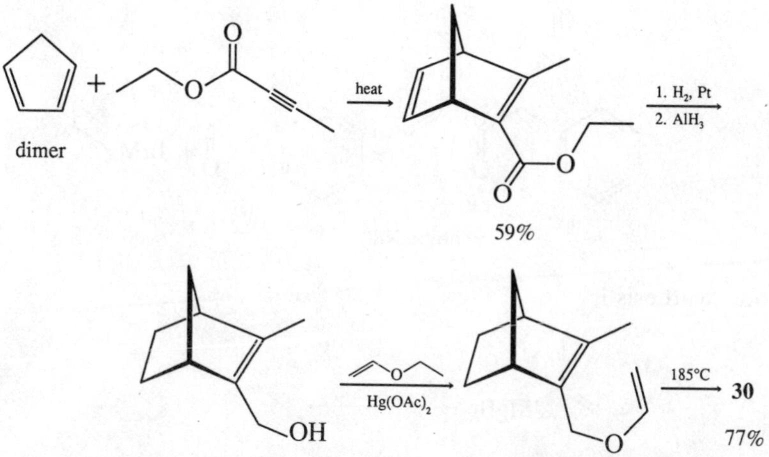

O

+ O

dimer

heat

1. H_2, Pt
2. AlH_3

O

O

59%

OH

O

$Hg(OAc)_2$

185°C

30 (30)

O

77%

Compound **31** contains unsaturation γ,δ to the bromo-functional carbon. Propose a carbonyl compound as a likely intermediate and then disconnect between the α and β carbons, write the rearranged allylic alcohol and the ester enol derivative (Scheme XXII). Continue with a disconnection at the alcohol.

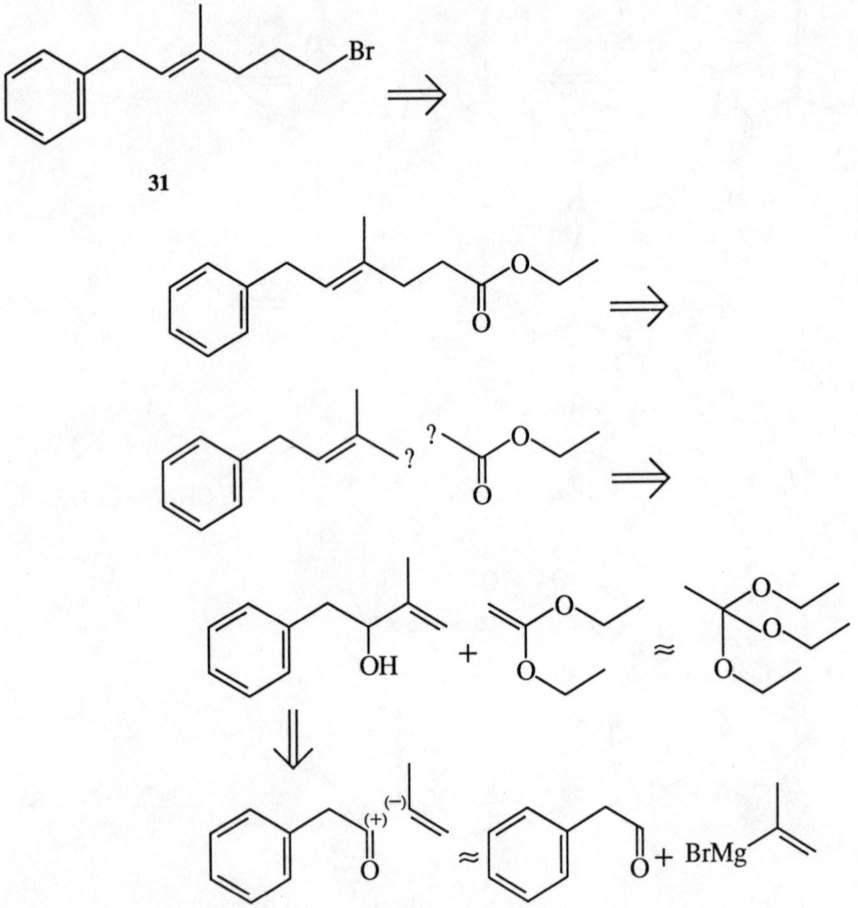

Br

31

O

O

?

?

O

O

O

OH

+

O

≈

O

O

O

⇓

(−)

(+)

O

≈

O

+ BrMg

Scheme XXII

The actual synthesis is[41]

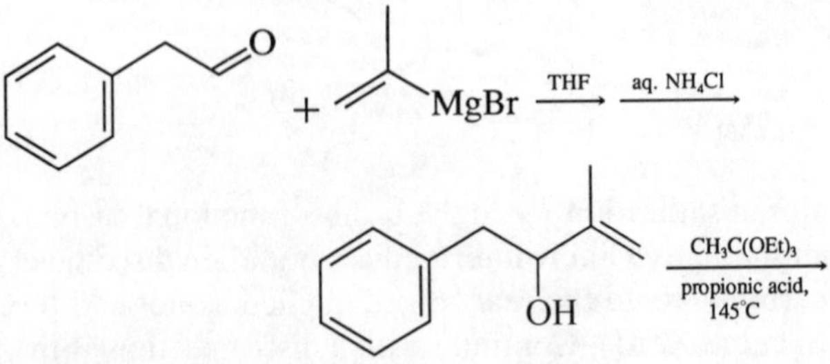

O

+

MgBr

THF

aq. NH_4Cl

OH

$CH_3C(OEt)_3$

propionic acid, 145°C

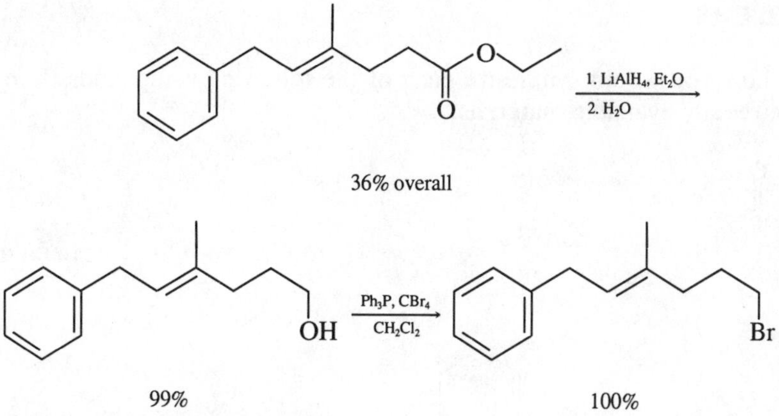

1. $LiAlH_4$, Et_2O
2. H_2O

36% overall

Ph_3P, CBr_4
CH_2Cl_2

OH Br (31)

99% 100%

6.8 FINAL NOTE

In longer syntheses it is not always routine to apply certain key steps. Some plans are highly novel creations that few persons would bring together. To pick one example among a great many, would Scheme XXIII seem obvious?[42]

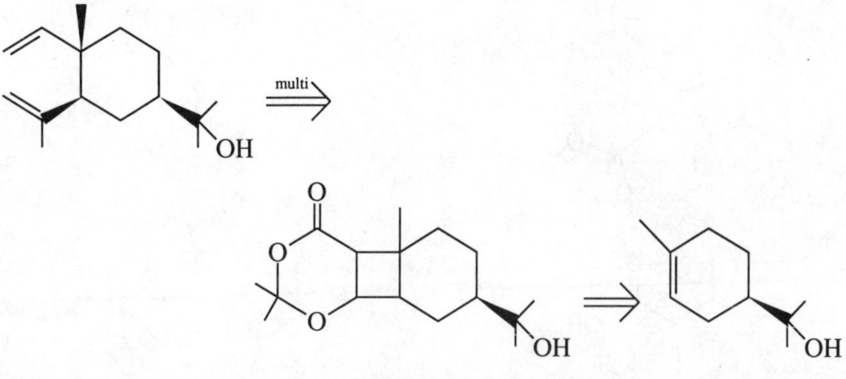

multi

OH

O O O OH OH

Scheme XXIII

On the other hand, bringing together reactions for a complex scheme from a list larger than a person could bring to mind can now be done by computer. Programs have been written, backed by large data collections, that use retrosynthetic analysis to provide reaction schemes for the synthesis of complex molecules.[43–45]

PROBLEMS

Show how you would synthesize each of the following compounds from simple readily available materials.

1. 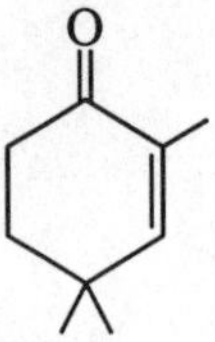 Ref. 46

2. 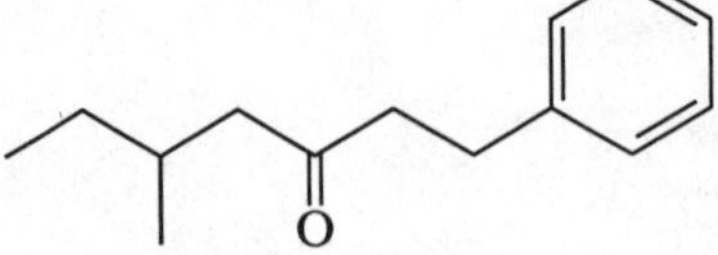 Ref. 5

3. 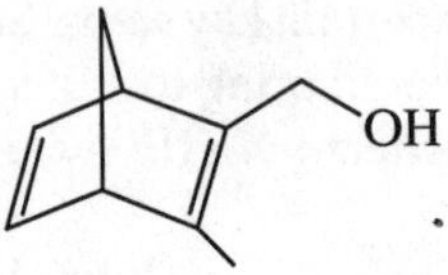 OH Ref. 47

4. 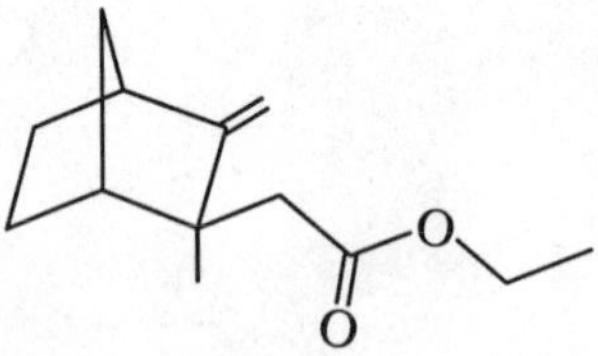 Ref. 47

5. 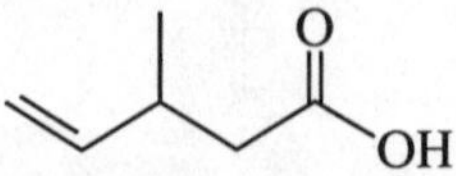 OH Ref. 48

6. Limonene ⟶ 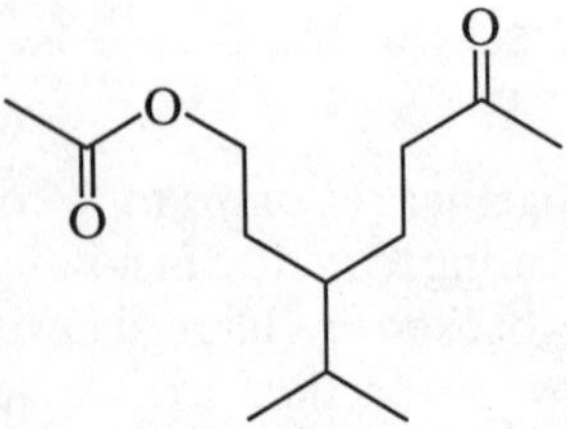 Ref. 49

7. 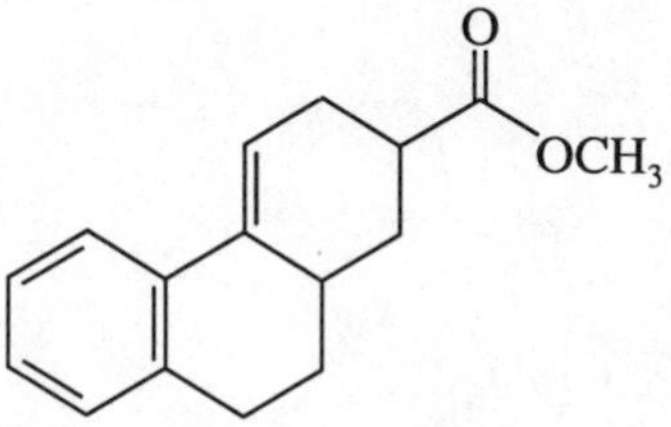 OCH_3 Ref. 50

8. 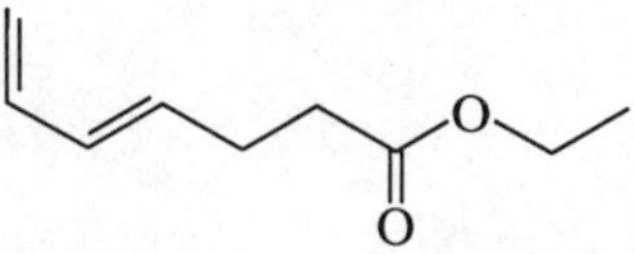 Ref. 51

9. 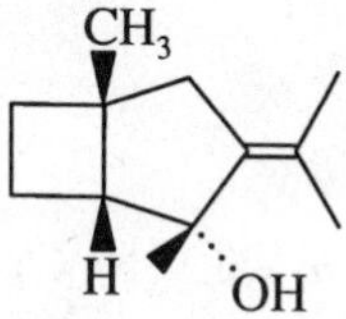 CH_3 H OH Ref. 52

10. 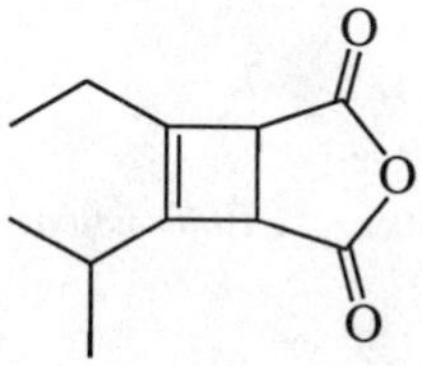 Ref. 53

11. 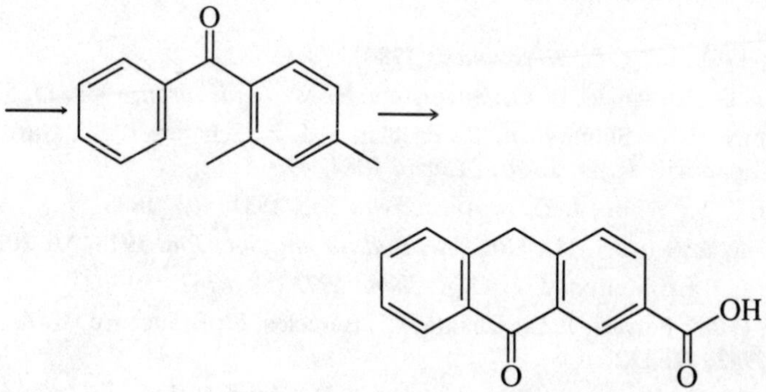 Ref. 54

12. 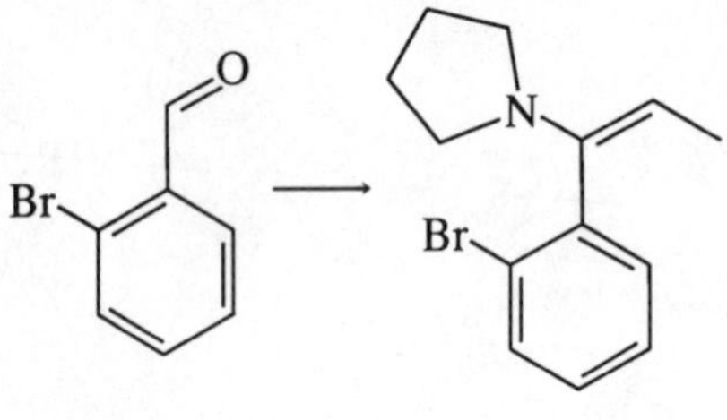 Ref. 55

13. 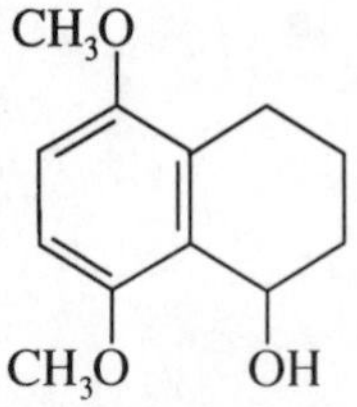 Ref. 56

REFERENCES

1. Warren, S. *Organic Synthesis: The Disconnection Approach,* Wiley, New York, 1982.
2. Ellison, R. A. *Synthesis* **1973,** 397; Ho, T. L. *Synth. Commun.* **1974,** 265; and many other papers.
3. Evans, D. A.; Andrews, G. C. *Acc. Chem. Res.* **1974,** *7,* 147.
4. Harding, K. E.; Burks, S. R. *J. Org. Chem.* **1984,** *49,* 40.
5. Yamashita, M.; Matsumiya, K.; Tanabe, M.; Suemitsu, R. *Bull. Chem Soc. Jpn.* **1985,** *58,* 407.
6. Seebach, D. *Angew. Chem. Internatl. Ed.* **1979,** *18,* 239.
7. Lever, O. W., Jr. *Tetrahedron* **1976,** *32,* 1943.
8. Richman, J. E.; Herrman, J. L.; Schlessinger, R. H. *Tetrahedron Lett.* **1973,** 3267.
9. Stowell, J. C.; King, B. K. *Synthesis* **1984,** 278.
10. Bellas, T. E.; Brownlee, R. G.; Silverstein, R. M. *Tetrahedron,* **1969,** *25,* 5149.
11. McCormick, J. P.; Shinmyozu, T.; Pachlatko, J. P.; Schafer, T. R., Gardner, J. W.; Stipanovic, R. D. *J. Org. Chem.* **1984,** *49,* 34.
12. Ruppert, J. F.; White, J. D. *J. Am. Chem. Soc.* **1981,** *103,* 1808.
13. Yamagiwa, S.; Kosugi, H.; Uda, H.; *Bull. Chem. Soc. Jpn.* **1978,** *51,* 3011.
14. McMurry, J. E.; Melton, J. *J. Org. Chem.* **1973,** *38,* 4367.
15. Zoretic, P. A.; Ferrari, J. L.; Bhakta, C.; Barcelos, F.; Branchard, B. *J. Org. Chem.* **1982,** *47,* 1327.
16. Kuehne, M. E.; Kirkemo, C. L.; Matsko, T. H.; Bohnert, J. C. *J. Org. Chem.*

1980, *45,* 3259; Stork, G.; Brizzollara, A.; Landesman, H. K.; Smuszkovicz, J.; Terrel, R. *J. Am. Chem. Soc.* **1963,** *85,* 207.

17. White, W. L.; Anzeveno, P. B.; Johnson, F. *J. Org. Chem.* **1982,** *47,* 2379.
18. Kane, V. V.; Doyle, D. L.; Ostrowski, P. C. *Tetrahedron Lett.* **1980,** *21,* 2643.
19. Plesnicar, B. In *Oxidation in Organic Chemistry,* Vol. 5-C, Trahanovsky, W. S., Ed., Academic, New York, p. 254.
20. Stowell, J. C. *J. Org. Chem.* **1976,** *41,* 560.
21. Gottschalk, F. J.; Weyerstahl, P. *Chem. Ber.* **1975,** *108,* 2799.
22. Bloomfield, J. J.; Owsley, D. C.; Nelke, J. M. *Org. React.* **1976,** *23,* 259.
23. Hart, H.; Curtis, O. E., Jr. *J. Am. Chem. Soc.* **1956,** *78,* 112.
24. Kumar, V. T. R.; Swaminathan, S.; Rajagopalan, K. *J. Org. Chem.* **1985,** *50,* 5867.
25. Bloomfield, J. J. *Tetrahedorn Lett.* **1968,** 587.
26. Ikeda, M.; Uno, T.; Homma, K.; Ohno, K.; Tamura, Y. *Synth. Commun.* **1980,** *10,* 438.
27. White, J. D.; Matsui, T.; Thomas, J. A. *J. Org. Chem.* **1981,** *46,* 3377.
28. Cargill, R. L.; Wright, B. W. *J. Org. Chem.* **1975,** *40,* 120.
29. Martin, S. F.; Benage, B. *Tetrahedron Lett.* **1984,** *25,* 4863.
30. Jakubowski, A. A.; Guziec, F. S., Jr.; Sugiura, M.; Tam, C. C.; Tishler, M.; Omura, S. *J. Org. Chem.* **1982,** *47,* 1221.
31. Ireland, R. E., Courtney, L.; Fitzimmons, B. J. *J. Org. Chem.* **1983,** *48,* 5186.
32. Callant, P.; Storme, P.; Van der Eycken, E.; Vanderwalle, M. *Tetrahedron Lett.* **1983,** *24,* 5797.
33. Yin, T.-K.; Lee, J.-G.; Borden, W. T. *J. Org. Chem.* **1985,** *50,* 531.
34. Bennett, G. B. *Synthesis* (review) **1977,** 589; Rhoads, S. J.; Raulins, N. R. *Org. React.* (review) **1975,** *22,* 1.
35. Crandall, J. K.; Magaha, H. S.; Henderson, M. A.; Widener, R. K.; Tharp, G. A. *J. Org. Chem.* **1982,** *47,* 5372.
36. Faulkner, D. J.; Petersen, M. R. *Tetrahedron Lett.* **1969,** 3243.
37. Johnson, W. S.; Yarnell, T. M.; Myers, R. F.; Morton, D. R.; Boots, S. G. *J. Org. Chem.* **1980,** *45,* 1254.
38. Majewski, M.; Snieckus, V. *Tetrahedron Lett.* **1982,** *23,* 1343.
39. Ireland, R. E.; Mueller, R. H. *J. Am. Chem. Soc.* **1972,** *94,* 5897.
40. Gibson, T.; Barnes, Z. J. *Tetrahedron Lett.* **1972,** 2207.
41. Guthrie, A. E.; Semple, J. E.; Joulie, M. M. *J. Org. Chem.* **1982,** *47,* 2369.
42. Baldwin, S. W.; Martin, G. F., Jr.; Nunn, D. S. *J. Org. Chem.* **1985,** *50,* 5720.
43. Corey, E. J. *Quart. Rev.* **1971,** *25,* 455.
44. Corey, E. J.; Johnson, A. P.; Long, A. K. *J. Org. Chem.* **1980,** *45,* 2051, and papers cited therein.

45. Hendrickson, J. B.; Grier, D. L.; Toczko, A. G. *J. Am. Chem. Soc.* **1985,** *107,* 5228.
46. Schuster, D. I.; Rao, J. M. *J. Org. Chem.* **1981,** *46,* 1515.
47. Monti, H.; Corriol, C.; Bertrand, M. *Tetrahedron Lett.* **1982,** *23,* 947.
48. Jager, V.; Gunther, H. J. *Tetrahedron Lett.* **1977,** 2543.
49. Kitahara, T.; Mori, K. *J. Org. Chem.* **1984,** *49,* 3281.
50. Ranu, B. C.; Sarkar, M.; Chakraborti, P. C.; Ghatak, U. R. *J. Chem. Soc. Perkin Trans. I* **1982,** 865.
51. Parker, K. A.; Iqbal, T. *J. Org. Chem.* **1982,** *47,* 337.
52. Cargill, R. L.; Wright, B. W. *J. Org. Chem.* **1975,** *40,* 120.
53. Smith, A. B., III; Boschelli, D. *J. Org. Chem.* **1983,** *48,* 1217.
54. Utz, C. G.; Shechter, H. *J. Org. Chem.* **1985,** *50,* 5705.
55. Boger, D. L.; Duff, S. R.; Panek, J. S.; Yasuda, M. *J. Org. Chem.* **1985,** *50,* 5782.
56. Chen, C.-P.; Swenton, J. S. *J. Org. Chem.* **1985,** *50,* 4569.

7

Mechanisms and Predictions

When planning a new reaction in organic chemistry, we look at the accumulated information on similar reactions in order to predict the best conditions for it. The more we know about the intimate details of the reaction process at the molecular level, the better will be our predictions. A particular reaction may be described as an ordered sequence of bond breaking and making and a series of structures that exist along the way from starting material to product. The description includes the concurrent changes in potential energy. Structures at energetic maxima are called *transition states,* and structures at minima are called *intermediates*. The complete description is called the *mechanism* of the reaction.

7.1 REACTION COORDINATE DIAGRAMS AND MECHANISMS

The energy–structure relationship is sometimes illustrated with a plot of potential energy versus progress along the pathway of lowest maximum potential energy. This is exemplified in Fig. I for the chain propagation steps of the familiar chlorination of methane. The first maximum is the transition state at which the C—H bond is partially broken and the H—Cl bond is partially formed. At the shallow minimum, the transient intermediate methyl radical exists. The second maximum is the transition state at which a Cl—Cl bond is partially broken and a C—Cl bond is partially

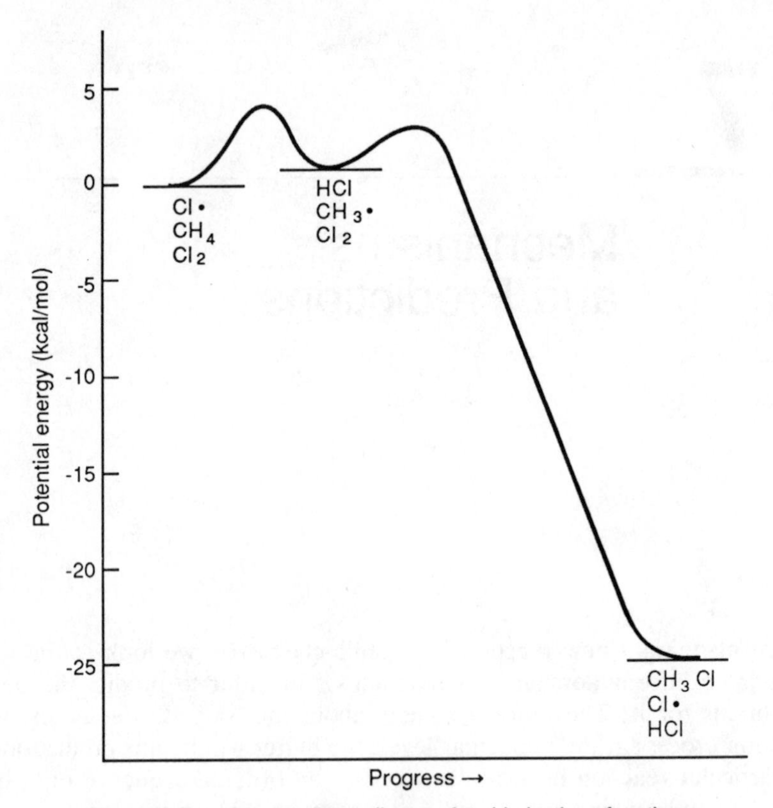

Figure I. Reaction coordinate diagram for chlorination of methane.

formed. Collisions interconvert kinetic and potential energy; thus the increase in potential energy required to reach a transition state corresponds to a decrease in kinetic energy, and a descent from a transition state corresponds to an increase in kinetic energy. Likewise, the net overall descent for this reaction corresponds to a net increase in kinetic energy, that is, the process is exothermal.

A reaction that requires a higher rise to a transition state (activation energy) will be slower than one requiring a lesser rise (if the probability factors are similar) because a smaller fraction of collisions will provide sufficient potential energy to make it.

The energy values for such plots are derived from measurements of overall exo- or endothermicity and from measurement of the effect of varying the temperature on the rate of the reaction (Section 7.3.6).

Many techniques have been developed for determining mechanisms, including complete product (and sometimes intermediate) identification, isotope labeling, stereochemistry, and kinetics,[1,2] as are covered in Section 7.3. In actuality two or more alternative mechanisms are proposed, and their differences are probed with these techniques. An observation is made that is incompatible with one of the proposed mechanisms, and that mechanism is eliminated from consideration, hopefully leaving one. Reasonable mechanisms that have withstood various experimental tests gain some acceptance and are then very useful in predicting the possible range of applications of the reaction and for suggesting changes in reaction conditions that will improve the yield and efficiency of the reaction.

Since molecules are individually too small and too fast for direct observation, our pictorial mechanisms are not the final word. Published mechanisms vary from unsupported conjecture to highly tested near certainty, and you should maintain a healthy skepticism in order to improve on what is available.

Even where the goal, a well-defined singular mechanism, is not yet attained for a given reaction, the observations can be used to make predictions. For example, using the Hammett equation (p. 170) we can predict from measured rates of some cases of a reaction how fast a new case with different substitution will occur.

7.2 THE HAMMOND POSTULATE

In a reaction coordinate diagram it is obvious that the potential energy content at a transition state is closer to that in the starting materials in an exothermal step and closer to the products in an endothermal step. Since potential energy is required to distort a molecule, the structure of the transition state will more closely resemble those molecules to which it is closer in potential energy. That is to say, a small vertical difference in a reaction coordinate diagram corresponds to a small horizontal difference. Transition states are late in endothermal steps and early in exothermal steps. This is the Hammond postulate,[3] and it is useful for predicting products where there is potentially close competition between two alternative steps.

In Fig. II we see a choice of two late transition states that are nearly as different in energy as the products are. We can predict that the thermodynamically more stable product will greatly predominate among the products (if the probability factors are similar). In Fig. III we see a choice of two early transition states, both resembling the starting material and thus resembling each other in structure and energy. We can predict that there may be little or no selectivity toward the more stable product. Comparisons

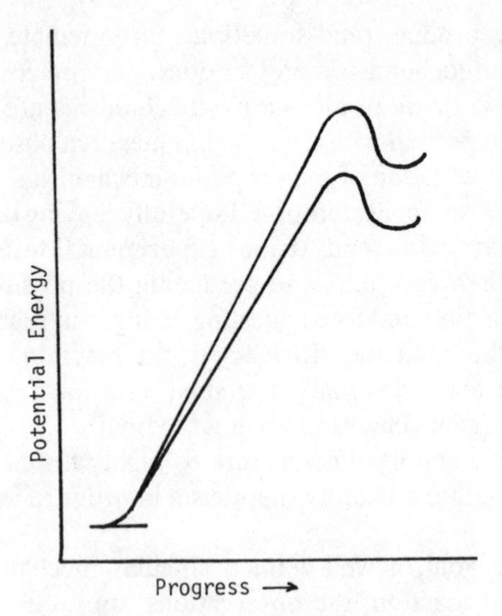

Figure II. Competing endothermal steps.

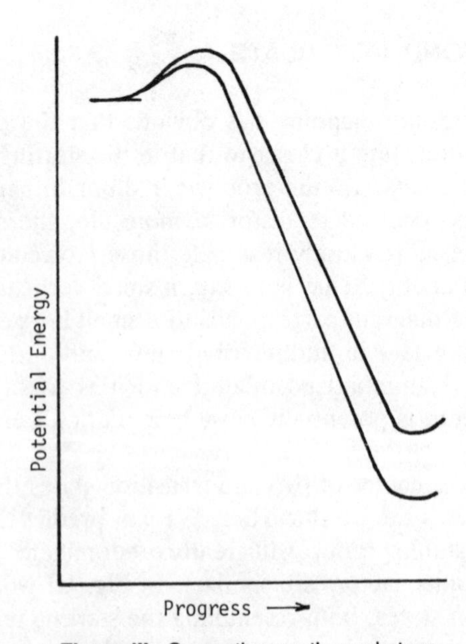

Figure III. Competing exothermal steps.

can be made for less extreme cases as well. Comparing two endothermal reactions, we can predict that the more endothermal one will be the more selective. For example, hydrogen atom abstraction from propane by Br atoms is more endothermal and more selective than by Cl atoms.

7.3 METHODS FOR DETERMINING MECHANISMS

7.3.1 Identification of Products and Intermediates

The primary information should be a thorough identification of the products of the reaction under investigation. If a mechanism proposal includes a temporary intermediate compound that may have some stability, attempts should be made to isolate some of it from the reaction by premature interruption. If it is not stable enough for this, it may yet be detectable spectroscopically in the reaction mixture while in progress. It may also be possible to divert the intermediate by adding a new reactant to the mixture. Finally, if the intermediate is a stable compound and can be prepared by another means, it can be used as a substitute starting material to determine whether it does give the same products overall, at least as rapidly.

7.3.2 Isotope Tracing

If competing mechanistic proposals differ in what atom in the starting material becomes a particular atom of the product, it can be traced by using an uncommon isotope in place of the usual one at one of the sites in the starting material. This technique was used to determine the site of proton removal in the isomerization of epoxides to allylic alcohols. One alternative (Eq. 1) allows that a ring hydrogen (or deuterium) be removed by base under the influence of the electronegative oxygen, followed by ring opening to a carbene and rearrangement.

O
D D
$^{-}$:$\ddot{N}Et_2$
:$\ddot{O}$:$^{-}$
D
:$\ddot{O}$:$^{-}$
D
(1)

Another alternative is the removal of a proton α to the ring with ring opening as in E2 eliminations from alkyl halides:

$$Et_2N:^- \quad \text{(trans-4,5-dideuterio-4,5-epoxyoctane)} \longrightarrow \text{(trans-5-octen-4-olate, } :\ddot{O}:^-\text{, D, D)} \tag{2}$$

The trans epoxide was prepared by reducing 4-octyne with sodium in deuterioammonia and oxidizing the dideutero-4-octene with monoperphthalic acid. Treating the trans epoxide (85% d_2 species) with lithium diethylamide gave *trans*-5-octen-4-ol that was 70% d_2 species by mass spectrometry. Thus almost all of the product likely formed by the process in Eq. 2 and not Eq. 1.[4]

Isotopes available for tracing include ^2H, ^3H, ^{13}C, ^{14}C, ^{15}N, and ^{18}O. The presence of the radioactive isotopes in the products or degradation derivatives of the products is determined by decay counting. NMR or mass spectra are used to locate ^{13}C or ^2H labels. If all the signals in a natural-abundance ^{13}C NMR spectrum of the products are identified (Chapter 10), the signal for the site labeled with enriched ^{13}C will be larger and identifiable. In ^2H-labeled compounds, the ^1H NMR signal for the labeled site will be absent or of reduced area.

7.3.3 Stereochemical Determination

If a reaction is carried out on a particular stereoisomer of starting material and the products are stereoisomerically identified (Chapter 3), a choice among mechanisms can often be made. In substitution and rearrangement reactions, inversion of configuration will indicate backside attack, retention will indicate frontside attack, and racemization will indicate formation of a flattened (achiral) intermediate such as a carbocation. If two steps have occurred between starting material and product, the interpretations will differ. Retention could be the result of two inversions, and racemization could be the result of some inverting attack by the former leaving group.

One may determine whether additions to alkenes are syn, anti, or mixed. For example, trifluoromethyl hypochlorite adds to alkenes, and one may propose several mechanisms:

$$CF_3OCl + \gt C{=}C\lt \longrightarrow \left[CF_3O^- + \gt \overset{+}{C}{-}\overset{Cl}{\overset{|}{C}}\lt \right] \longrightarrow \gt \underset{OCF_3}{\underset{|}{C}}{-}\overset{Cl}{\overset{|}{C}}\lt \tag{3}$$

$$CF_3OCl + \rangle C{=}C\langle \longrightarrow \left[CF_3O^- + \rangle \overset{Cl^+}{C{-}C} \langle \right] \longrightarrow \rangle \underset{OCF_3}{C}{-}\overset{Cl}{C}\langle \quad (4)$$

$$CF_3OCl + \rangle C{=}C\langle \longrightarrow \left[\begin{array}{c} CF_3O\cdots Cl \\ \rangle C{\cdots}C\langle \end{array} \right] \longrightarrow \rangle \overset{CF_3O}{C}{-}\overset{Cl}{C}\langle \quad (5)$$

When this addition was carried out with pure *cis*-2-butene, only the erythro product was obtained, and with pure *trans*-2-butene, only the threo product was obtained.[5] If Eq. 3 were the mechanism, a mixture would be expected; if Eq. 4 were the mechanism, the results would have been the opposite. Only Eq. 5 is in accord with the stereochemical results, that is, a concerted (or nearly so) syn addition.

In elimination reactions, a similar comparison of the stereochemistry of starting materials and products can indicate syn, anti, or mixed processes.

7.3.4 Concentration Dependence of Kinetics

The measurement of rates of reactions under various conditions gives several different kinds of mechanistic information.[6] The concentration dependence of rates can give information on the number of steps in a mechanism and on the species involved in reaction collisions.

For kinetic purposes the components in a homogeneous solution are measured in concentration terms, usually moles solute per liter of solvent. When a reaction is in progress concentration(s) of starting material(s) is (are) decreasing while the concentration(s) of product(s) is (are) increasing. At any one moment a certain number of moles of a product are appearing per liter of solution per second. This is called the *rate* of the reaction. It generally decreases as the supply of starting material is depleted. This is shown graphically in Fig. IV for the reaction $A \rightarrow B + C$. The rate at any one moment is the slope of the graph. The slope of the curve for [A] is equal but opposite in sign to that for [B]; thus there are two alternative rate expressions:

$$\text{Rate} = \frac{d[B]}{dt} \quad \text{or} \quad \text{rate} = \frac{-d[A]}{dt} \qquad (6)$$

(rate is in units of moles per liter per second).

Experimentally, a solution of known concentration of starting material A is prepared, and then as the reaction proceeds, the concentration of A

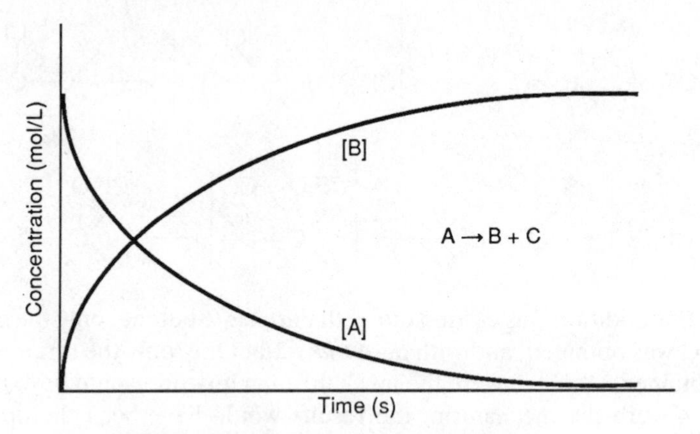

Figure IV. Change in concentration with time.

and/or B is measured repeatedly. In the simple reaction A → B + C, compound A continually decomposes and requires no coreactants or catalysts. When the supply of A reaches half the original concentration, the rate should be half the initial rate. If the rate at any point is divided by the [A] at that point in time, the quotient should be a constant k, sometimes called the *specific rate:*

$$\frac{\text{Rate}}{[\text{A}]} = k \quad \text{or} \quad \frac{-d[\text{A}]}{dt} = k[\text{A}] \tag{7}$$

The value of k is thus independent of the concentration of A, but it does vary with changes in temperature or with structural variation in A. Since by Eq. 7 the rate is proportional only to the concentration of one component to the first power, this is called a *first-order reaction.* The measurement of k values (rate constants) for various related compounds at various temperatures leads to other mechanistic information, as will be shown later in this chapter.

Rather than repeatedly approximating the slopes along a curve, the k value is more readily extracted from a linear plot that results from the integrated form of Eq. 7:

$$-\int_{[\text{A}]_0}^{[\text{A}]_t} \frac{d[\text{A}]}{[\text{A}]} = k\int_0^t dt; \qquad \ln\frac{[\text{A}]_0}{[\text{A}]_t} = kt \tag{8}$$

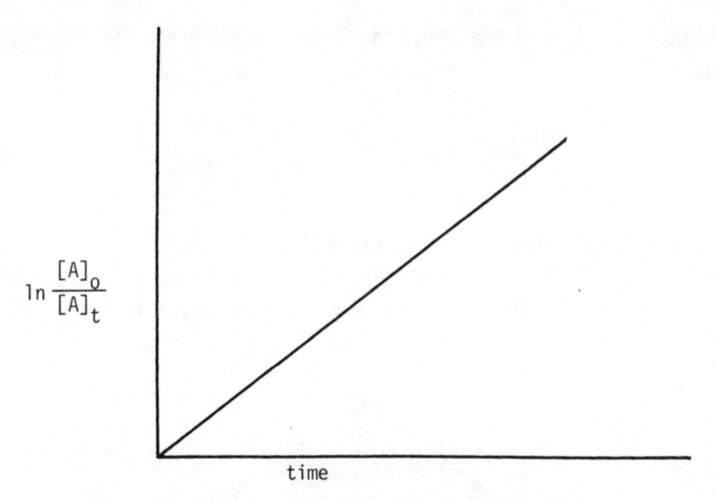

Figure V. Linear plot of kinetic data for a first-order reaction.

Plotting the logarithm of the ratio of an initial [A] to the [A] at time t versus time gives a straight line of slope k (Fig. V). The value of k from the linear plot is based on the whole set of points graphically or least-squares optimized. Notice also that the actual value of [A] does not need to be measured; only a ratio of values of [A] is needed. This sort of ratio is readily obtained spectroscopically as ratios of areas of diminishing peaks in a series of NMR spectra or ratios of absorbance from a spectrophotometer. Finally, if the data do not give a straight line, the reaction is not a first-order reaction.

Dimerization reactions (Eq. 9) give a different kinetic result. Here a collision between two A molecules is necessary; therefore, as the concentration of A decreases, the rate drops faster than found for Eq. 7.

$$\mathrm{A} + \mathrm{A} \longrightarrow \mathrm{B} \tag{9}$$

At half the initial concentration, collisions would be only one-fourth as frequent. Division of the rate by $[\mathrm{A}]^2$ gives a constant in this case (Eq. 10). Since the rate is proportional to the concentration of A squared, it is called a *second-order reaction.*

$$\frac{\mathrm{Rate}}{[\mathrm{A}]^2} = k \quad \text{or} \quad \frac{-d[\mathrm{A}]}{dt} = k[\mathrm{A}]^2 \tag{10}$$

Once again, the data give a linear plot with a slope of k if the rate equation is integrated:

$$-\int_{[A]_0}^{[A]_t} \frac{d[A]}{[A]^2} = k \int_0^t dt; \qquad \frac{1}{[A]_t} - \frac{1}{[A]_0} = kt \tag{11}$$

Another simple bimolecular reaction (Eq. 12) gives a similar rate equation (Eq. 13), which can be integrated to give a linear plot (Eq. 14). This reaction is said to be first order in A and first order in B or second order overall.

$$A + B \longrightarrow C \tag{12}$$

$$\frac{-d[A]}{dt} = k[A][B] \tag{13}$$

$$\frac{1}{[A]_0 - [B]_0} \ln \frac{[B]_0[A]_t}{[A]_0[B]_t} = kt \tag{14}$$

For this type of reaction it is common[7] to simply plot the variables $\ln([A]_t/[B]_t)$ versus t to check for linearity and then calculate k, using Eq. 14.

The kinetics for reactions of the Eq. 12 type are simplified if a large excess of B is used so that as the $[A]_t$ diminishes, [B] changes to only a negligible extent, that is, $[B]_0 \approx [B]_t$. In this case a plot of $-\ln([A]_t/[A]_0)$ versus time is linear as it was for the first-order reaction (Eq. 7). This is thus called *pseudo-first-order conditions*. The observed first-order rate constant may be converted to the actual second-order rate constant by multiplying by $-1/[B]_0$, as can be deduced from Eq. 14. One must repeat the experiment with one or more other concentrations of excess B to be certain that the reaction is first order in [B] (e.g., not zero or second order, which would still give pseudo-first-order data).

Many reactions involve multistep mechanisms. Consider the two-step process shown in Eq. 15, where an intermediate compound I is formed in the first step and consumed in the second.

$$\begin{aligned} A + B &\xrightarrow{k_1} I \\ I &\xrightarrow{k_2} C \end{aligned} \tag{15}$$

If the first step is relatively slow and the second step fast, the I will be consumed as rapidly as it is formed. The concentration of I will remain very low and practically constant. Assuming that constancy (known as the *steady-state approximation*), we may equate the expressions for the rate of formation of I and the rate of decomposition of I:

$$k_1[\mathrm{A}][\mathrm{B}] = k_2[\mathrm{I}] \tag{16}$$

Substituting from Eq. 16, we can express the rate of formation of C in terms of measureable concentrations:

$$\begin{aligned} \frac{d[\mathrm{C}]}{dt} &= k_2[\mathrm{I}] \\ \frac{d[\mathrm{C}]}{dt} &= k_1[\mathrm{A}][\mathrm{B}] \end{aligned} \tag{17}$$

This rate expression is the same as that for Eq. 12, meaning that this two-step mechanism is not kinetically distinguishable from the one-step mechanism. The slow first step is rate determining, and subsequent events are not indicated in the kinetics. Detection or diversion of the intermediate would be necessary.

A similar result occurs when the first step is rapid and reversible and the second step is slow:

$$\begin{aligned} \mathrm{A} + \mathrm{B} &\underset{k_{-1}}{\overset{k_1}{\rightleftharpoons}} \mathrm{I} \\ \mathrm{I} &\xrightarrow{k_2} \mathrm{C} \end{aligned} \tag{18}$$

The first step provides a low concentration of I that is related by an equilibrium constant to [A] and [B]:

$$K = \frac{[\mathrm{I}]}{[\mathrm{A}][\mathrm{B}]} \tag{19}$$

the rate of formation of C is proportional to [I], which from Eq. 19 is proportional to [A][B]:

$$\begin{aligned} \frac{d[\mathrm{C}]}{dt} &= k_2[\mathrm{I}] \\ \frac{d[\mathrm{C}]}{dt} &= k_2K[\mathrm{A}][\mathrm{B}] \end{aligned} \tag{20}$$

Once again we observe simple second-order kinetics indistinguishable from

Eqs. 12 and 15. Detection of I and determination of the equilibrium constant would distinguish these.

In the following example reaction of this sort,[8] the prior equilibrium became apparent in the curvature of a plot as will be seen below. Treatment of 1,1-diphenyltrichloroethanol with sodium hydroxide caused elimination of chloroform. The proposed mechanism is

$$\mathrm{Ph_2C(OH)CCl_3 + OH^- \rightleftharpoons Ph_2C(\ddot{\underset{..}{O}}{:}^-)CCl_3 + H_2O}$$

$$\mathrm{Ph_2C(\ddot{\underset{..}{O}}{:}^-)CCl_3 \longrightarrow Ph_2C{=}\ddot{\underset{..}{O}} + {:}\bar{C}Cl_3}$$

$$\mathrm{{:}\bar{C}Cl_3 + H_2O \longrightarrow HCCl_3 + {:}\ddot{\underset{..}{O}}H^-} \tag{21}$$

In this reaction the anionic intermediates were present in low concentrations and not detected directly. A unique feature here is that the concentration of OH^- does not change because it is regenerated in the third step. Measurements showed that the reaction is pseudo first order in the chloroalcohol and also first order in $[OH^-]$.

The reaction was run in the sample cell of a spectrometer, where the temperature was controlled to 0.1°C. The absorbance at 258 nm for benzophenone was followed and plotted versus time as in Fig. VI. Absorbance

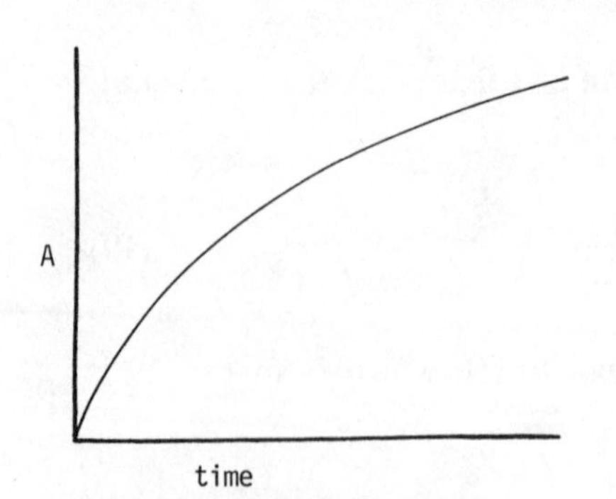

Figure VI. Change in absorbance with progress of the reaction.

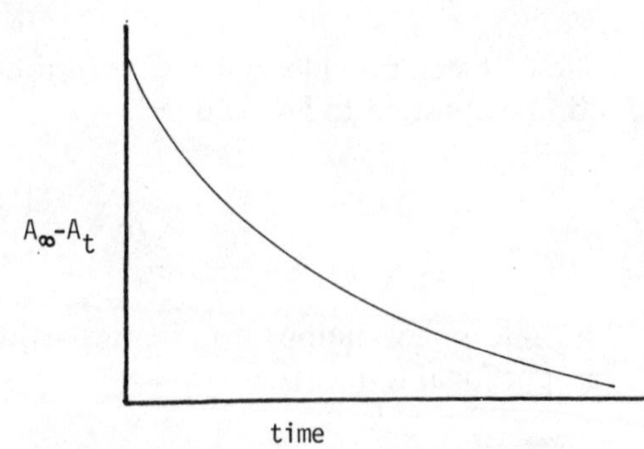

Figure VII. Change in $A_\infty - A_t$ with progress of the reaction.

is proportional to concentration. Figure VII shows a plot of $A_\infty - A_t$, which is proportional to the concentration of the chloroalcohol, assuming 100% conversion:

$$[\mathrm{ROH}]_t = j(A_\infty - A_t) \tag{22}$$

Rewriting Eq. 8 for this case (Eq. 23) and substituting from Eq. 22, we find Eq. 24. Since A_0 is zero, we have Eq. 25.

$$\ln \frac{[\mathrm{ROH}]_0}{[\mathrm{ROH}]_t} = kt \tag{23}$$

$$\ln \frac{j(A_\infty - A_0)}{j(A_\infty - A_t)} = kt \tag{24}$$

$$\ln A_\infty - \ln(A_\infty - A_t) = kt \tag{25}$$

A plot of $\ln A_\infty - \ln(A_\infty - A_t)$ versus t is linear and the slope is the pseudo-first-order rate constant k_{obs}. Since $\ln A_\infty$ is a constant, that sets the intercept; it does not affect the slope and may be omitted.

Numerous runs were followed, beginning with $1.5 \times 10^{-5}M$ chloroalcohol and different concentrations of OH^-. The rate was proportional to the $[OH^-]$ in the pH range of 10–12, indicating a first-order dependence on $[OH^-]$, but at higher $[OH^-]$ the rate increase was less than proportional. This is apparent in a plot of the log of the pseudo-first-order rate constants against pH (Fig. VIII). The linear portion of the plot is in accord with Eq. 21 but would just as well fit a reaction of the sort in Eq. 12. However, since the chloroalcohol has a pK_a of 12, it is largely converted to alkoxide

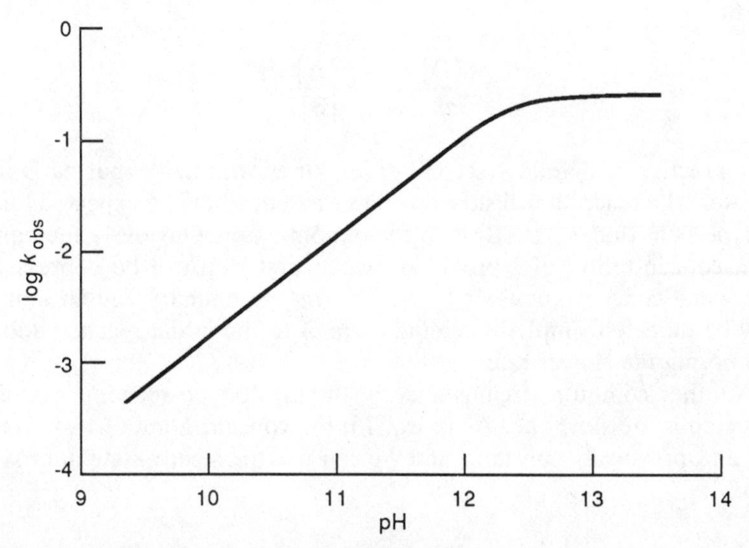

Figure VIII. Logarithm of pseudo-first-order rate constants versus pH for the reaction of Eq. 21 at 25°C. Reprinted by permission from Nome, F.; Erbs, W.; Correia, V. R. *J. Org. Chem.* **1981,** *46,* 3802. Copyright 1981 American Chemical Society.

near pH 14 and additional OH^- would do little to change this; therefore, the slope levels off. This is in accord with the prior equilibrium in Eq. 21 but not Eq. 12. Thus the slow step follows the preequilibrium.

A more complex expression can be written that will predict the linear and curved portions[9] but is beyond our scope of coverage here.

The mechanism in Eq. 21 is called an E1cB because it is an elimination (of $CHCl_3$) from the conjugate base of the chloroalcohol. The 1 signifies that the reaction is first order in the conjugate base (step 2).

Another sort of preliminary equilibrium is illustrated in Eq. 26, where the starting material A dissociates rapidly and reversibly to give a low concentration of intermediate I and by-product B.

$$\begin{aligned} \mathrm{A} &\underset{k_{-1}}{\overset{k_1}{\rightleftharpoons}} \mathrm{I} + \mathrm{B} \\ \mathrm{I} + \mathrm{C} &\xrightarrow{k_2} \mathrm{D} \end{aligned} \tag{26}$$

The rate expression is first stated for the slow step, the consumption of I by C:

$$\frac{d[\mathrm{D}]}{dt} = k_2[\mathrm{I}][\mathrm{C}] \tag{27}$$

The unmeasurable [I] is replaced by using the equilibrium expression for the first step:

$$\frac{d[\mathrm{D}]}{dt} = \frac{k_2K[\mathrm{A}][\mathrm{C}]}{[\mathrm{B}]} \tag{28}$$

This reaction will follow second-order kinetics initially, but as B accumulates, the reaction will slow down more than would be expected for Eq. 13-type behavior, since [B] is in the denominator. Obviously the equilibrium concentration of I provided by the first step will be depressed by increasing concentrations of B. In fact, the preliminary equilibrium step may be detected simply by adding extra B to the initial reaction solution and noting the slower rate.

Another common circumstance is an Eq. 26-type reaction where the first step is the slow one. As in Eq. 15, the concentration of I will remain low and practically constant, and we can use the steady-state approximation:

$$\frac{d[\mathrm{I}]}{dt} = 0 = k_1[\mathrm{A}] - k_{-1}[\mathrm{I}][\mathrm{B}] - k_2[\mathrm{I}][\mathrm{C}] \tag{29}$$

The rate of product formation depends on the [I] and [C] (Eq. 30), but substituting from Eq. 29, we find Eq. 31. Early in the reaction the concentrations of I and B are very small, and the reaction follows first-order kinetics.

$$\frac{d[\mathrm{D}]}{dt} = k_2[\mathrm{I}][\mathrm{C}] \tag{30}$$

$$\frac{d[\mathrm{D}]}{dt} = k_1[\mathrm{A}] - k_{-1}[\mathrm{I}][\mathrm{B}] \tag{31}$$

Added extra B may slow the reaction, but from the kinetics, it is obvious that there are two steps because reactant C, which is involved in the product formation, is absent from the rate expression. It must come in later in the fast step. The familiar S_N1 reactions are of this type.

Many other reaction sequences lead to a wide variety of rate equations, some of which are easily analyzed in ways analogous to those developed above, and many of which require more complicated treatments.[6]

7.3.5 Isotope Effects in Kinetics

The ground-state vibrational energy of a bond is lower for a bond to a heavier isotope than it is for a lighter isotope. Therefore, the activation energy required to break the bond with the heavier isotope will be greater than that required for the lighter isotope. The higher activation energy process is slower. The largest differences are found with hydrogen, where the mass ratios of the isotopes are greatest. For deuterium, rate constant ratios k_H/k_D range up to about 10. The smaller mass ratio for ^{12}C to ^{13}C gives $k_{^{12}C}/k_{^{13}C}$ up to about 1.1. These ratios are called *primary kinetic isotope effects*.[1,10] Heavier isotopes located near a reaction site but not involved in bond breaking give a smaller secondary kinetic isotope effect.

If the rate-determining step of a reaction mechanism involves the breakage of a C—H bond, the corresponding C—D compound will be substantially slower. The maximum effect occurs if the transition state occurs at the midpoint of hydrogen transfer, and the former bonding partner, the coming bonding partner, and the hydrogen are colinear. If the C—H bond is broken in a step subsequent to the rate determining one, k_H/k_D will be zero.

A distinction was made between two proposed mechanisms for the fragmentation of an oxaziridine, using a combination of the primary kinetic isotope effect and stereochemistry.[11] Amines can behave as bases and attack at hydrogen, or they can be nucleophiles and attack at carbon or nitrogen;

E2 and S_N2 reactions are examples. In Eq. 32 a tertiary amine (brucine) is shown in the role of base in a step resembling an E2 reaction.

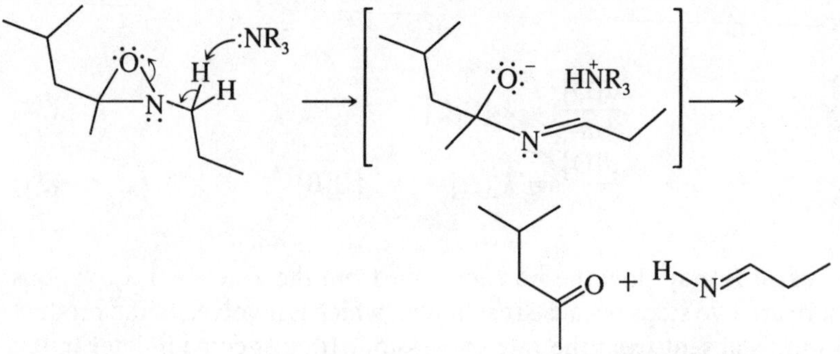

:NR$_3$ H H HN$^+$R$_3$ (32)

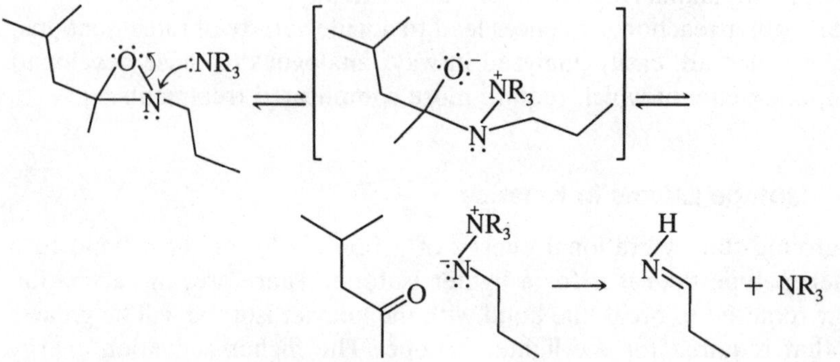

:NR$_3$ N$^+$R$_3$ N$^+$R$_3$ H + NR$_3$ (33)

In Eq. 33 the amine is shown attacking the nitrogen as in an S_N2 reaction. In order to disprove one of these alternatives, the dideutero analog **1** was prepared and the rates were measured for the fragmentation of both the hydrogen and deutero compounds with various concentrations of brucine in refluxing acetonitrile.

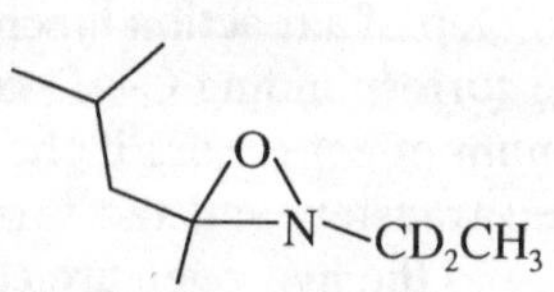

CD_2CH_3

1

The ratio of the pseudo-first-order rate constants k_H/k_D was found to be 4.25. Thus the removal of the H or D is in or before the rate-determining step. This is in accord with Eq. 32, with the first step rate-determining. It would also be in accord with Eq. 33 if the last step were rate-determining.

If the last step were the slow one, either intermediates would accumulate (they do not in this case) or the early steps must be fast and reversible, giving only a low concentration of intermediate. The fact that ring opening is not reversible was proven stereochemically. Inversion of configuration on nitrogen does not readily occur when the nitrogen is part of a three-membered ring; therefore, the *Z* and *E* isomers of the oxaziridine could be prepared and isolated. One isomer was treated with brucine until 43% of it was converted to products, and the remaining oxaziridine was examined by ^{1}H NMR. This showed complete absence of the other diastereomer, which would have been present if some of the acyclic intermediate had reclosed. Thus Eq. 33 is not in accord with the results, and Eq. 32 remains as the most reasonable mechanism.

7.3.6 Temperature Effects on Kinetics

In any reaction step involving some bond breaking, there will be a potential energy high point called a *transition state* as described in Section 7.1. The higher the temperature, the larger the proportion of molecules with sufficient kinetic energy to successfully reach the transition state condition. Determining how steeply the rate of a reaction increases with increasing temperature gives two kinds of information about the transition state: (1) the enthalpy of activation $\Delta H,^{\ddagger}$ that is, the difference in ΔH (and, therefore, structure) between starting material and the activated complex at the transition state; and (2) an indication of the change in the extent of organization (entropy, ΔS) of the atoms and solvent molecules as they proceed to the transition state. Reactions with a larger $\Delta H^{\ddagger}$ are accelerated to a greater extent by a temperature increase.

The change in enthalpy and entropy from starting materials to products in a reaction may be calculated if an equilibrium constant can be measured for the reaction at several temperatures. If we could set up a simple equilibrium between starting materials and the transition state for a reaction and measure the amounts of each present, we could calculate ΔG, ΔH, and ΔS for that change. However, a transition state is too short-lived, and we cannot measure the molar concentration of it to calculate the equilibrium constant for the formation of it.

Transition-state theory affords us a way to determine a "concentration" of activated complexes from rate measurements. At a certain temperature, a certain proportion of activated complexes will proceed on to product formation. Knowing the rate of product formation and that proportionality, we can calculate the concentration of the activated complex, [AB‡] in Eq. 34.

$$A + B \longrightarrow AB^{\ddagger} \longrightarrow C \qquad (34)$$

The proportionality constant is given in Eq. 35. It is the same for any reaction and consists of Boltzmann's constant k_B, absolute temperature T, and Planck's constant h.

$$\frac{-d[A]}{dt} = \frac{k_B T}{h}[AB^{\ddagger}] \tag{35}$$

The rate of the reaction is measurable and the rate constant relating it to known concentration of starting materials may be calculated:

$$\frac{-d[A]}{dt} = k_{rate}[A][B] \tag{36}$$

Substituting from Eq. 36 into Eq. 35, we obtain Eq. 37, from which we may extract the equilibrium constant $K^{\ddagger}$ for the formation of the activated complex (Eq. 38).

$$k_{rate}[A][B] = \frac{k_B T}{h}[AB^{\ddagger}] \tag{37}$$

$$K^{\ddagger} = \frac{[AB^{\ddagger}]}{[A][B]} = \frac{k_{rate} h}{T k_B} \tag{38}$$

An equilibrium constant* is a reflection of the change in free-energy ΔG in the process (Eq. 39).[12] The free energy is composed of an enthalpy portion and an entropy portion (Eq. 40).

$$\Delta G^{\ddagger} = -RT \ln K^{\ddagger} \tag{39}$$

$$\Delta G^{\ddagger} = \Delta H^{\ddagger} - T\,\Delta S^{\ddagger} \tag{40}$$

Substituting the proportioned rate constant for $K^{\ddagger}$ (Eq. 38 into Eq. 39) and taking this expression for ΔG into Eq. 40, we find Eq. 41. Isolation of $\ln k_{rate}/T$ gives Eq. 42.

$$-RT \ln \frac{k_{rate} h}{T k_B} = \Delta H^{\ddagger} - T\,\Delta S^{\ddagger} \tag{41}$$

$$\ln \frac{k_{rate}}{T} = -\frac{\Delta H^{\ddagger}}{RT} + \frac{\Delta S^{\ddagger}}{R} + \ln \frac{k_B}{h} \tag{42}$$

* $K^{\ddagger}$ must be rendered dimensionless.

Since $\Delta H^{\ddagger}$ and $\Delta S^{\ddagger}$ are found to be nearly constant over a range of temperature, the only variables are k_{rate} and T. Furthermore, there is a linear relationship between $\ln k_{rate}/T$ and $1/T$ where the slope is $-\Delta H^{\ddagger}/R$ and the intercept is $\Delta S^{\ddagger}/R + \ln k_B/h$. Plotting these from a set of rate constants at several temperatures should give a straight line (Eyring plot), from which $\Delta H^{\ddagger}$ and $\Delta S^{\ddagger}$ may be obtained. The line also indicates the quality of the data and allows extrapolations to predict rates at new temperatures.

Very similar results are obtained by simply plotting $\ln k_{rate}$ versus $1/T$ (Arrhenius plot). In this case the slope is $-E_a/R$, where E_a is called the *Arrhenius activation energy*. For a reaction in solution, E_a may be converted to $\Delta H^{\ddagger}$ by subtracting RT[12] (about 0.6 kcal at room temperature).

For unimolecular reactions, $\Delta S^{\ddagger}$ indicates whether the transition state has either more or less freedom of motion than the starting molecule. For example, the large positive $\Delta S^{\ddagger}$ for the thermal decomposition of the azo compound in Eq. 43[13] is in accord with a simple fragmentation mechanism where the transition state is a more loosely bound, more freely moving structure than the starting molecule.

N=N $\xrightarrow[\text{heat}]{Ph_2O}$ [···N≡N···]‡ ⟶ 2 · + N_2 (43)

$\Delta S^{\ddagger} = 16.3$ cal/mol K

In contrast with this, the large negative $\Delta S^{\ddagger}$ for the reaction in Eq. 44[14] requires that some freedom of motion in the starting molecule, such as rotation of the ring substituent bonds, is restricted in the transition state.

(44)

$\Delta S^{\ddagger} = -11.7$ cal/mol K

The $\Delta S^{\ddagger}$ values calculated from second-order rate constants are not independent of the concentration units and cannot be individually evaluated as above.[15] They can, however, be used to compare similar reactions.[12]

7.3.7 Substituent Effects on Kinetics

The order of bond breaking or bond forming in most reactions leads to temporary development of a positive or negative electrostatic charge at a particular site in the transition-state structure. Determining whether the charge is positive or negative and whether it is relatively large or small allows postulation of the sequence of bond changes. How can we determine this?

Certain substituents are known to withdraw electron density from a reaction site, thus rendering a developing negative charge at that site less intense and lessening the energy required to develop that negative charge. If a negative charge is partially formed on progressing from starting material to the transition state of a reaction, the substance with the electron withdrawing substituent will react at a faster rate than one with a hydrogen atom. Other substituents are known to donate electron density to a reaction site, and these would intensify a developing negative charge, slowing the reaction. In other reactions a positive charge is developed at the reaction site. Here the substituents would have the opposite effect; that is, the electron donating substituents will give faster reactions. For those reactions where the substituents are on a benzene ring and far enough from the reaction site to be sterically out of the way, the substituent effects have been correlated quantitatively. The Hammett equation[16] (Eq. 45) states that the $\log_{10}$ of the ratio of the rate constant for some reaction with a substituent present (k) to the rate constant with no substituent (k_0) is equal to the product of a factor indicating the sensitivity of that reaction to substituent effects (ρ) and the characteristic electronic influence of the substituent (σ).

$$\log_{10} \frac{k}{k_0} = \rho\sigma \tag{45}$$

Table I lists the σ values for a selection of common substituents in meta and para positions. The ρ value for a reaction is determined by first measuring rate constants for a series of examples of the reaction with various substituents present. These are used to plot $\log_{10} k/k_0$ versus σ. The slope of this plot is ρ. With the value of ρ and the table of σ values, one may predict k values of other examples with different substituents.

Where did the σ values come from? Hammett chose to define them from a particular reaction, the ionization of benzoic acid. In this case, and many others, the equilibrium constants rather than rate constants were measured. For equilibria, the Hammett equation is written in terms of equilibrium constants:

$$\log_{10} \frac{K}{K_0} = \rho\sigma \tag{46}$$

Since the ionization of a series of substituted benzoic acids is used to define

TABLE I. Selected Hammett σ Values

Substituent	σ_{meta}	σ_{para}	σ^+_{para}
NMe_2	−0.21	−0.83	−1.7
O^-	−0.17	−0.52	—
NH_2	−0.16	−0.66	−1.3
CO_2	−0.10	0.00	−0.03
CH_3	−0.07	−0.17	−0.31
C_2H_5	−0.07	−0.15	−0.30
H	0	0	0
Ph	0.06	−0.01	−0.17
OH	0.12	−0.37	−0.92
OCH_3	0.12	−0.27	−0.78
SCH_3	0.15	0.00	−0.60
$NHCOCH_3$	0.21	0.00	0.00
OPh	0.25	−0.32	−0.5
SH	0.25	0.15	—
$CONH_2$	0.28	—	—
F	0.34	0.06	−0.07
I	0.35	0.18	0.13
CHO	0.35	0.22	—
Cl	0.37	0.23	0.11
COOH	0.37	0.45	0.42
CO_2R	0.37	0.45	0.48
$COCH_3$	0.38	0.50	—
Br	0.39	0.23	0.15
$OCOCH_3$	0.39	0.31	—
CF_3	0.43	0.54	—
CN	0.56	0.66	0.66
NH_3^+	0.63	—	—
NO_2	0.71	0.78	0.79
NMe_3^+	0.88	0.82	0.41

Source: Ritchie, C. D.; Sager, W. F. *Progr. Phys. Org. Chem.* **1964,** 2, 323.

σ values, the ρ for this reaction was defined as 1.000. To find the σ_m for the chloro substituent, the K_a values for *m*-chlorobenzoic acid, 1.48×10^{-4}, and benzoic acid, 6.27×10^{-5}, are entered in Eq. 46 and it is solved for σ, which equals 0.373, the table value.

In the ionization of benzoic acid, a negative charge develops on the organic molecule. Any other reaction in which a negative charge develops would be affected in the same direction by the substituents, and it would be found to have a positive value for ρ. In kinetic data, the charge development in the transition state is indicated by the positive ρ. In reactions where a positive charge develops in the transition state or the product, the ρ values will be negative. In reactions where the sensitivity to substituents is low, the ρ can approach zero, and where it is high, the ρ value can be as high as ±5 or more. This sensitivity is related to the distance separating the charge and the substituent, the transmitting ability of the intervening atoms, and also to the size of the charge, usually a fraction of a unit charge in transition states.

Estimates of the size of the charge in a transition state have been made by taking a ratio of ρ values for a rate and an equilibrium in very similar reactions.[18]

In some reactions the electron donating substituents in the para position can have a greater effect than their σ values would predict. These are cases where a direct resonance contribution can be made as in benzylic carbocation formation:

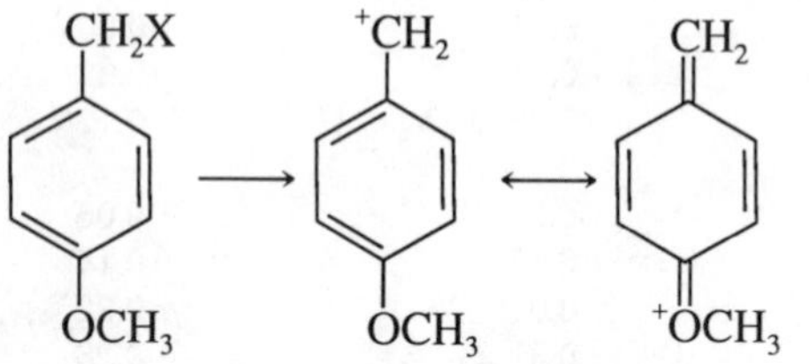

$$p\text{-}CH_3O\text{-}C_6H_4\text{-}CH_2X \longrightarrow p\text{-}CH_3O\text{-}C_6H_4\text{-}{}^{+}CH_2 \longleftrightarrow CH_2{=}C_6H_4{=}{}^{+}OCH_3 \tag{47}$$

For such reactions, another set of substituent values, σ^+, was developed.[19] The rates of solvolysis of meta- and para-substituted cumyl chlorides in 90% aqueous acetone (Eq. 48) were measured.

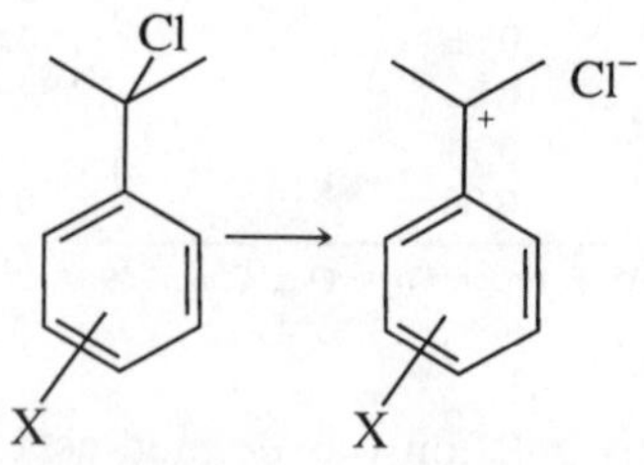

$$X\text{-}C_6H_4\text{-}C(CH_3)_2Cl \longrightarrow X\text{-}C_6H_4\text{-}{}^{+}C(CH_3)_2 \quad Cl^- \tag{48}$$

The ρ for the reaction was found to be −4.54 by plotting only the meta isomers against σ. This was then used to obtain σ^+ values according to Eq. 49 with the para isomers. These are found in Table I also.

$$\log \frac{k}{k_0} = -4.54\sigma^+ \tag{49}$$

An investigation of the thermal decomposition of oxetanones is presented here to illustrate the use of the Hammett equation. 2-Oxetanones decompose with first-order kinetics to give carbon dioxide and an alkene

(Eq. 50).[20] A series of substituted 3- and 4-aryl-2-oxetanones was prepared and the rate constants determined for the decomposition at 150°C.

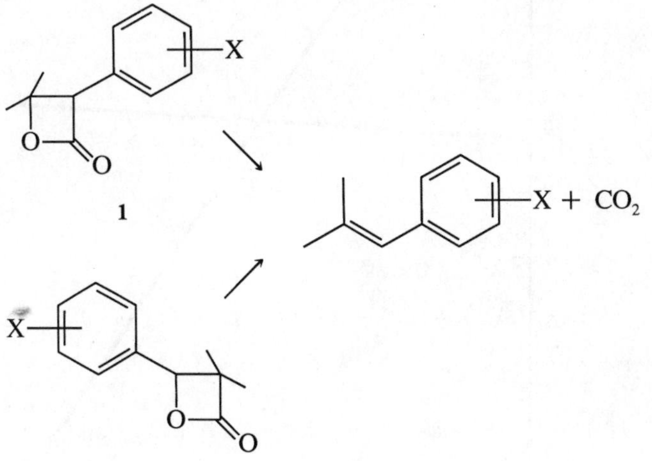

$$\mathbf{1}, \mathbf{2} \longrightarrow \text{ArX-alkene} + CO_2 \qquad (50)$$

A plot of log k/k_0 for each of these versus σ^+ is shown in Figure IX. The line for compound series **1** is almost flat ($\rho = +0.03$) indicating no significant charge develops at ring atom 3 in the transition state. The line for compound series **2** shows a substantial negative slope ($\rho = -1.52$), indicating that a partial positive charge develops at ring atom 4 in the transition state as indicated in **3**.

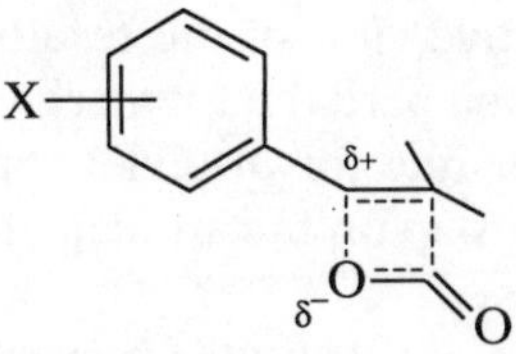

3

The partial plus charge is at a benzylic site, which is the reason for using σ^+ instead of σ. Thermolysis of *cis*-3-*tert*-butyl-4-phenyl-2-oxetanone gives pure *cis*-alkene, which suggests a concerted mechanism with some dipolar character, because a completely formed carbocation would be detached from the oxygen and likely allow rotation of the σ bond to give a mixture of *cis*-alkene and *trans*-alkene.

It is further possible that without a 4-aryl group, no significant charge would develop on ring atom 4 and the reaction would be more cleanly

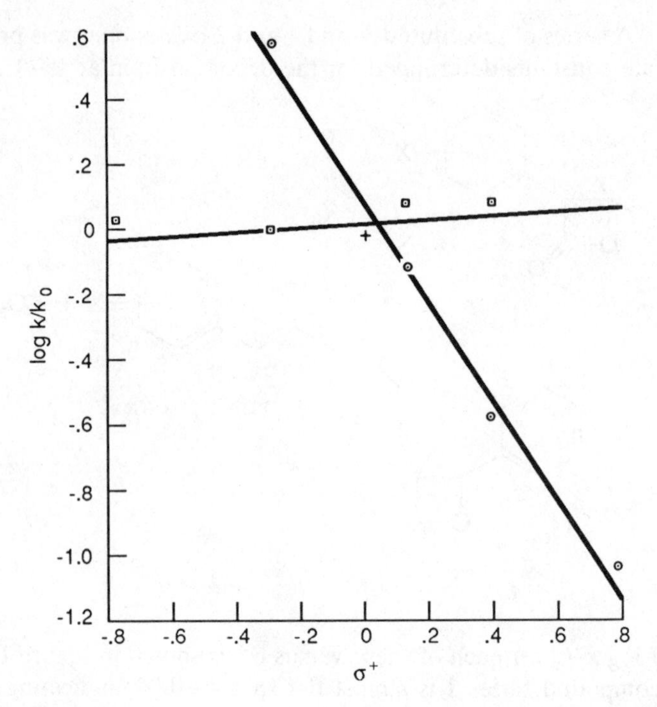

Figure IX. Plot of log k/k_0 versus σ^+ for the reactions in Eq. 50 (3-aryloxetanones ⊡; 4-aryloxetanones ⊙).

concerted, as allowed by the fact that compounds **1** all react faster than compounds **2.**

One may gain a qualitative view of the results without measuring ρ or using a graph by noting that a partial positive charge develops at ring atom 4 as indicated by the faster rate for *p*-tolyl-**2** compared to phenyl-**2.** The electron-donating methyl group on the aromatic ring stabilizes the δ+ and renders it more easily formed.

The familiar bromination of substituted benzenes with Br_2 in acetic acid also correlates with σ^+, and the ρ is -12, indicating substantial plus-charge development very close to the substituents as shown for anisole[19] in Eq. 51. Notice also that the activating ortho, para-directing groups generally have negative σ^+ values and the deactivating meta-directing groups have positive σ^+ values.

:OCH$_3$ $\xrightarrow{Br_2}$ H, Br / $^+$OCH$_3$ ⟷ H, Br / + / :OCH$_3$ ⟷ H, Br / + / :OCH$_3$ ⟷ H, Br / + / :OCH$_3$ (51)

The ρ values for an enormous number of reactions have been determined, and a lengthy list of σ values is available. Beyond this, many variations on the original equation have been formulated.[17]

The Hammett equation is called a *linear free-energy relationship*. The log K and log k values are proportional to ΔG and $\Delta G^{\ddagger}$ values, respectively (Eq. 39) for equilibria and rates. The change in ΔG or $\Delta G^{\ddagger}$ with changing substituents for many reactions is linearly related to the σ scale. Therefore, the change in ΔG or $\Delta G^{\ddagger}$ with changes in substituents in one reaction is linearly related to the change in ΔG or $\Delta G^{\ddagger}$ in another reaction for the same changes in substituents:

$$\frac{\Delta\,\Delta G_1}{\rho_1} = \frac{\Delta\,\Delta G_2}{\rho_2} \qquad (52)$$

where $\Delta\,\Delta G_1$ is the change in free energy for reaction 1 for a change in substituents and $\Delta\,\Delta G_2$ is the change in free energy for reaction 2 for the same change in substituents.

7.4 REPRESENTATIVE MECHANISMS

After decades of mechanistic investigations and the discerning of generalities, it is now common practice to propose reasonable mechanisms for new reactions, considering their relationships to ones that have been investigated mechanistically. At the simplest level, these proposals consist of a succession of reaction intermediates without great attention to transition states. A sampling of such proposals is gathered in this section without supporting mechanistic data. An accounting of atoms and electrons must be included. For simplicity, only one resonance form of each ion or radical is given in most cases.

Some working principles to keep in mind in writing such proposals include:

Sites of unlike charge attract one another and often become bonded together.

Elements of different electronegativity that are bonded together carry partial charges.

In most ionic and concerted reactions, electrons remain paired throughout the process.

An odd, unpaired electron exists on radical species temporarily.

Complete electron octets are maintained on all C, N, O, or F atoms except at carbocation, radical, carbene, and nitrene sites where fewer than eight electrons reside temporarily.

In most steps, reactivity should decrease on progressing to product. For example, highly basic compounds give products of lower basicity, and unstable 1° carbocations rearrange to 2° to 3° where possible.

7.4.1 Reactions in Basic Solution

A. Overall reaction:[21]

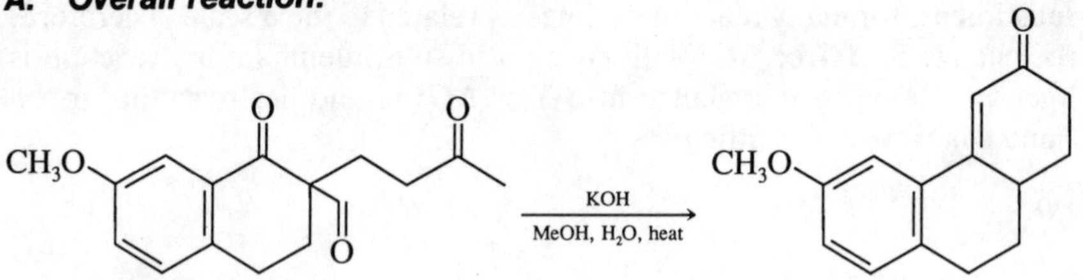

Intermediates:

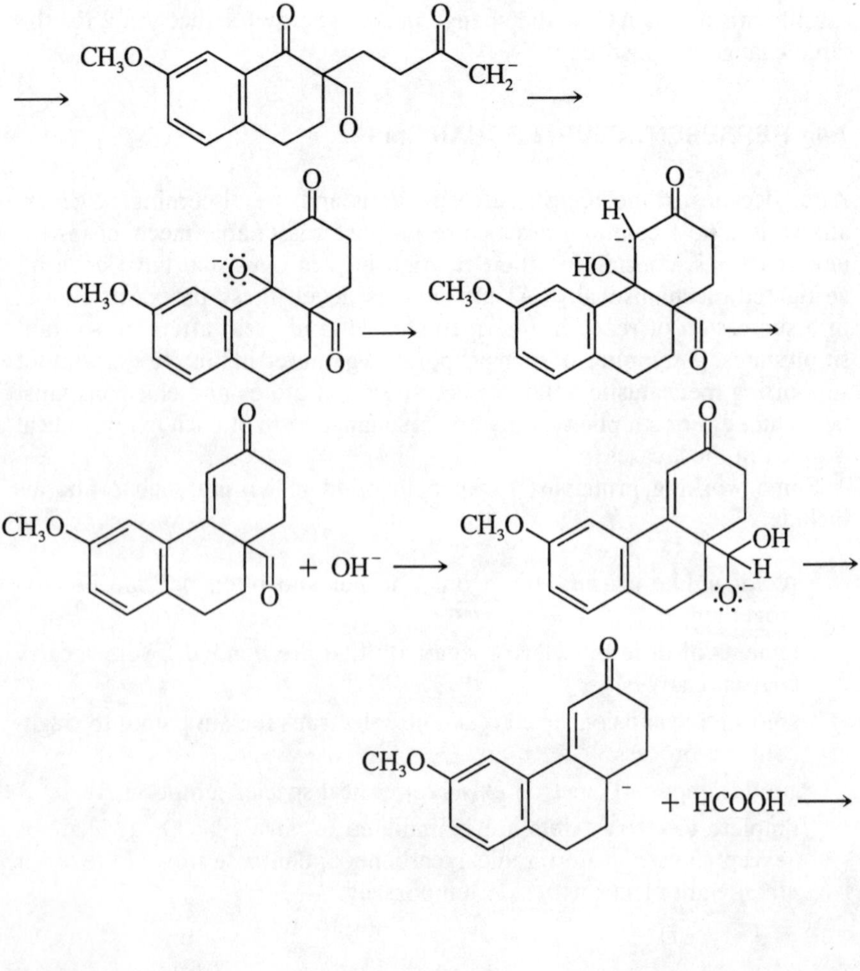

Treatment of a ketone with the relatively weak base OH^- generates small equilibrium amounts of enolate anions at sites α to the carbonyl group. The nonbonding electron pair on the carbanion site is attracted to the partial plus on another ketone site and becomes a bond. The final weak base present is formate.

B. *Overall reaction:*[22]

$NNHSO_2Ph$ — 1. 2 *n*-BuLi, $Me_2NCH_2CH_2NMe_2$, −50°C to rt; 2. D_2O → D

Intermediates:

→ :N–N̈⁻–SO_2Ph → $^-\ddot{C}H_2$:N–N̈–SO_2Ph →

:N=N̈⁻ + $PhSO_2^-$ → (vinyl carbanion) + N_2 $\xrightarrow{D_2O}$

Two equivalents of base remove two relatively acidic hydrogens, and then the weak base phenylsulfenate is formed, followed by the very stable molecule N_2. The vinyl carbanion is then deuterated.

C. *Overall Reaction:*[23]

O, O, N, CH_3, Ph + $(CH_3)_3S^+I^-$ + KOH $\xrightarrow{DMSO}$

O, O, N, CH_3, Ph + CH_3SCH_3

Intermediates:

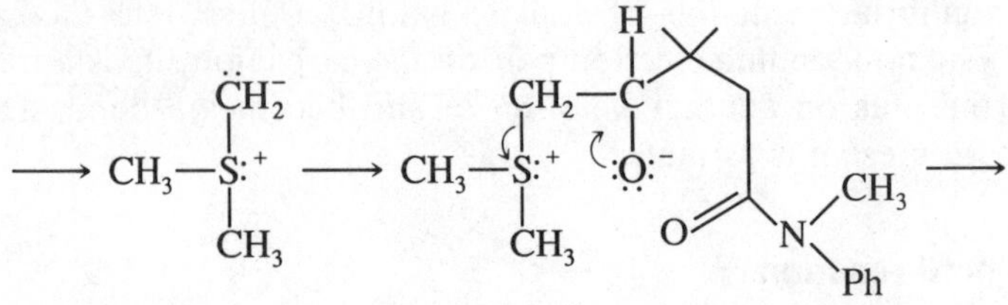

$$\longrightarrow CH_3-\ddot{S}^{+}(\ddot{C}H_2^-)(CH_3) \longrightarrow CH_3-\ddot{S}^{+}(CH_2-C(H)-\ddot{O}:^-)(CH_3) \longrightarrow$$

The hydroxide removes a proton from the trimethylsulfonium ion to give a 1,2-dipolar species called an *ylide,* which has nucleophilic character on the CH_2 group. This adds to an aldehyde to give an alkoxide that displaces the dimethyl sulfide leaving group.

D. Overall Reaction:[24]

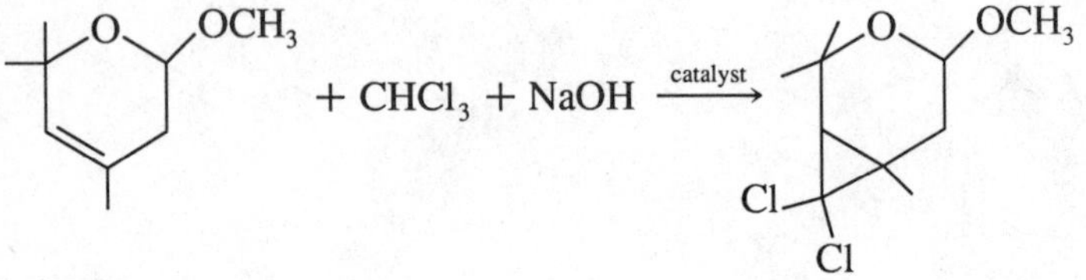

$$+ CHCl_3 + NaOH \xrightarrow{\text{catalyst}}$$

Intermediates:

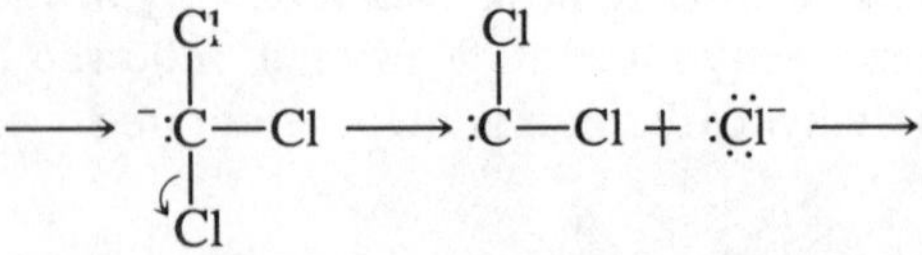

$$\longrightarrow {}^-:C(Cl)_2-Cl \longrightarrow :C(Cl)-Cl + :\ddot{Cl}^- \longrightarrow$$

The hydroxide removes a proton from chloroform to give the trichlorocarbanion which loses a chloride ion to give the neutral dichlorocarbene. This is electrophilic and forms two σ bonds to the alkene site.

Some other electrophiles that convert alkenes to cyclopropanes are not free carbenes but have metals coordinated with their electrophilic site. These are called *carbenoids*. The familiar Simmons–Smith reaction (Section 5.3) is an example. The electrophile is a zinc organometallic derived from a dihalomethane and zinc–copper couple. Another frequently used example is the copper-, rhodium-, or palladium-catalyzed decomposition of diazoketones and esters[25,26] (Eq. 53).[27] The presence of the metal atom in the electrophile is shown in the variation of the stereoselectivity of the reaction with changes in the other ligands on metal.

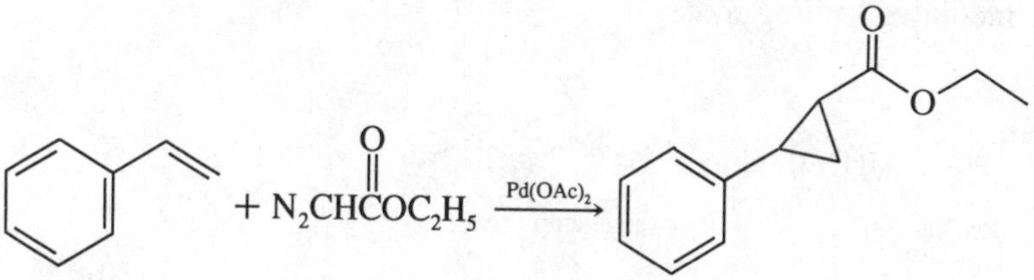

$$+ N_2CHCOC_2H_5 \xrightarrow{Pd(OAc)_2} \quad (53)$$

98%

E. Overall Reaction:[28]

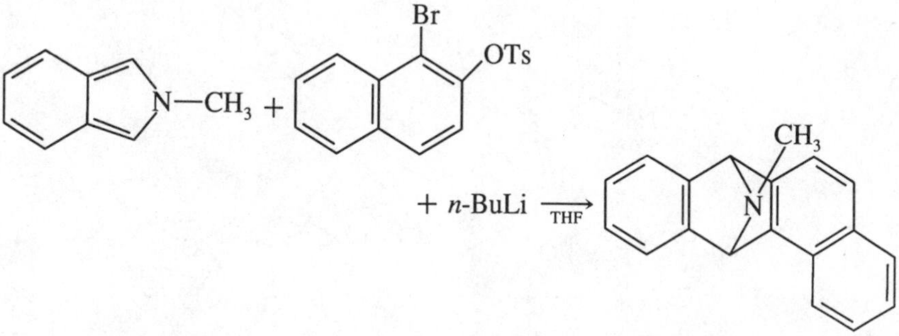

Br, OTs

$N{-}CH_3$ + ... + n-BuLi $\xrightarrow{THF}$ (CH_3, N)

Intermediates:

→ (Li, OTs) → →

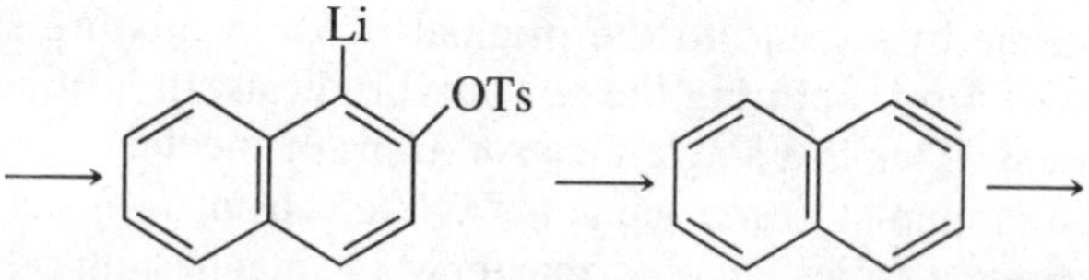

The n-butyllithium exchanges with the bromo compound to give n-butyl bromide and an aryllithium. Elimination of lithium tosylate gives the naphthalyne, which combines with the isoindole in a concerted process.

F. Overall Reaction:[29]

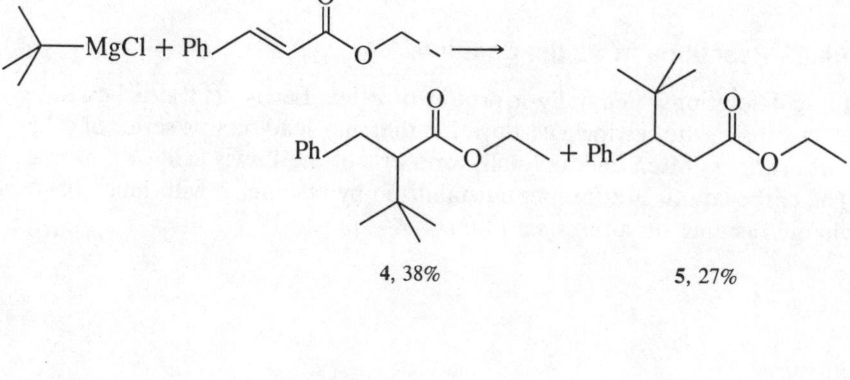

—MgCl + Ph (O, O) ⟶

Ph (O, O) + Ph (O, O)

4, 38% **5**, 27%

Intermediates leading to **4:**

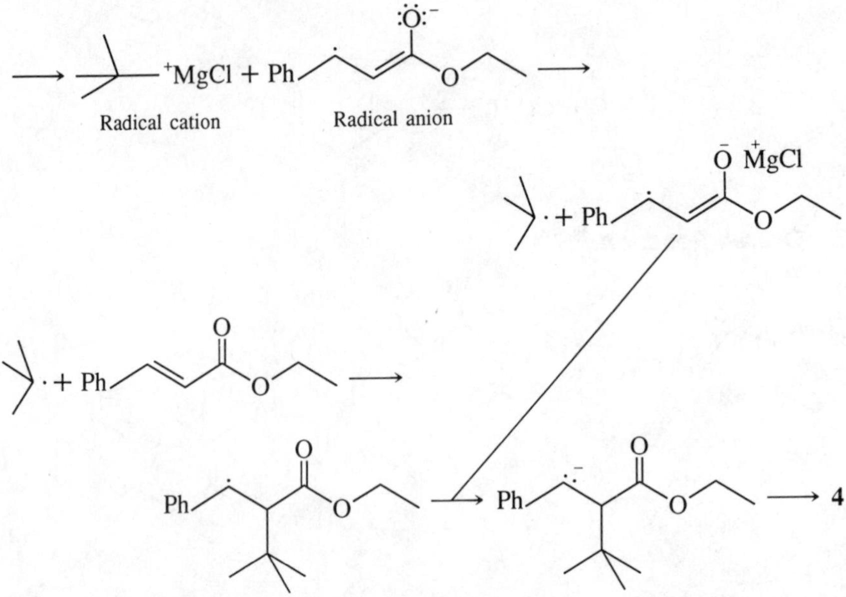

Radical cation

Radical anion

Addition of α-methylstyrene to the original reaction mixture suppresses the formation of **4** by capturing the *tert*-butyl radicals, indicating that it is a radical process. This is a single-electron-transfer mechanism where the easily reduced cinnamate ester gains a single electron from the Grignard reagent to become a radical anion. Transfer of the magnesium cation leaves a *tert*-butyl radical that attacks another ethyl cinnamate molecule to give a benzylic radical. This radical then takes back the electron originally transferred, giving the benzylic carbanion which upon protonation gives product **4.** Product **5** may arise from conventional carbanion conjugate addition or from the collapse of the radical pair.

7.4.2 Reactions in Acidic Solution

In acid solutions, generally a proton or other Lewis acid attaches to an electron-rich site, giving a carbocation that may lead on to a series of other carbocations, often ending finally with loss of the Lewis acid or a proton. The carbocations are frequently stabilized by resonance with much of the charge residing on an oxygen atom.

A. Overall Reaction:[30]

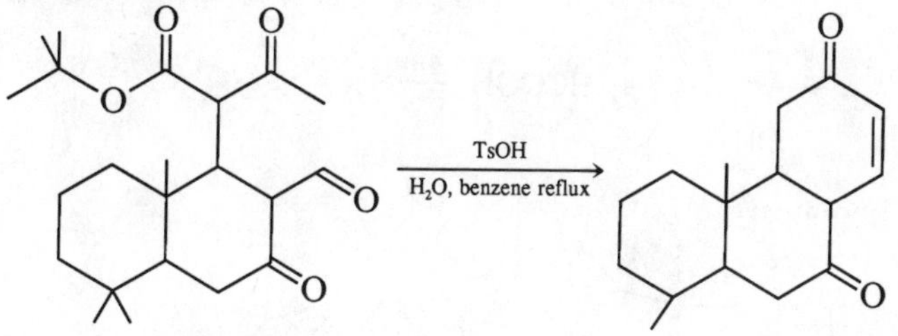

TsOH

H_2O, benzene reflux

Intermediates:

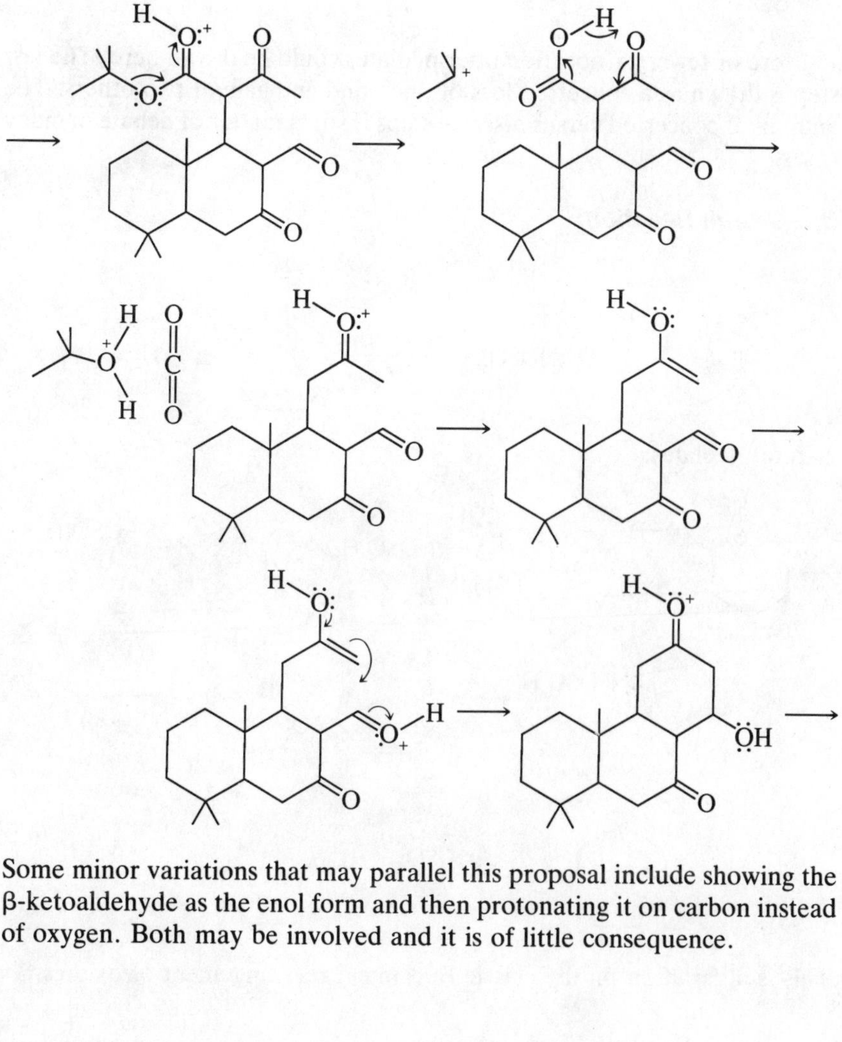

Some minor variations that may parallel this proposal include showing the β-ketoaldehyde as the enol form and then protonating it on carbon instead of oxygen. Both may be involved and it is of little consequence.

B. Overall Reaction:[31]

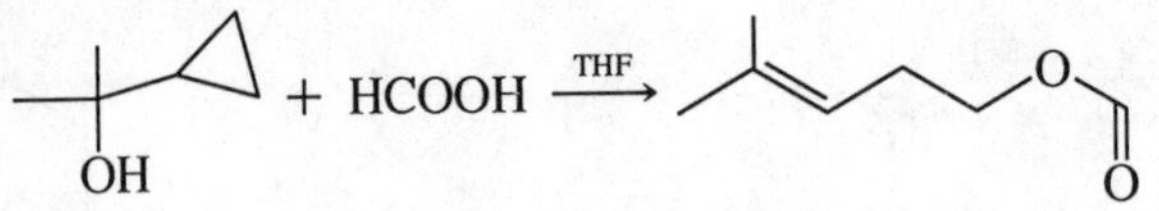

+ HCOOH $\xrightarrow{\text{THF}}$

OH

O

O

Intermediates:

H_2O

:Ö—CH

O

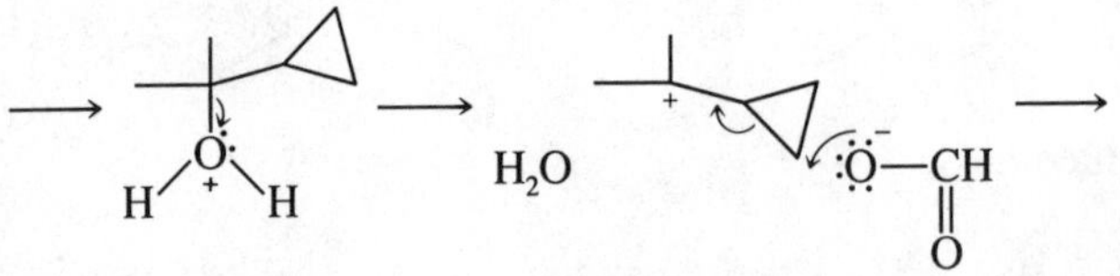

More or fewer carbocation intermediates could be drawn here. The last step is drawn as a concerted loss of one bond and gain of two others. The amount of concerted versus discrete steps is still a matter of debate in many cases.

C. Overall Reaction:[32]

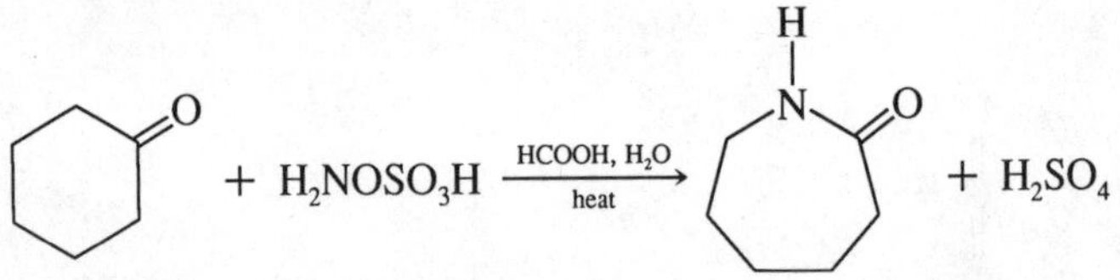

+ H_2NOSO_3H $\xrightarrow[\text{heat}]{\text{HCOOH, } H_2O}$ + H_2SO_4

Intermediates:

$\ddot{O}^+$—H

OH H NOSO$_3$H H

N—OSO$_3$H

N $\bar{O}SO_3H$

N H :ÖSO$_3$H O H

N O—H H_2SO_4

H N$^+$ $^-OSO_3H$ OH

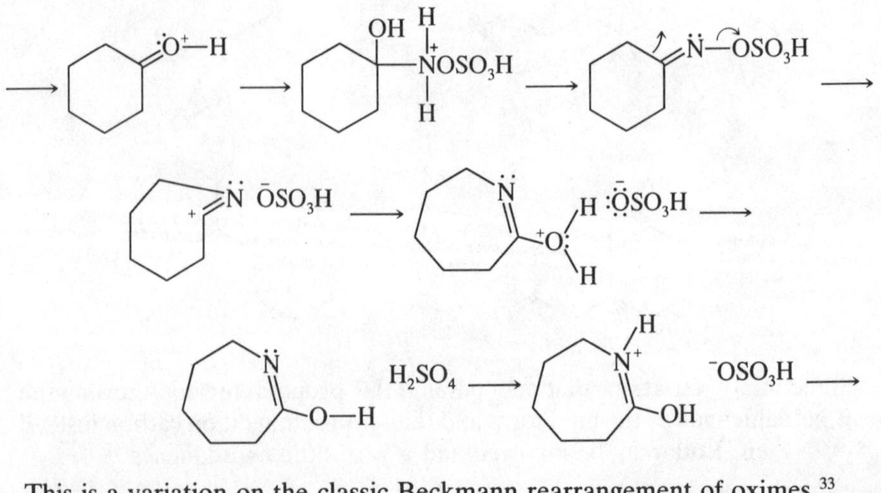

This is a variation on the classic Beckmann rearrangement of oximes.[33]

7.4.3 Free-Radical Reactions

Certain classes of compounds will decompose thermally or photochemically in a homolytic way; that is, a bonding pair of electrons will unpair. Aliphatic azo compounds and peroxides commonly do so and can be used in catalytic amounts to initiate radical chain reactions of other molecules.

A. Overall Reaction:[34]

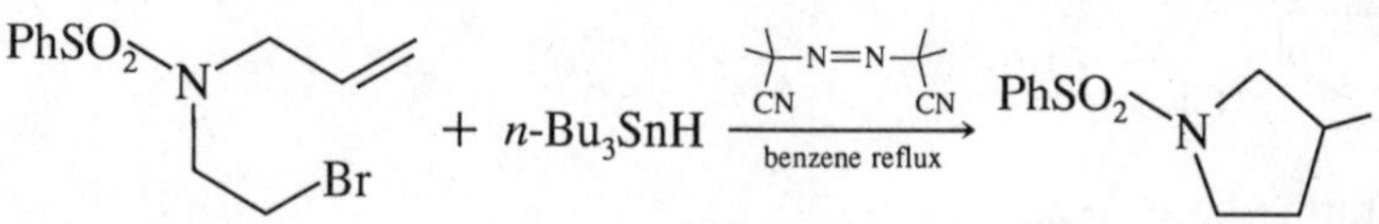

$PhSO_2$–N(–CH$_2$CH=CH$_2$)(–CH$_2$CH$_2$Br) + n-Bu_3SnH —(Me$_2$C(CN)–N=N–C(CN)Me$_2$; benzene reflux)→ $PhSO_2$–N(3-methylpyrrolidine)

Intermediate steps:

1. Me$_2$C(CN)–N=N–C(CN)Me$_2$ ⟶ 2 Me$_2\dot{C}$(CN) + N_2 (CN, CN, CN)

2. Me$_2\dot{C}$(CN) + n-Bu_3SnH ⟶ Me$_2$C(CN)–H + n-$Bu_3Sn\cdot$

} initiation

1. n-$Bu_3Sn\cdot$ + $PhSO_2$–N(–CH$_2$CH=CH$_2$)(–CH$_2$CH$_2$Br) ⟶ n-Bu_3SnBr + $PhSO_2$–N(–CH$_2$CH=CH$_2$)(–CH$_2$$CH_2\cdot$)

2. $PhSO_2$–N(–CH$_2$CH=CH$_2$)(–CH$_2$$CH_2\cdot$) ⟶ $PhSO_2$–N(pyrrolidine)–3-$CH_2\cdot$

3. $PhSO_2$–N(pyrrolidine)–3-$CH_2\cdot$ + n-Bu_3SnH ⟶ $PhSO_2$–N(pyrrolidine)–3-CH_3 + n-$Bu_3Sn\cdot$

} propagation

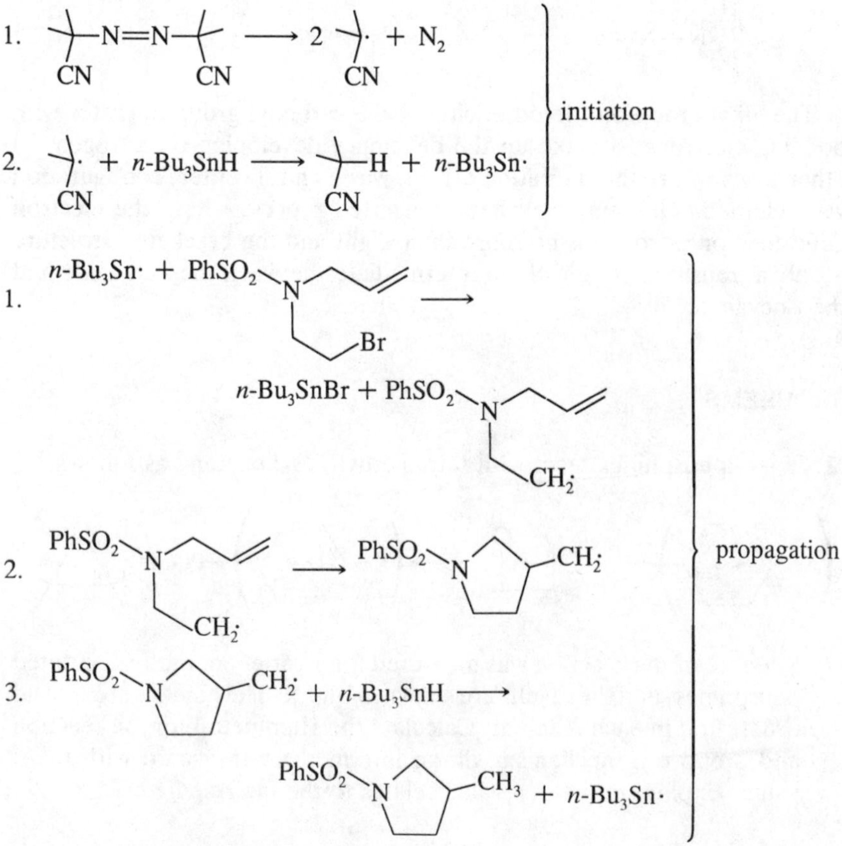

7.4.4 Rearrangement to Electron-Deficient Nitrogen

The Lossen and Hoffmann rearrangements involve an α elimination from an amide nitrogen to leave the nitrogen neutral but short of an octet of electrons. The loss of N_2 in the Curtius and Schmidt rearrangements gives a similar circumstance on nitrogen.

A. Overall Reaction:[35]

$$\text{propanoyl chloride} + n\text{-Bu}_3\text{SnN}_3 \xrightarrow[50\text{–}80^\circ\text{C}]{} \text{CH}_3\text{CH}_2\text{N=C=O} + n\text{-Bu}_3\text{SnCl} + \text{N}_2$$

Intermediate steps:

$$\text{CH}_3\text{CH}_2\text{C(=O)}\ddot{\text{N}}\text{=}\overset{+}{\text{N}}\text{=}\bar{\text{N}}\text{:} \longrightarrow \left[\ \text{transition state: C(=O)}\cdots\text{N}\cdots\text{N}\equiv\text{N}\ \right]^{\ddagger} \longrightarrow \text{O=C=N–CH}_2\text{CH}_3 + \text{N}_2$$

The alkyl group on the other side of the carbonyl group migrates with bonding electrons to make up the deficiency developing on nitrogen. In other cases where the migrating carbon was a chiral center, configuration was retained. This may well be a concerted process where the electron deficiency on nitrogen is no more than slight and the bracketed structure is only a transition state with no intermediates between the acyl azide and the isocyanate.

PROBLEMS

1. Triarylphosphines react with tetramethyl-1,2-dioxetane as follows:[36]

$$(X\text{–C}_6\text{H}_4)_3\text{P} + \text{tetramethyl-1,2-dioxetane} \longrightarrow (X\text{–C}_6\text{H}_4)_3\text{PO} + \text{tetramethyloxirane}$$

The rate of the reaction was measured for a variety of para-substituted phosphines, and the results are shown in Fig. X. The kinetics are second order, first in each reactant. Calculate the Hammett ρ for the reaction and propose a mechanism via an intermediate in accord with the ρ value. Explain how the ρ value relates to the intermediate.

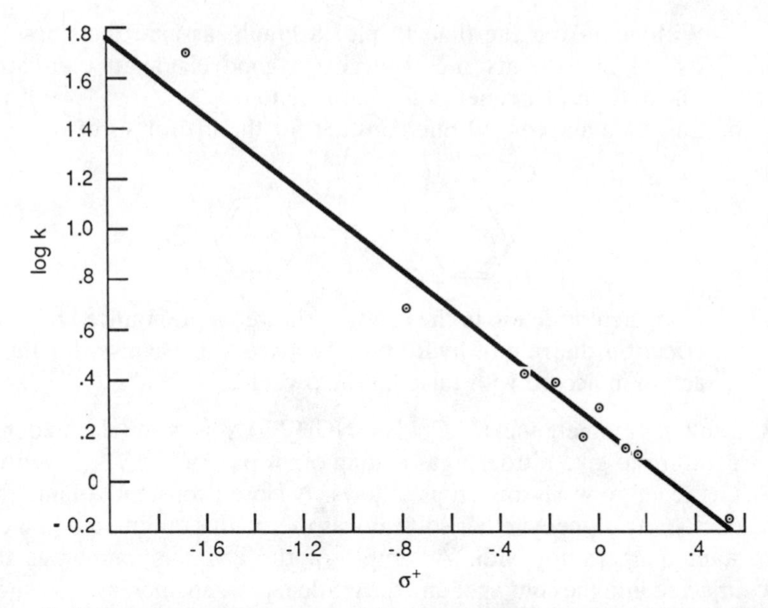

Figure X. Plot of rate constants versus σ^+ for the reaction of triarylphosphines with tetramethyl-1,2-dioxetane. Reprinted with permission from Baumstark, A. L.; McCloskey, C. J.; Williams, T. E.; Chrisope, D. R. *J. Org. Chem.* **1980,** *45,* 3593. Copyright 1980 American Chemical Society.

2. Imidoyl chlorides hydrolyze to give amides as you might expect from their structural similarity to acid chlorides.[37]

$$X-C_6H_4-N=C(Cl)-C_6H_5 + H_2O \longrightarrow X-C_6H_4-NH-C(=O)-C_6H_5 + HCl$$

The pseudo-first-order rate constants for hydrolysis in excess aqueous dioxane at 25°C for three examples are tabulated:

$X = -NO_2$	$3.5 \times 10^{-4}\ s^{-1}$
$X = -Cl$	$112. \times 10^{-4}\ s^{-1}$
$X = -H$	$460. \times 10^{-4}\ s^{-1}$

a. Without taking the time to plot a graph, assume that these are typical substituents and they give a good Hammett correlation; calculate the Hammett ρ for the reaction.
b. Calculate a predicted rate constant for the hydrolysis of

c. Added chloride ion in the solution showed a pronounced retarding effect on the rate of hydrolysis. Propose a mechanism for the reaction in accord with this and the ρ value.

3. Diphenylazomethane ($PhCH_2N{=}NCH_2Ph$) reacts at elevated temperature to give nitrogen gas among other products.[38] The kinetics of this reaction were studied as follows. A closed constant volume flask containing diphenyl ether solvent was thermally equilibrated in a constant temperature bath. A sample of the azo compound was then injected into the container and the gradual pressure increase measured. The cumulative increase in pressure ΔP at a list of times is tabulated below for two reaction temperatures.

150°C		**175°C**	
Time (min)	**ΔP (mm Hg)**	**Time (min)**	**ΔP (mm Hg)**
0	0.0	0	0.0
8	5.5	2	18.3
16	10.2	4	31.7
24	14.5	6	40.7
32	18.0	8	47.2
40	21.1	10	51.6
48	24.0	12	54.6
56	26.2	∞	61.8
64	28.3		
72	30.0		
∞	40.4		

a. How can you determine whether this is a first- or second-order reaction from these data?
b. Find the value of the rate constant k graphically at each temperature (or write a computer program that graphically displays the data for evaluation of linearity and calculates k).

Hint: The total amount of A ($PhCH_2N{=}NCH_2Ph$) or the initial concentration of A is proportional to the total amount of N_2 obtained and to ΔP_∞. The amount of A remaining at time t is proportional to $\Delta P_\infty - P_t$. The ratio of $[A]_0/[A]_t$ then equals $\Delta P_\infty/\Delta P_\infty - \Delta P_t$.

c. Calculate the activation enthalpy and entropy for the reaction.

Draw likely intermediates in the mechanism of each of reactions 4–8.

4.

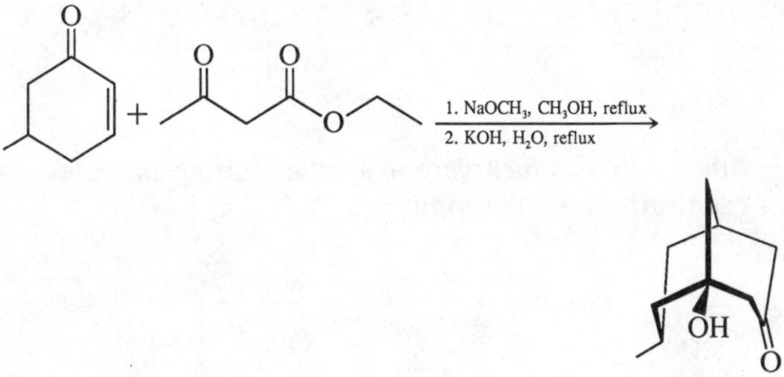

1. $NaOCH_3$, CH_3OH, reflux
2. KOH, H_2O, reflux

OH
O

Ref. 39

5.

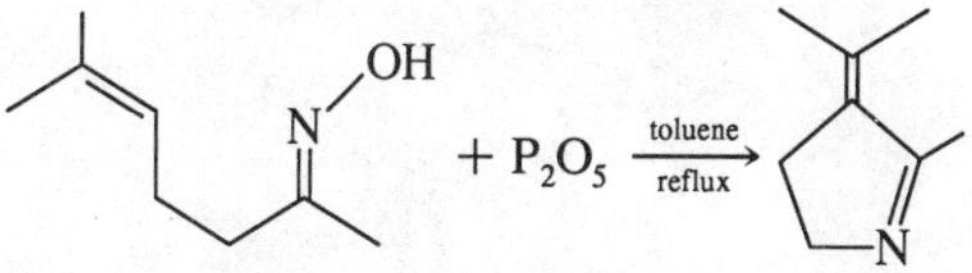

OH
N

+ P_2O_5 toluene reflux

N

Ref. 40

6.

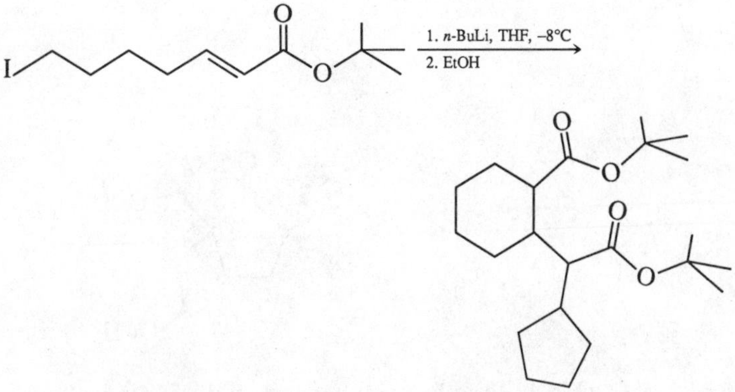

O
I
O

1. *n*-BuLi, THF, –8°C
2. EtOH

O
O
O
O

Ref. 41

7.

O

BF_3
Et_2O

O

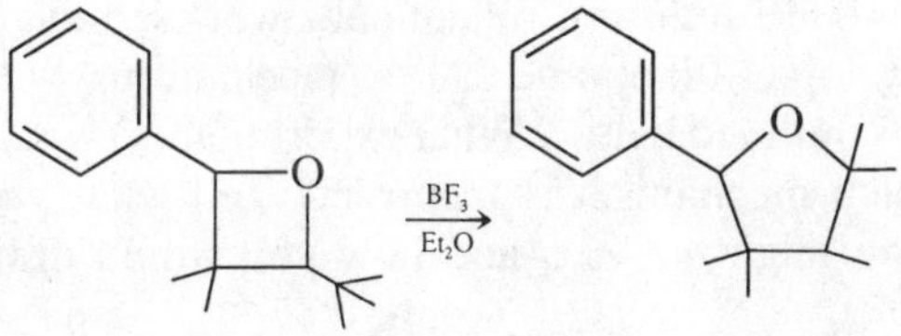

Ref. 42

8.

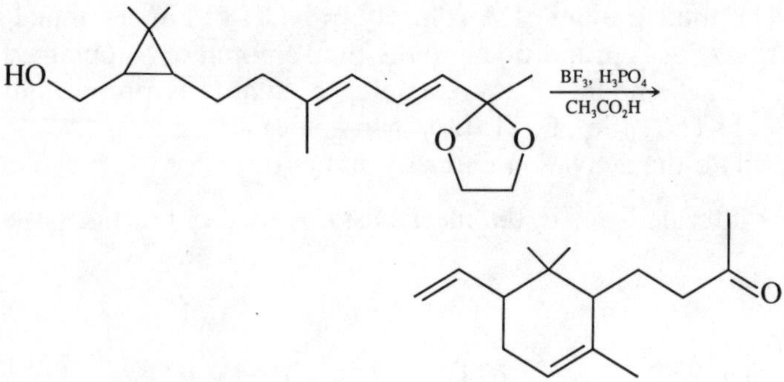

HO … BF_3, H_3PO_4 / CH_3CO_2H →

Ref. 43

9. Indicate with numbers which carbon in the starting material is the source of each carbon in the product:

a. CH_3O … OH … OH … OCH_3 —aq. HCl / THF→ Ref. 44

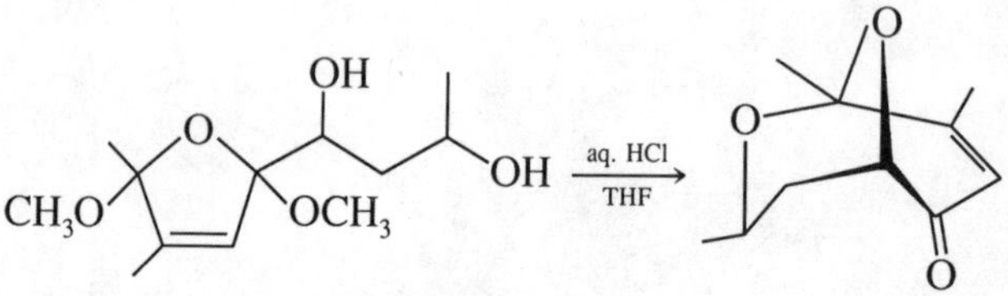

b. OCH_3 — 1. $NaOCH_3$ 2. CH_3OH, reflux 3. toluene reflux 4. $BrCH_2CO_2CH_3$ →

OCH_3 … OCH_3 Ref. 45

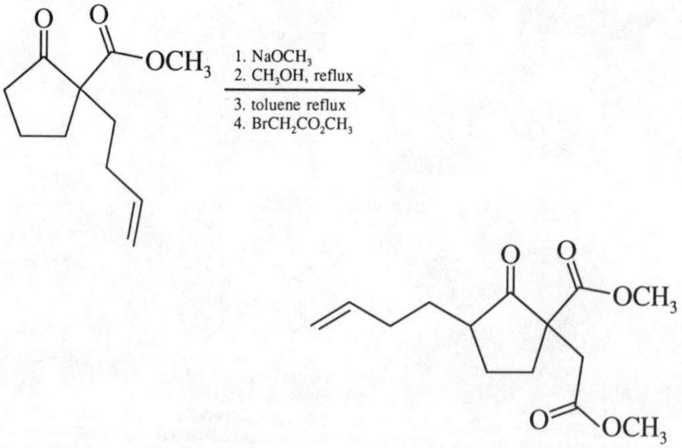

10. Treatment of 1-chloro-2-methylcyclohexene with 5 equivalents of methyllithium in TMEDA–THF followed by aqueous workup gave a 47% yield of 1-methylbicyclo[4.1.0]heptane.[46] Two mechanisms were proposed for this process as outlined below. What experiment(s) would you use to determine which mechanism is incorrect? Tell what you would do, what results you might expect, and how you would draw your conclusions.

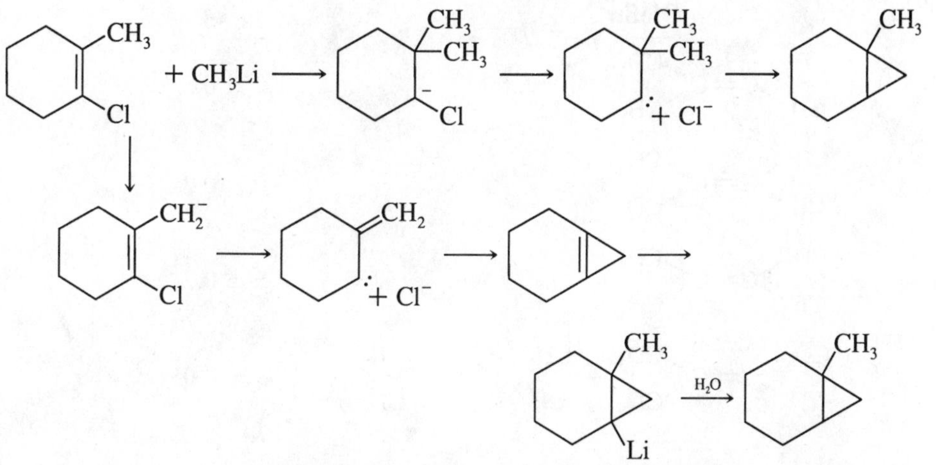

CH_3 Cl + CH_3Li ⟶ CH_3 CH_3 Cl ⟶ CH_3 CH_3 + Cl^- ⟶ CH_3

↓

CH_2^- Cl ⟶ CH_2 + Cl^- ⟶ ⟶

CH_3 Li $\xrightarrow{H_2O}$ CH_3

11. The reaction of monoolefins with nitroso compounds gives hydroxylamine products.[47] The process was thought to occur by either a one-step ene reaction (Eq. 1) or a two-step reaction via a transient aziridine oxide intermediate (Eq. 2). The following deuterium isotope effects were measured. Explain the similarities and differences among these results in terms of one of these mechanisms. Use the results to exclude one mechanism.

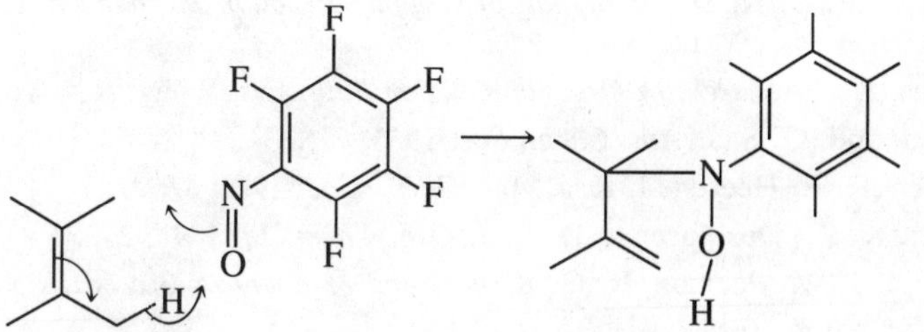

H N O F F F F F ⟶ N O H (1)

+ F F F F F N=O ⟶ [O N F F F F F] ⟶ HO N F F F F F

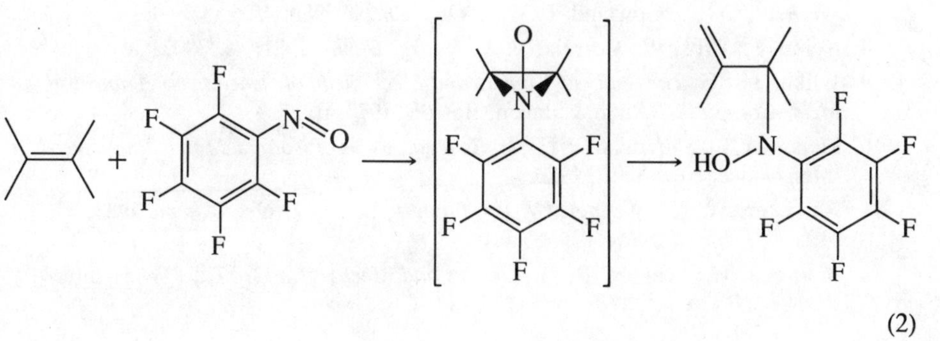

(2)

Olefin	k_H/k_D
$(CD_3)(CH_3)C{=}C(CD_3)(CH_3)$	1.2 ± 0.2
$(CD_3)(CH_3)C{=}C(CH_3)(CD_3)$	3.0 ± 0.2
$(CD_3)_2C{=}C(CH_3)_2$	4.5 ± 0.2
$(CD_3)_2C{=}C(CD_3)_2$ + $(CH_3)_2C{=}C(CH_3)_2$ (1 : 1)	1.03 ± 0.05

The first three cases are intramolecular competitions where k_H is the rate of formation of the $C{=}CH_2$ product and k_D is the rate of formation of the $C{=}CD_2$ alternative product. The last case is an intermolecular competition. The results were determined by ^1H NMR analysis of the products.

REFERENCES

1. Carpenter, B. K. *Determination of Organic Reaction Mechanisms,* Wiley-Interscience, New York, 1984.
2. Sykes, P. *The Search for Organic Reaction Pathways,* Wiley, New York, 1972.
3. Hammond, G. S. *J. Am. Chem. Soc.* **1953,** *77,* 334.
4. Cope, A. C.; Heeren, J. K. *J. Am. Chem. Soc.* **1965,** *87,* 3125.
5. Johri, K. K.; DesMarteau, D. D. *J. Org. Chem.* **1983,** *48,* 242.
6. Moore, J. W.; Pearson, R. G. *Kinetics and Mechanism,* 3rd ed., Wiley-Interscience, New York, 1981.
7. Wigfield, D. C.; Gowland, F. W. *J. Org. Chem.* **1980,** *45,* 653.
8. Nome, F.; Erbs, W.; Correia, V. R. *J. Org. Chem.* **1981,** *46,* 3802.
9. Wilkins, R. G. *The Study of Kinetics and Mechanism of Reactions of Transition Metal Complexes,* Allyn & Bacon, Boston, 1974, p. 27.
10. Melander, L.; Saunders, W. H., Jr. *Reaction Rates of Isotopic Molecules,* Wiley-Interscience, New York, 1980.
11. Rastetter, W. H.; Wagner, W. R.; Findeis, M. A. *J. Org. Chem.* **1982,** *47,* 419.
12. Moore, J. W.; Pearson, R. G. *Kinetics and Mechanism,* 3rd Ed., Wiley-Interscience, New York, 1981, pp. 178–181.

13. Martin, J. C.; Timberlake, J. W. *J. Am. Chem. Soc.* **1970,** *92,* 975.
14. Hammond, G. S.; DeBoer, C. *J. Am. Chem. Soc.* **1964,** *86,* 899.
15. Robinson, P. J. *J. Chem. Ed.* **1978,** *55,* 509.
16. Hammett, L. P. *J. Am. Chem. Soc.* **1937,** *59,* 96.
17. Ritchie, C. D.; Sager, W. F. *Progr. Phys. Org. Chem.* **1964,** *2,* 323.
18. Poh, B.-L. *Can. J. Chem.* **1979,** *57,* 255.
19. Brown, H. C.; Okamoto, Y. *J. Am. Chem. Soc.* **1958,** *80,* 4979.
20. Imai, T.; Nishida, S. *J. Org. Chem.* **1979,** *44,* 3574.
21. Turner, R. B.; Nettleton, D. E., Jr.; Ferebee, R. *J. Am. Chem. Soc.* **1956,** *78,* 5923.
22. Stemke, J. E.; Bond, F. T. *Tetrahedron Lett.* **1975,** 1815.
23. Majewski, M.; Snieckus, V. *J. Org. Chem.* **1984,** *49,* 2684.
24. Slessor, K.; Oehlschlager, A. C.; Johnston, B. D.; Pierce, H. D., Jr.; Grewal, S. K.; Wickremesinghe, L. K. G. *J. Org. Chem.* **1980,** *45,* 2290.
25. Dave, V.; Warnhoff, E. W. *Org. React.* **1970,** *18,* 217.
26. Barton, W. R.; DeCamp, M. R.; Hendrick, M. E.; Jones, M. Jr.; Levin, R. H.; Sohn, M. B. In *Carbenes,* Jones, M. Jr.; Moss, R. A. Eds., Wiley-Interscience, New York, 1973; Marchand, A. P.; MacBrockway, N. *Chem. Rev.* **1974,** *74,* 431.
27. Anciaux, A. J.; Hubert, A. J.; Noels, A. F.; Petiniot, N.; Teyssie, P. *J. Org. Chem.* **1980,** *45,* 695.
28. Gribble, G. W.; LeHoullier, C. S.; Sibi, M. P.; Allen, R. W. *J. Org. Chem.* **1985,** *50,* 1611.
29. Holm, T.; Crossland, A.; Madsen, J. O. *Acta Chem. Scand. B* **1978,** *32,* 754.
30. Meyer, W. L.; Manning, R. A.; Schroeder, P. G.; Shew, D. C. *J. Org. Chem.* **1977,** *42,* 2754.
31. Moiseenkov, A. M.; Czeskis, B. A.; Nefedov, O. M. *Synthesis* **1985,** 932.
32. Olah, G. A.; Fung, A. P. *Synthesis* **1979,** 537.
33. Donaruma, L. G.; Heldt, W. Z. *Org. React.* **1960,** *11,* 1.
34. Padwa, A.; Nimmesgern, H.; Wong, G. S. K. *J. Org. Chem.* **1985,** *50,* 5620.
35. Kricheldorf, H. R.; Leppert, E. *Synthesis* **1976,** 329.
36. Baumstark, A. L.; McCloskey, C. J.; Williams, T. E.; Chrisope, D. R. *J. Org. Chem.* **1980,** *45,* 3593.
37. Hegarty, A. F.; Cronin, J. D.; Scott, F. L. *J. Chem. Soc. Perkin Trans. II,* **1975,** 429.
38. Bandlish, B. K.; Garner, A. W.; Hodges, M. L.; Timberlake, J. W. *J. Am. Chem. Soc.* **1975,** *97,* 586.
39. Kraus, G. A.; Hon, Y.-S. *J. Org. Chem.* **1985,** *50,* 4605.
40. Gawley, R. E.; Termine, E. J. *J. Org. Chem.* **1984,** *49,* 1946.
41. Cooke, M. P., Jr., *J. Org. Chem.* **1984,** *49,* 1144.

42. Carless, H. A. J.; Trivedi, H. S. *J. Chem. Soc. Chem. Commun.* **1979,** 382.
43. Chopra, A. K.; Khambay, B. P. S.; Madden, H.; Moss, G. P.; Weedon, B. C. L. *J. Chem. Soc. Chem. Commun.* **1977,** 357.
44. DeShong, P.; Ramesh, S.; Perez, J. J.; Bodish, C. *Tetrahedron Lett.* **1982,** *23,* 2243.
45. Dauben, W. G.; Walker, D. M. *Tetrahedron Lett.* **1982,** *23,* 711.
46. Gassman, P. G.; Valcho, J. J.; Proehl, G. S.; Cooper, C. F. *J. Am. Chem. Soc.* **1980,** *102,* 6519.
47. Seymour, C. A.; Greene, F. D. *J. Org. Chem.* **1982,** *47,* 5226.

8

Electron Delocalization, Aromatic Character, and Concerted Reactions

Alternating behavior has been observed in many different kinds of reactions. Protonation of a conjugated polyene gives a carbocation with the positive charge specifically delocalized to alternating carbons of the chain. For example, protonation of hexatriene gives positive charge on carbons 1, 3, and 5 as manifest in final product distributions and in spectra. Similar distributions exist in conjugated carbanions.

Another alternation appears in the series of cyclic molecules that show aromatic character. If the ring contains a conjugated cycle of $4n + 2$ π electrons, we find peculiar spectroscopic and reaction characteristics that are absent in the $4n$ set (n is the series 0, 1, 2, 3, . . .).

In this chapter you will see reactions that occur with a certain stereochemistry when the reactant(s) contain(s) $4n + 2$ π electrons but do not occur or occur with a different stereochemistry when there are $4n$ π electrons.

Recognition of these alternating patterns has allowed many predictions. In this chapter these patterns are rationalized in terms of molecular orbitals.[1]

8.1 MOLECULAR ORBITALS

Electrons in atoms are in constant motion around the nuclei. For an electron there is a continual trade-off between potential and kinetic energy as it varies in distance from the nucleus but the total remains exactly the same. The motion cannot be followed, but a probability of finding it at any and all points around the nucleus can be defined. The mathematical forms that give this probability distribution are the same as those used to define waves. An atomic orbital is a wavefunction that has a numerical value for each point in space around the nucleus. The square of that number is directly related to the probability of the electron being at that point at any moment in time. For a *p* orbital the wavefunction values are high at two points on opposite sides of the nucleus and diminish with distance from these points, to zero at a plane through the nucleus or at infinite distance from the atom. The sign of the function (phase of the wave) is opposite on each side of that zero (nodal) plane through the nucleus. On paper we can show a cross section of this with contour lines as shown in Figure I.

The square of this is the electron density function. A plot of the amplitude of the $2p_y$ orbital of fluorine along the *y* axis is shown in Figure II. The square of this is shown also to indicate the relative electron density along that axis. In the fluorine molecule the molecular orbitals are conveniently described as combinations (overlapping) of the two atomic *p* orbitals. They overlap in two ways: with reinforcement between the nuclei (σ orbital) or with interference between the nuclei (σ^* orbital). This is illustrated in Figure III, where you can see that the electron density between nuclei in the σ^* orbital would be very small and the nuclei would repel each other. However, the electron density is largely concentrated between the nuclei in the σ orbital, and the nuclei are attracted to it and, therefore,

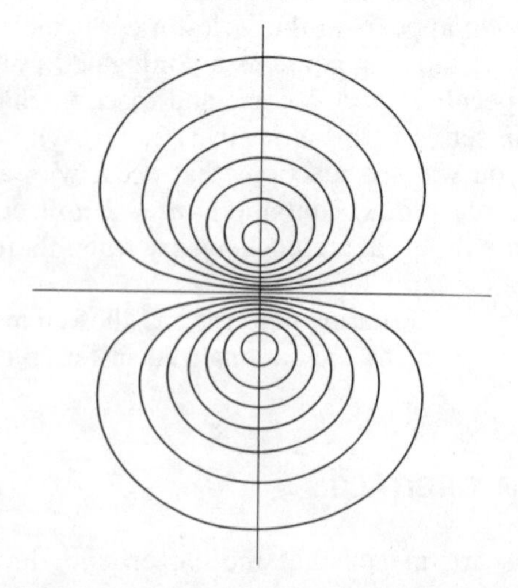

Figure I. A $2p_y$ atomic orbital at the *xy* plane.

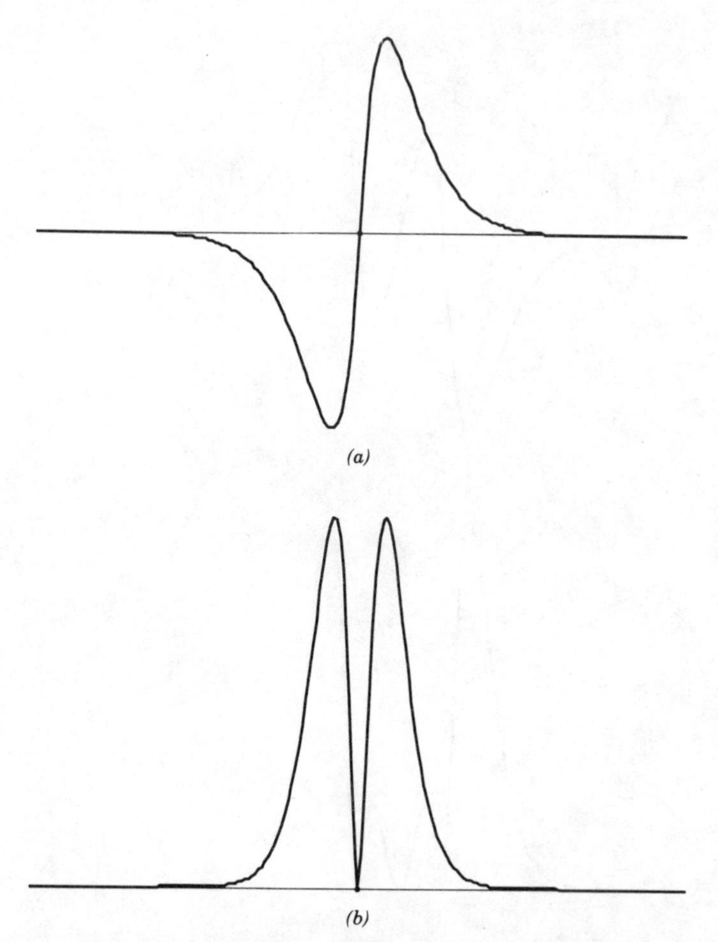

Figure II. (*a*) The value of the $2p_y$ atomic orbital of a fluorine atom (ψ_{2p_y}) along the *y* axis and (*b*) the square of that function ($\psi_{2p_y}{}^2$) indicating electron density along that axis. The nucleus is at the center and the full trace width is 10 Å.

together. These are the bonding and antibonding orbitals. Of course the electron density in the *volume* between the nuclei is important, but for simplicity we have sampled that along the axis in these plots.

The bonding electron pair in F_2 occupies the lower energy σ orbital and the σ* is unoccupied in the ground state.

In ethylene, besides the σ orbitals, there are π molecular orbitals described as laterally overlapping *p* orbitals. Here again there are two ways of overlapping: with reinforcement (π) and with interference (π*) in the space above and below the nodal plane. A conventional cross-sectional

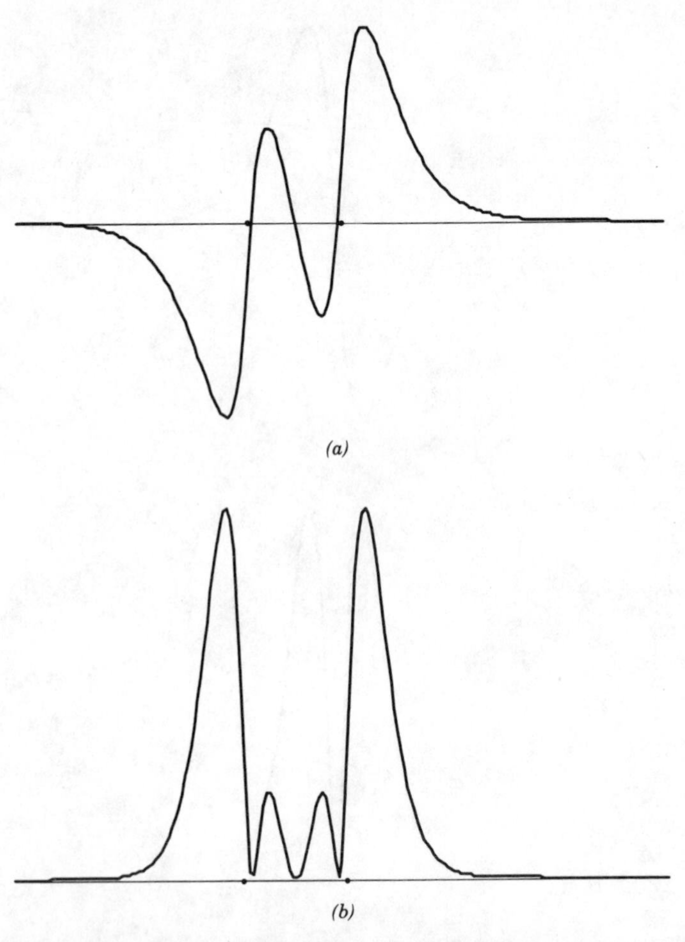

Figure III. Molecular orbital values and electron densities along the internuclear axis in the F_2 molecule, based on overlap of ψ_{2p_y} at the normal bonding distance. (*a*) σ^*, (*b*) σ^{*2}, (*c*) σ, (*d*) σ^2. These plots were provided by S. L. Whittenburg, University of New Orleans.

representation of these is shown in Fig. IV with single contour lines where opposite phases are indicated by shaded and unshaded lobes. As in σ bonds, the bonding pair of electrons resides in the lower energy orbital where the attractive forces effectively bond the carbon atoms. The difference in energy between one electron in an isolated *p* orbital and that electron in an ethylene π orbital is called β (18 kcal/mol). the π-bonding energy of ethylene with two electrons is 2β. As more *p* orbitals are mixed in a linear array a matching number of π molecular orbitals are formed and the lowest energy orbital is lower, approaching a minimum of -2β per electron (Fig.

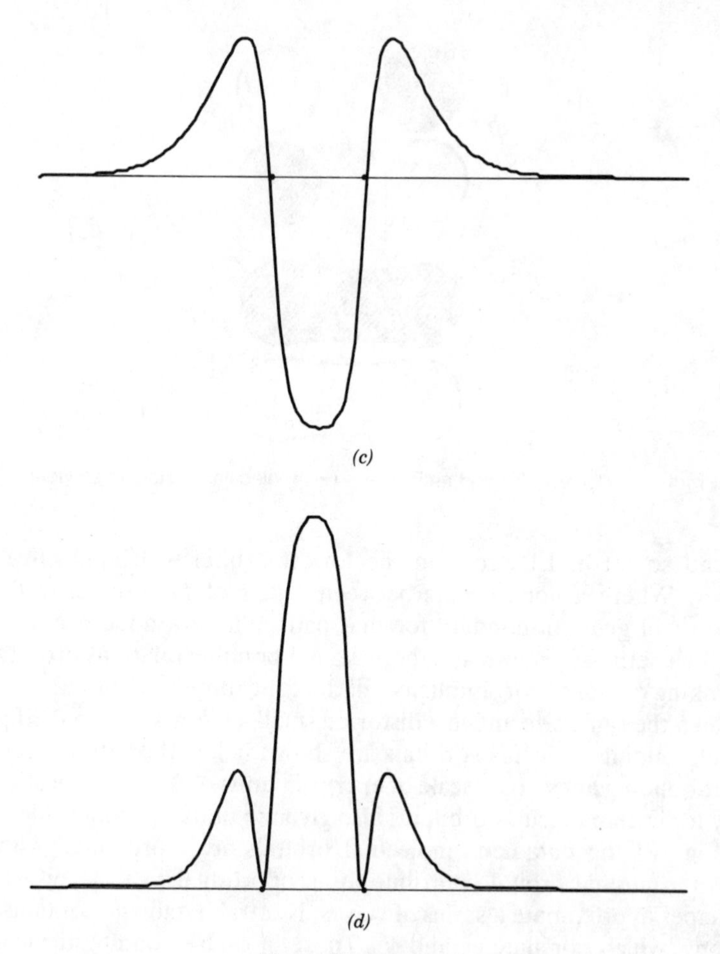

Figure III. *Continued*

V). Each molecular orbital is composed of portions of the atomic *p* orbitals, and the extent to which each atomic orbital contributes to each molecular orbital is given as a coefficient (LCAO–MO).[1] The square of the coefficient is the electron density at that atom if one electron occupies the molecular orbital with that coefficient. In butadiene the sum of the squares of all the coefficients on carbon 1 (or any other carbon in Fig. V) is one atomic orbital. Likewise, the sum of the squares of the coefficients of the four atoms contributing to one molecular orbital is also 1. In each set of molecular orbitals shown in Fig. V, the lowest (ψ_1) has no nodes (beyond the one in the plane of the nuclei), the next higher (ψ_2) has one node, ψ_3 has

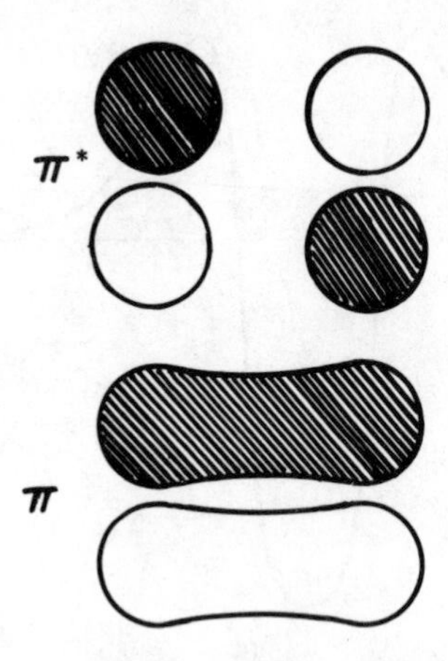

Figure IV. Bonding (π) and antibonding (π*) molecular orbitals of ethylene.

two, and so forth. Electrons in the lowest orbital will bond all atoms together. Where a node occurs between a pair of carbons, electron occupation will give antibonding for that pair. Where a node occurs at an atom, the coefficient is zero and there is a nonbonding relationship between the flanking carbons. For simplicity, all the contributions of atomic orbitals are drawn the same size and in a distorted small narrow shape in the figure. The net bonding molecular orbitals are shown below the zero energy line and antibonding above on a scale of energy in units of β. The delocalization energy for each molecular orbital is also given in units of β alongside each.

In Fig. VI the butadiene molecular orbitals are represented with the size of the atomic orbital contributions proportional to the coefficients. The shapes approximate a series of waves. Neutral butadiene contains four electrons, which populate ψ_1 and ψ_2. Those in ψ_1 are bonding throughout and particularly so between C-2 and C-3, where the coefficients are large. Those in ψ_2 bond C-1 to C-2 and C-3 to C-4 but give antibonding between C-2 and C-3. This antibonding is relatively weak owing to the small coefficients at C-2 and C-3. The net bonding in butadiene is greater than in isolated ethylenes, which is in accord with the observation of greater thermodynamic stability in conjugated systems compared to nonconjugated. This is observed experimentally in the lower heat of hydrogenation of conjugated dienes compared to nonconjugated analogs and in the selective formation of conjugated dienes in elimination reactions. Using the bonding energy in Fig. V, we find that two ethylenes with two electrons each (4β) are less than four electrons in ψ_1 and ψ_2 of butadiene ($1.618 \times 2 + 0.618 \times 2 = 4.472\beta$). The orbitals above the zero line are net antibonding, and if they were occupied, it would weaken the molecule.

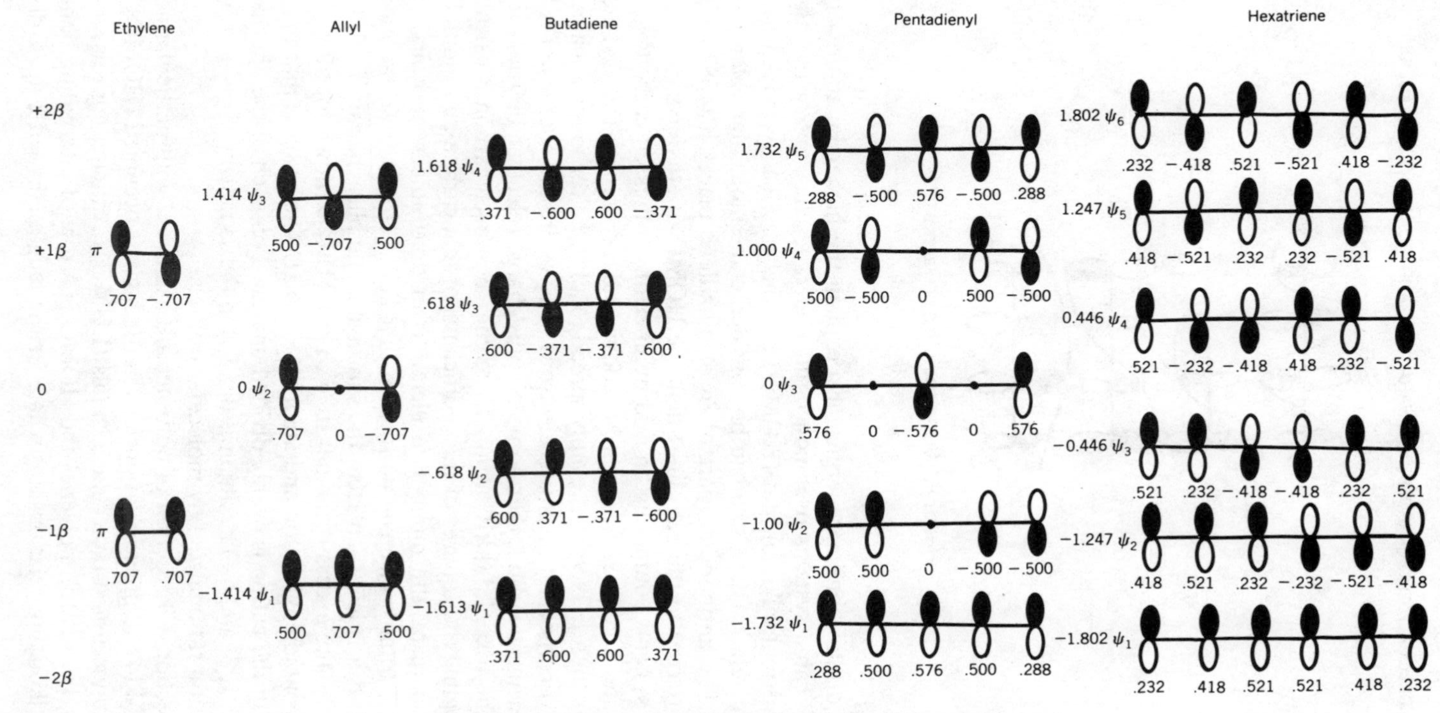

Figure V. π Molecular orbitals of linear C-2–C-6 systems.

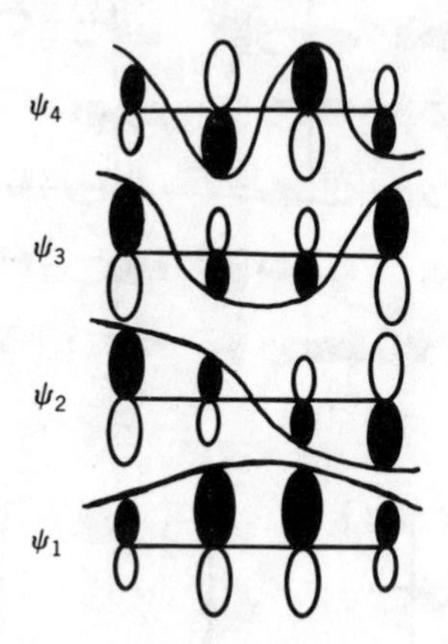

Figure VI. Butadiene π molecular orbitals.

The orbital bonding energies also explain the higher kinetic reactivity of conjugated systems over nonconjugated. The electrons in the highest occupied molecular orbital (HOMO) are the most easily moved to new bonding relationships and can be considered the molecular valence electrons.[2] The antibonding part of ψ_2 of butadiene places it 0.382β higher energy (closer to the zero line) than the HOMO of ethylene.

The systems containing an odd number of conjugated carbons are reactive intermediate radicals, carbocations, and carbanions. The allyl radical contains three π electrons evenly spread to all three carbons (no charges). The allyl carbocation has two electrons in ψ_1 where the coefficient is high on the central carbon; therefore, the plus charge (electron deficiency) will be on the first and third carbons. The charge at each carbon is calculable by summing the squares of the coefficients of that carbon and multiplying each square by the number of electrons in that molecular orbital. This gives the total π-electron density on that atom. If it is 1.00, the atom is neutral; if it is less than 1.00, the shortage is the plus charge; if it is more than 1.00, the excess is the negative charge. This is shown for all carbons in the pentadienyl cation and anion in Fig. VII. The result is that the plus charge in the cation is evenly divided one-third on each of the alternating carbons 1, 3, and 5. The anion is divided in the same way. This is the first alternation rationalized by molecular orbitals.

In Section 8.3 reaction of polyenes are examined with particular attention paid to the phase relationships in the highest occupied (HOMO) and lowest unoccupied molecular orbitals (LUMO). Anticipating this, notice that for the neutral polyenes the HOMO is a row of alternating bonding and antibonding relationships whatever the length (shaded 2 up, 2 down,

	C-1	C-2	C-3	C-4	C-5
ψ_3 coefficient	0.576	0	−0.576	0	0.576
Coefficient^2	0.333	0	0.333	0	0.333
2 coefficient^2	0.666	0	0.666	0	0.666
ψ_2 coefficient	0.500	0.500	0	−0.500	−0.500
Coefficient^2	0.250	0.250	0	0.250	0.250
2 coefficient^2	0.500	0.500	0	0.500	0.500
ψ_1 coefficient	0.288	0.500	0.576	0.500	0.288
Coefficient^2	0.083	0.250	0.333	0.250	0.083
2 coefficient^2	0.166	0.500	0.666	0.500	0.166

The electron density at each carbon in the cation is $2\psi_1^2 + 2\psi_2^2$, and the charge is $1.00 - (2\psi_1^2 + 2\psi_2^2)$:

	C-1	C-2	C-3	C-4	C-5
	0.666	1.00	0.666	1.00	0.666
	+0.333	0	+0.333	0	+0.333

The electron density at each carbon in the anion is $2\psi_1^2 + 2\psi_2^2 + 2\psi_3^2$, and the charge is $1.00 - (2\psi_1^2 + 2\psi_2^2 + 2\psi_3^2)$

	C-1	C-2	C-3	C-4	C-5
	1.332	1.00	1.332	1.00	1.332
	−0.33	0	−0.33	0	−0.33

Figure VII. Calculation of the charge on each atom of the pentadienyl cation and anion.

2 up, etc.), and that the C_{4n} have the opposite phase at the first and last carbons while the C_{4n+2} have the same phase at the first and last carbons.

8.2 AROMATIC CHARACTER

Cyclic compounds in which there is overlap of a *p* orbital on every ring atom have π molecular orbitals as exemplified in Fig. VIII. As with acyclic cases there are as many molecular orbitals as there were contributing atomic orbitals. Unlike the acyclic cases, there are degenerate (equal energy) pairs of molecular orbitals. An energy-level diagram for four of these rings is shown in Fig. IX. In the ground state the electrons would populate the lower-energy orbitals following Hund's rule as marked with arrows. In cyclobutadiene the lowest pair is bonding throughout, but the next two are in the nonbonding orbitals ψ_2 and ψ_3. This is unstable compared to the dimer compound where all electrons are bonding; therefore, the monomer is unisolable and the dimer forms readily. In the cyclopentadienide or the isoelectronic furan, all the electrons are in orbitals of some net bonding, giving a stable system. The same is true for benzene. In cyclooctatetraene (if it were conjugated) two electrons would occupy nonbinding orbitals. This again would be unstable and it is relieved by not conjugating. The molecule is tub shaped and the electrons are all in bonding "ethylene"

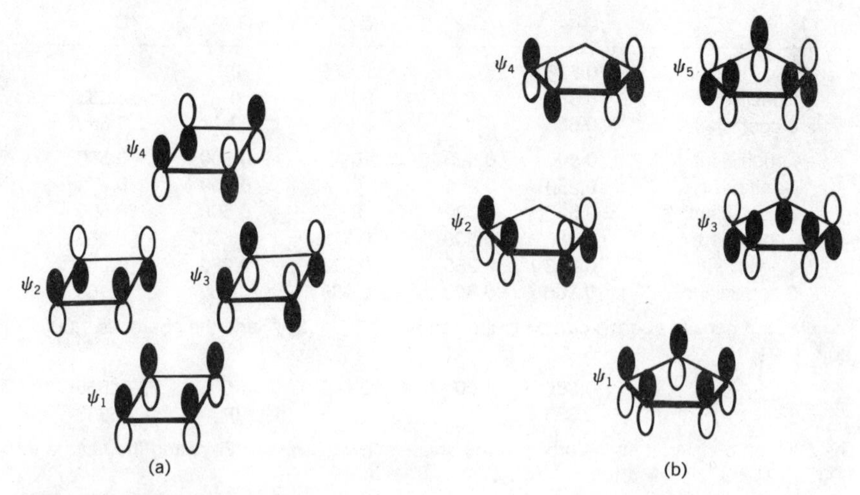

Figure VIII. (*a*) Cyclobutadiene π molecular orbitals; (*b*) cyclopentadienyl π molcular orbitals.

orbitals. These circumstances alternate through the cyclic vinylogous series continuing to larger rings. Those with $4n + 2$ π electrons have closed-shell occupation (the highest occupied energy level filled with four electrons) and have stability, in fact, greater stability than found in the acyclic cases. The $4n$ series is less stable than acyclic cases and will either rapidly react or twist out of conjugation.

The greater stability in the $4n + 2$ series is called "aromatic character"

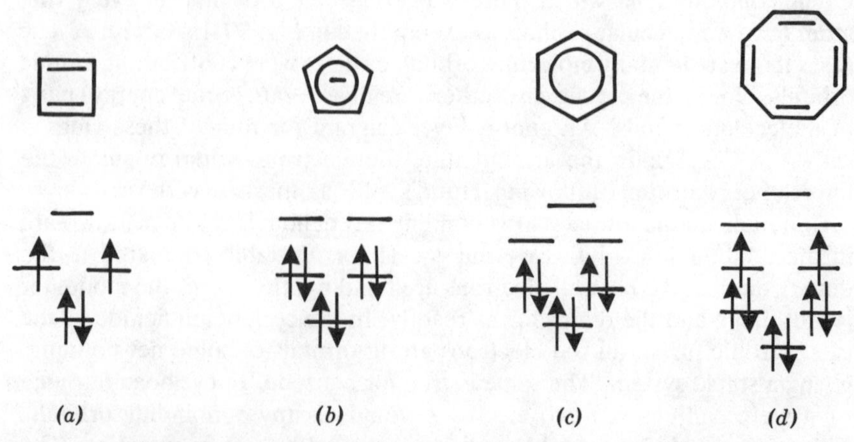

Figure IX. Energy levels of cyclic π molecular orbitals: (*a*) cyclobutadiene, (*b*) cyclopentadienyl, (*c*) benzene, (*d*) conjugated cyclooctatetraene.

and is manifest in numerous physical and chemical properties. They give a lower combustion energy compared to acyclic analogs and resist addition reactions that would interrupt the ring of π bonding. The ^1H NMR spectra show differences also. Hydrogens projecting outward in the plane of an aromatic ring appear several ppm downfield of the vinyl hydrogens in acyclic analogs. In contrast, those hydrogens extending inward or above the ring appear considerably upfield.

Molecules with $4n$ π electrons are called *antiaromatic* (if near planar) and promptly react or are easily distorted from planarity. The $(CH)_{16}$ and $(CH)_{24}$ rings have been prepared, and they do show the opposite ^1H NMR chemical-shift effects from those found for the $4n + 2$ compounds.[3]

There is some stability in closed-shell electron occupation even if four of the electrons are nonbonding. Cyclooctatetraene can be reduced to the dianion by alkali metals, and the ring becomes aromatic. X-ray crystallography of 1,3,5,7-tetramethylcyclootatetraene dianion shows that the ring carbons form an equilateral planar octagon.[4]

Those monocyclic aromatic systems where n is greater than 1 are not nearly as chemically resistant as benzene, although some of them undergo electrophilic substitution. The heterocycles furan and pyrrole have a ring of 6 π electrons and show aromatic character, but higher homologues such as 1,4-dioxocin, which has a ring of 10 π electrons shows olefinic, not aromatic character.[5]

Aromatic systems with fused six-membered rings have $4n + 2$ π electrons and aromatic character but less stabilization per ring than in benzene.[6]

This is the second alternation, rationalized as having closed-shell occupation of degenerate molecular orbitals.

8.3 CONCERTED REACTIONS

Numerous organic reactions proceed through a cyclic transition state in a single step involving no intermediates. These are called *pericyclic reactions*. Certain bonds break, while others form in concert. These reactions are divided into two groups taking stereochemical choices into account. Examples from one group have been observed under thermal conditions, while examples from the other group are found only under photochemical conditions. In the thermally observable reactions the phase relationships of the combining orbitals are such that the bonding electrons maintain bonding character from starting materials, through the transition state, and in the products.

Selection rules were put forth by Woodward and Hoffmann based on the conservation of orbital symmetry.[7] Molecular orbitals in the starting material are correlated with orbitals in the product according to their symmetry elements. If all the ground-state occupied molecular orbitals in the starting material correlate with ground-state molecular orbitals in the product, the reactions will be allowed thermally. If, instead, the lowest excited state of the starting material correlates with the lowest excited state of the product, the reaction is allowed photochemically. It is the HOMO that is pivotal in these choices.

Fukui examined the interaction of the HOMO and LUMO alone (the frontier orbitals) and rationalized the same rules.[8,9] Basically each reaction is viewed as the coming together (or dissociation) of two sets of molecular orbitals intra- or intermolecularly. The HOMO of one is matched with the LUMO of the other, and if the overlap at both sites of projected new bond formation between them is in-phase (a bonding overlap), the reaction is allowed. This method of analysis is detailed with examples in Sections 8.3.1–8.3.3.

Another approach, concentrating on the transition state is given in Section 8.3.4.

All these analyses view the reactions as reversible and allowedness applies both ways. Other factors such as ring strain, steric hindrance, and bond energies determine the actual direction.

8.3.1 Electrocyclic Reactions

An electrocyclic reaction[10] is the closure of a conjugated polyene to give a cyclic compound with one less π bond, or the reverse. For the first example, consider the opening of a cyclohexadiene to give a hexatriene (Eq. 1). A σ bond breaks, and new π bonding develops.

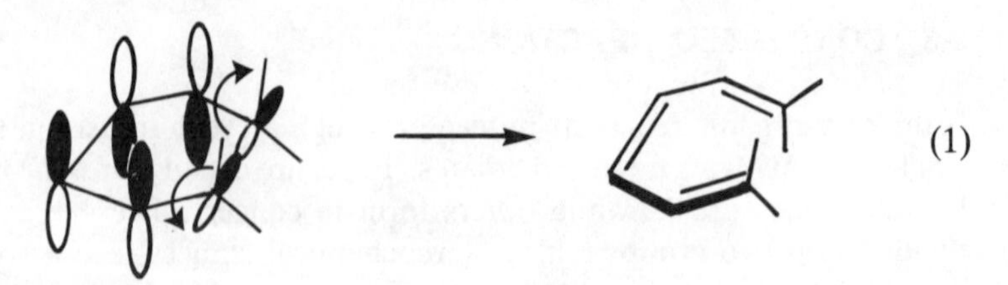

(1)

The highest-energy electrons available are in the HOMO of the diene, the phase of which is indicated. This may donate electron density to the LUMO of the sigma bond, the phase of which is also shown. It can be seen that a particular rotating motion will bring the orbitals into the new bonding relationship with bonding overlap. Viewed from the side, one appears to rotate about 90° counterclockwise while the other goes clockwise. This is called disrotatory motion. The opposite, where both rotate clockwise or both counterclockwise is called conrotatory motion. These are illustrated in a general way in Fig. X. The other arbitrary choice of phase indication in the initially independent molecular orbitals is as shown in Eq. 2.

(2)

This difference has no physical meaning and both choices should be kept in mind in all pericyclic reactions. Which of the two disrotatory motions occurs will be determined in most cases by steric effects among the substituents, or both may occur. Conrotation is forbidden for this reaction because it leads to antibonding overlap at one end.

In this analysis we looked only at the closest-lying LUMO and HOMO, which are called the *frontier orbitals*. If we had instead examined the HOMO of the σ bond and the LUMO of the diene, we would have again concluded that the disrotatory process was the allowed one.

The conclusion applies to the reverse of the reaction as well, and actually the closure of the six-membered ring is very favorable, and we see it generally in that direction. We can analyze that reverse reaction by arbitrarily viewing the hexatriene as a diene and ethylene system about to interact:

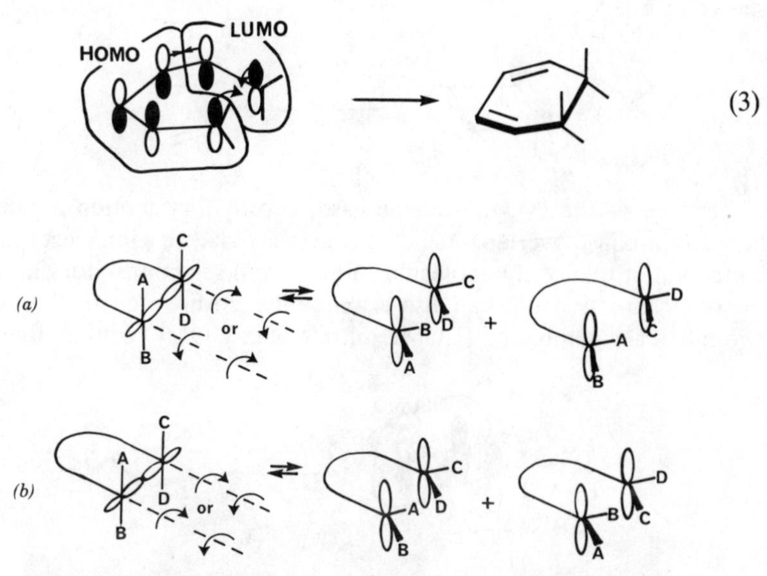

Figure X. (*a*) Disrotatory and (*b*) conrotatory ring, opening or closing.

The HOMO of the diene and the LUMO of the ethylene can overlap in phase (bonding) if the motion is disrotatory (just as we concluded for the closure). The stereochemistry was demonstrated in the cyclization of the isomers of 2,4,6-octatriene.[11] The *E,Z,E* isomer gave only the cis ring compound. The *Z,Z,E* and *Z,Z,Z* isomers were in thermal equilibrium, but rate measurements indicate that the *Z,Z,E* isomer likely gave the trans ring compound:

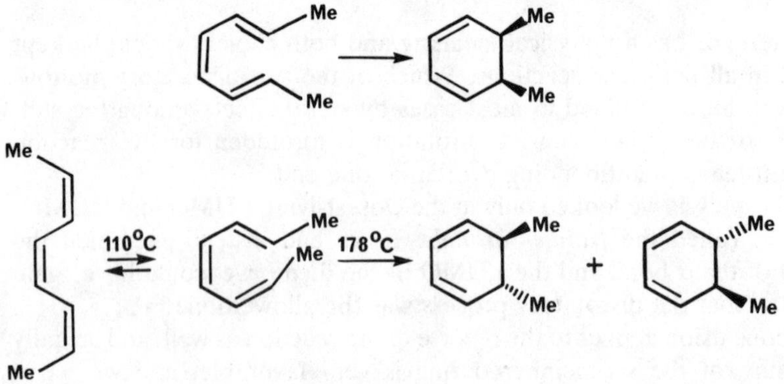

Me Me → Me Me

Me Me 110°C ⇌ Me Me 178°C → Me Me + Me Me (4)

The ring opening of cyclobutenes can be analyzed by the frontier orbital theory also. The LUMO of the σ bond and the HOMO of the π bond are shown in Eq. 5.

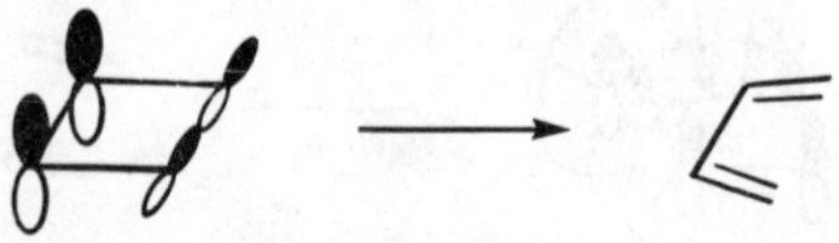

(5)

In contrast to the cyclohexadiene case, conrotatory motion is necessary here for in-phase overlap. As in the previous case, we can invert the phase indication of one of the molecular orbitals and see conrotatory motion in the other direction. We can also examine the closure viewing the diene as two ethylenes combining (LUMO of one with HOMO of the other):

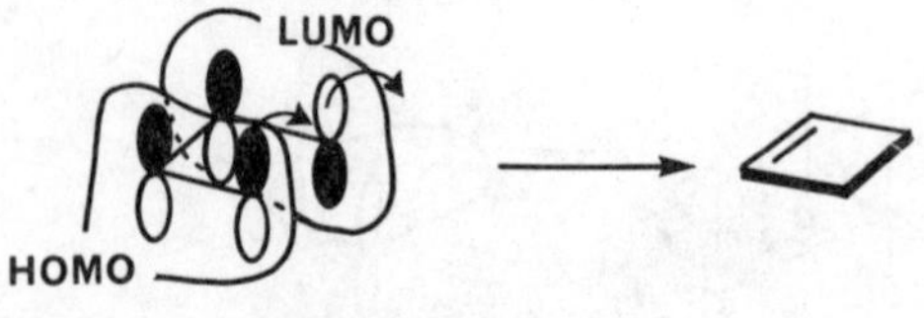

LUMO HOMO (6)

Both directions of conrotation in the opening of *cis*-3,4-dimethylcyclobutene lead to *Z,E*-2,4-hexadiene (Eq. 7).[12] The *trans*-3,4-dimethylcyclobutene should lead to the *E,E* and *Z,Z* diene, but the steric strain developed in rotation toward the *Z,Z* diene was prohibitive.

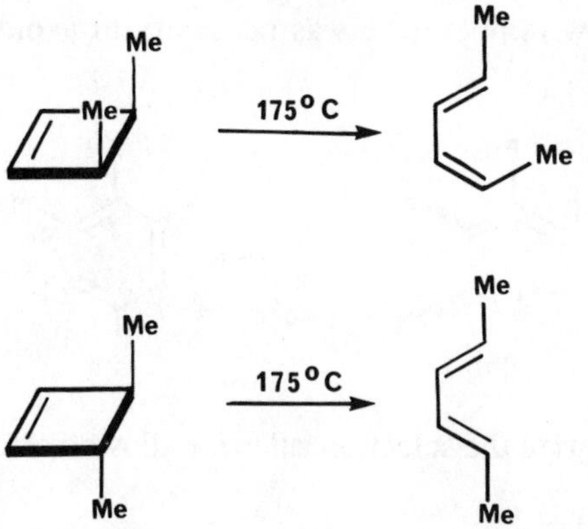

Me Me 175°C Me Me

(7)

Me Me 175°C Me Me

Another experiment showed both directions of conrotation for both opening and closure, and the complete absence of disrotation:[13]

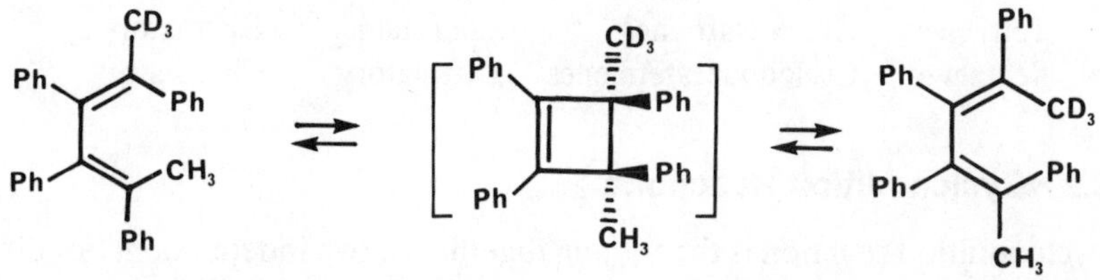

CD_3 Ph Ph Ph CH_3 Ph ⇄ [CD_3 Ph Ph Ph Ph CH_3] ⇄ Ph Ph CD_3 Ph Ph CH_3

(8)

Each isomer of the deuterium labeled diene was heated at 124°C and a 1:1 mixture of them was obtained, containing none of the cis,cis isomer. In another experiment the trans,trans and the cis,cis dienes (no deuterium) were thermally equilibrated and none of the cis,trans diene was formed.

Higher vinylogs continue the alternation. The closure of a tetraene to a cyclootatriene (or the reverse) can be analyzed as above and one finds that conrotation is the thermally allowed process.

Under photochemical conditions we find the opposite results. Photons generally excite an electron from the HOMO to the next higher molecular orbital. This higher orbital was the LUMO, but it becomes the higher singly occupied molecular orbital (HSOMO). This orbital can then interact with the LUMO of the other reacting system to give the product in the excited state, which subsequently descends to the ground state. The

HSOMO will have the phase relationship opposite to that of the former HOMO and therefore give the opposite stereochemistry compared to the ground state thermal reactions. For example, irradiation of *cis*-5,6-dimethyl-1,4-diphenylcyclohexadiene gave the conrotatory triene product (Eq. 9).[14] The low temperature was necessary to avoid a subsequent sigmatropic shift.

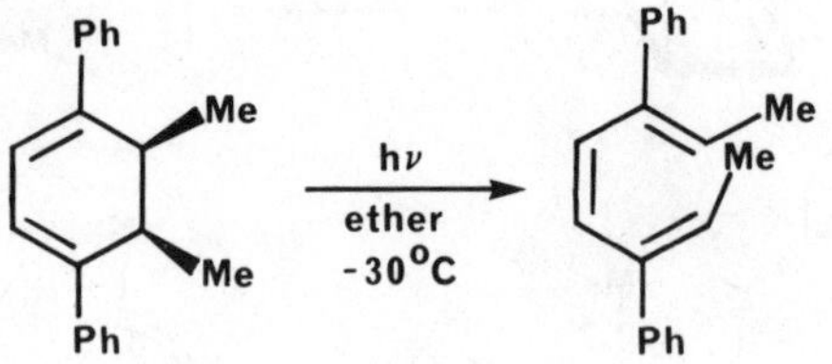

Ph Me Me Ph —hν, ether, -30°C→ Ph Me Me Ph (9)

We can summarize the selection rules for allowed electrocyclic reactions as follows:

Closures	Openings	Thermal	Photochemical
Dienes	Cyclobutenes	Conrotatory	Disrotatory
Trienes	Cyclohexadienes	Disrotatory	Conrotatory
Tetraenes	Cyclooctatrienes	Conrotatory	Disrotatory
Pentaenes	Cyclododecatetraenes	Disrotatory	Conrotatory

8.3.2 Cycloaddition Reactions

A cycloaddition reaction is the joining together of two independent bonding systems to form a ring with two new σ bonds. The reverse is called a *retro cycloaddition* reaction, and the selection rules again apply in both directions of a given reaction.

If butadiene and an appropriately substituted ethylene approach and begin to overlap as in Eq. 10, we find that there is a favorable phase relationship between the HOMO of one and the LUMO of the other for a face-to-face joining. This is, of course, the familiar Diels–Alder reaction, and it is thermally allowed.

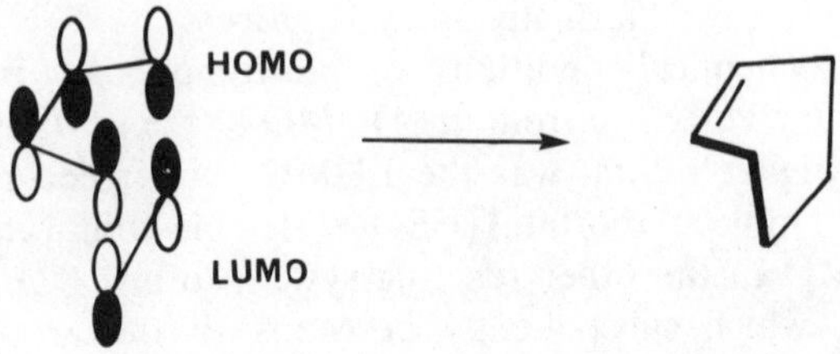

HOMO LUMO → (10)

The stereochemical consequences of this approach are already illustrated in Section 6.6.

If two ethylenes are similarly brought together face-to-face, the HOMO–LUMO phase relationship is unfavorable for formation of a cyclobutane (Eq. 11). Under photochemical conditions, however, this is a valuable route to cyclobutanes (Section 6.4).

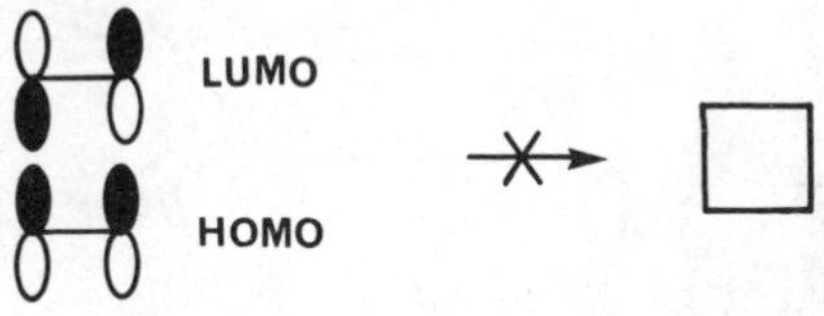

LUMO

HOMO

(11)

The allowedness may be determined by considering the HSOMO with the LUMO as in Eq. 12. Those substituents that were cis on an ethylene will be cis on the cyclobutane. Regioisomers are possible also.

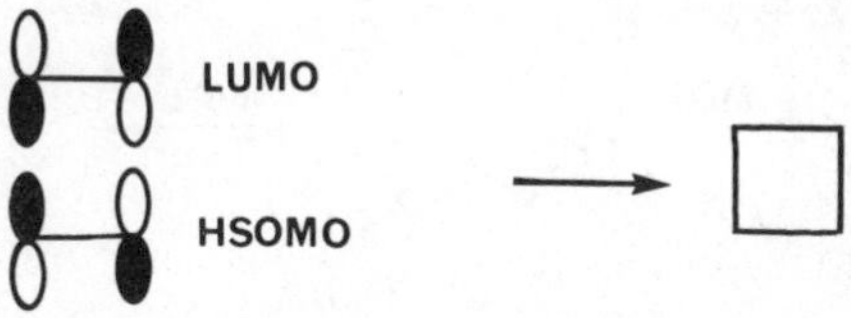

LUMO

HSOMO

(12)

A triene and an ethylene combining face to face at their ends is again not thermally favorable (Eq. 13), as is also a diene combining with a diene (Eq. 14) since the phase relations are unfavorable to bonding. These components may instead combine in a Diels–Alder reaction to give a vinylcyclohexene.

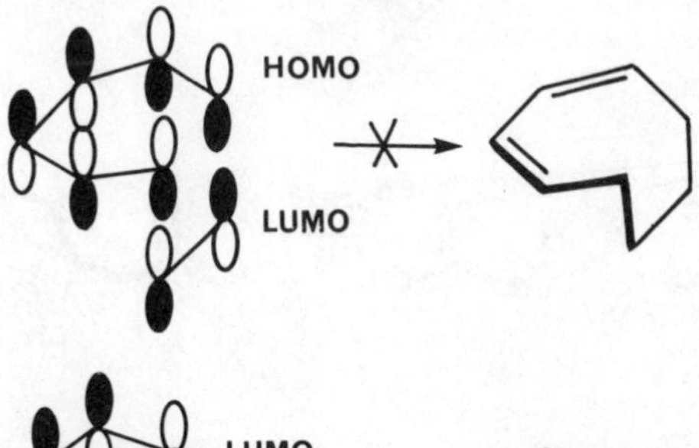

HOMO

LUMO

(13)

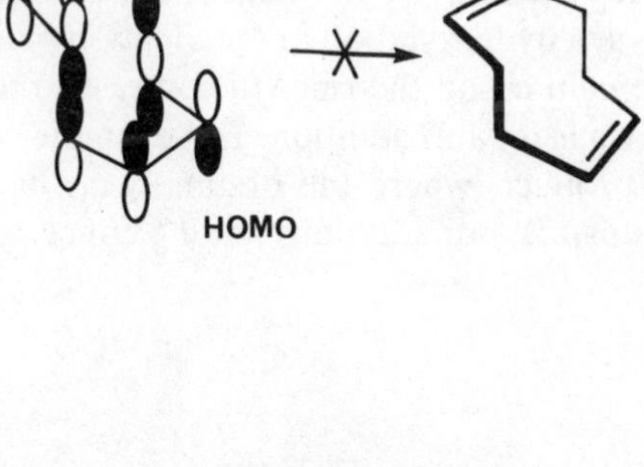

LUMO

HOMO

(14)

Going one step further, the combination of a diene and a triene (Eq. 15), or a tetraene and an ethylene, is favorable face to face. Specific examples are shown in Eq. 16[15] and Eq. 17.[16]

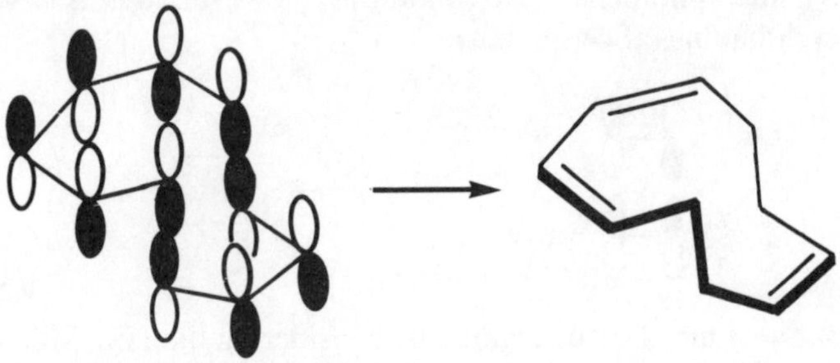

(15)

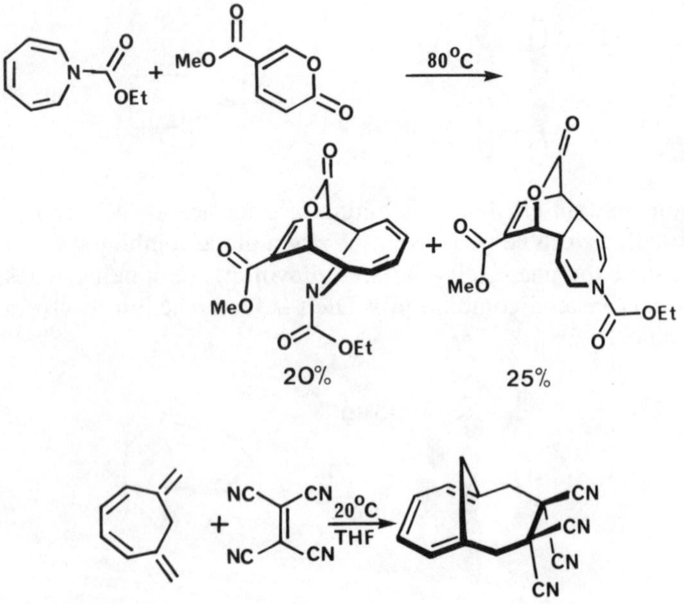

80°C

(16)

20% 25%

20°C THF

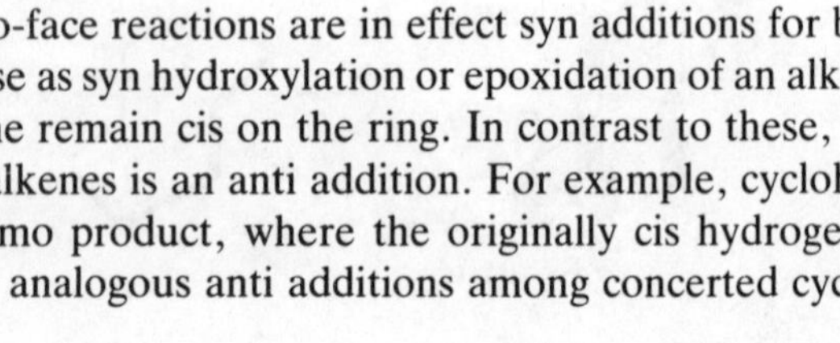

(17)

These face-to-face reactions are in effect syn additions for both systems in the same sense as syn hydroxylation or epoxidation of an alkene. Groups cis on the alkene remain cis on the ring. In contrast to these, the addition of bromine to alkenes is an anti addition. For example, cyclohexene gives the trans dibromo product, where the originally cis hydrogens are trans also. There are analogous anti additions among concerted cycloadditions.

For example, heptafulvalene combines with tetracyanoethylene this way (Eq. 18).

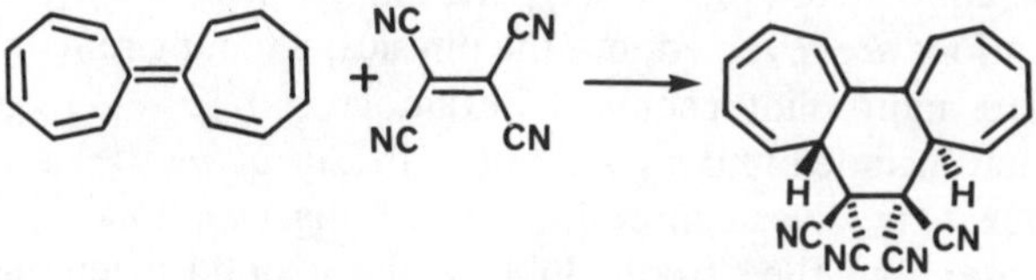

(18)

Heptafulvalene has 14 π electrons and the face-to-face reaction with an ethylene at the carbons shown is thermally forbidden, but by attaching at the top face at one end and the bottom at the other, the opposite phase interaction is reached and the reaction is thermally allowed. On the tetracyanoethylene side it is a syn reaction. Anti additions are called *antarafacial,* and syn additions are called *suprafacial.* The process of Eq. 18 is abbreviated $[\pi 14_a + \pi 2_s]$, where the numbers indicate the number of electrons reorganizing in each piece and the subscripts a and s indicate antarafacial and suprafacial for each piece. Antarafacial cycloadditions occur in polyenes that are highly strained or twisted to allow such access.

The bracketed abbreviations may be applied to all the concerted reactions; that is, Eq. 10 is $[\pi 4_s + \pi 2_s]$, Eq. 12 is $[\pi 2_s + \pi 2_s]$, Eq. 15 is $[\pi 6_s + \pi 4_s]$, and Eq. 17 is $[\pi 8_s + \pi 2_s]$.

An "addition" to a σ bond also offers the choice of antarafacial or suprafacial. Going back to an electrocyclic case, the thermal conrotatory opening of a cyclobutene has one face of a π bond attaching to a front lobe from a σ bond at one end and a back lobe at the other end. This is a $\sigma 2_a$ process. Equation 19 indicates the new bonding with dashed lines.

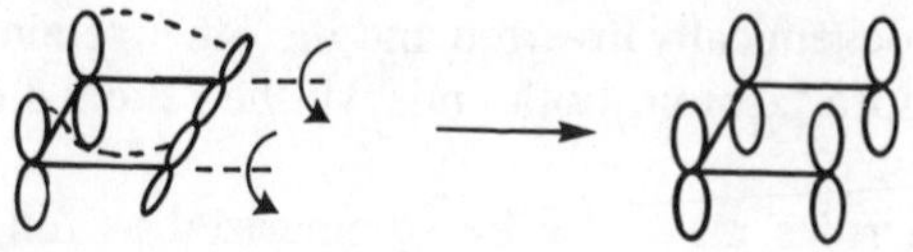

(19)

$[\sigma 2_a + \pi 2_s]$

The disrotatory opening of a cyclohexadiene has the front lobes on both ends (or back lobes on both ends) of the σ bond attaching to the same face of the system, and this is a $\sigma 2_s$ process (Eq. 20).

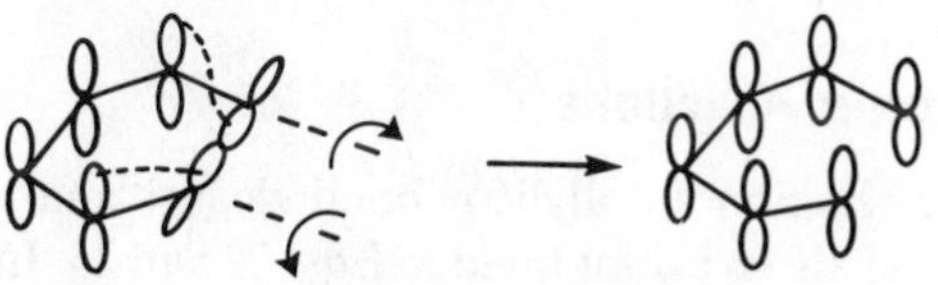

(20)

$[\sigma 2_s + \pi 4_s]$

In electrocyclic reactions there are two ways to abbreviate each process because the π system could be considered antarafacially instead. Equation 19 could be abbreviated [$\sigma 2_s + \pi 2_a$] and Eq. 20, [$\sigma 2_a + \pi 4_a$]. Both a and/or s designations are reversed, but the physical meaning remains the same.

There are many photochemical intramolecular [2 + 2] cycloaddition reactions that include breaking a σ bond. In Eq. 21 we see that irradiation of a cyclohexenone gives a three-membered ring. A σ bond is broken from a chiral center, but the "front" lobe of the σ bond extending from the chiral center was utilized in forming a new σ bond, and the configuration is retained.[17] This is the expected result for a concerted reaction as indicated by the phase relationships for the π HSOMO and the σ LUMO. The new bonding is indicated by dashed lines.

LUMO CH$_3$ O HSOMO $h\nu$ CH$_3$ O

[$\sigma 2_a + \pi 2_a$]

(21)

Look again at the breaking σ bond. The CH_2 end of it is undergoing stereochemical inversion. Although this is not detectable here, it was found by means of deuterium labeling in a similar molecule.[18]

Whenever a σ bond is broken in an antara fashion in a concerted reaction wherein both carbons are sp^3-hybridized at the finish (as in Eq. 21), one end will be stereochemically inverted and the other retained. If the σ bond is broken in a supra fashion, both ends will be retained or both ends will be inverted.

The selection rules may now be summarized as follows. If the total number of electrons reorganizing in the two combining systems is $4n$ ($n = 1, 2, 3, \ldots$), [a + s] reactions are thermally allowed and [a + a] and [s + s] are photochemically allowed. If the total number of electrons is $4n + 2$, [s + s] and [a + a] are thermally allowed and [a + s] is photochemically allowed. The reverse of allowed reactions is also allowed.

8.3.3 Sigmatropic Reactions

In a sigmatropic reaction an allylic σ bond cleaves and a new one forms further along the chain as exemplified in Eqs. 22 and 23. In Eq. 22 a carbon three positions along from the detaching one attaches to a site three positions along in the other chain. This is called a *shift of order* [3,3].

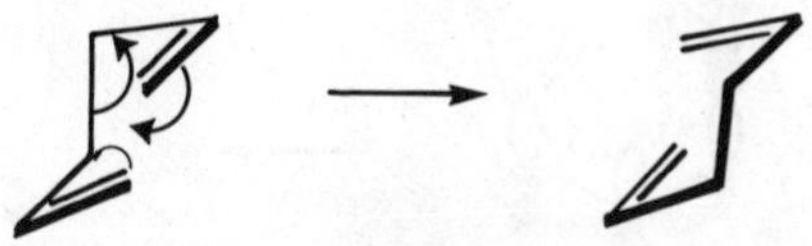

(22)

In Eq. 23 the same atom, a hydrogen, detaches and reattaches on the fifth position along the chain. This is called a *shift of order* [1,5].

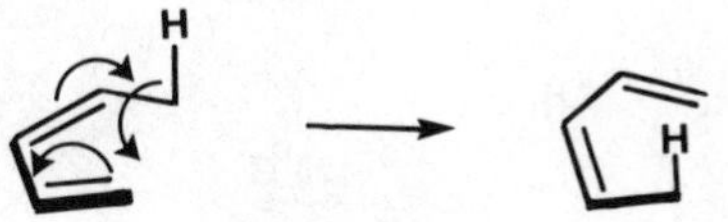

(23)

$[\sigma 2_s + \pi 4_s]$

For frontier orbital analysis this is viewed as a σ bond combining with a diene bond. Examining the phase relationships of the diene HOMO with the sigma LUMO, we find an in phase overlap (Eq. 24) when the hydrogen reattaches to the same face of the π system, that is, a suprafacial shift.

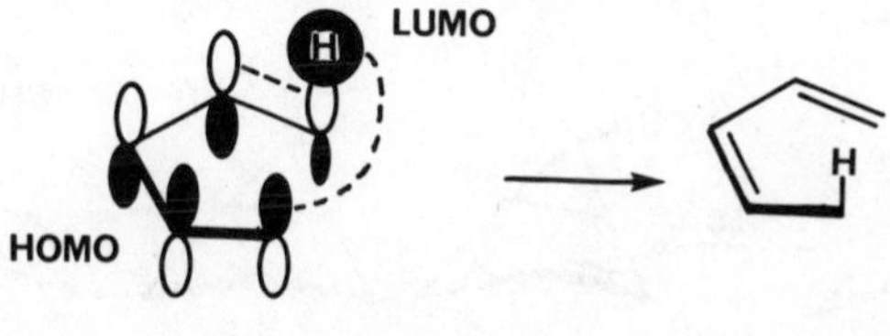

LUMO

HOMO

(24)

This is a thermally allowed reaction. In contrast, a [1,3] suprafacial shift of a hydrogen is not allowed, as indicated by the antibonding relationship shown in Eq. 25.

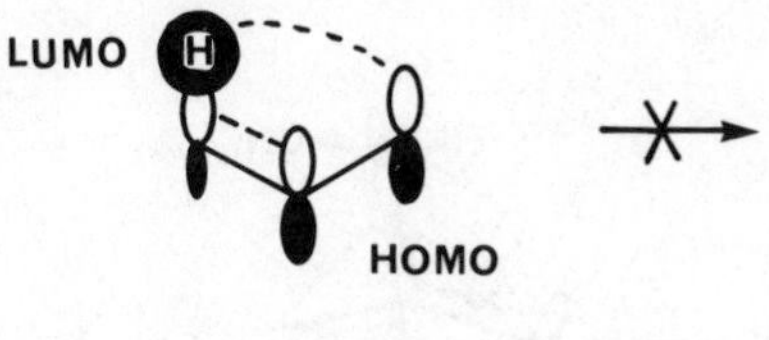

LUMO

HOMO

(25)

If the [1,3] migrating group is a carbon instead of a hydrogen, there is a back lobe of opposite phase and the σ bond can be used antarafacially to give in phase overlap, but the migrating carbon will undergo inversion of configuration:

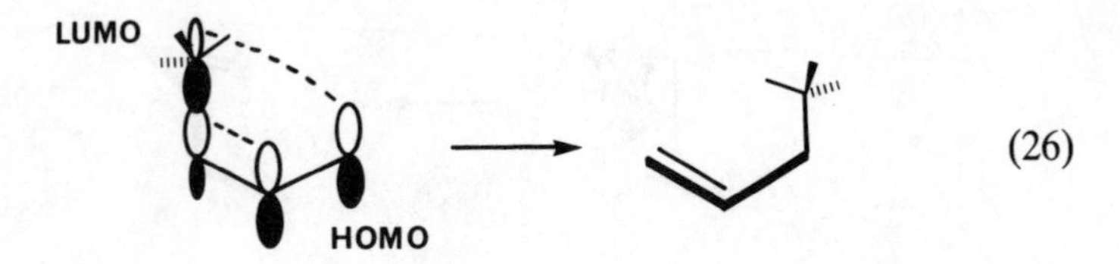

(26)

An example of this is the thermal rearrangement of the exo-bicyclo [2.1.1] compound in Eq. 27.[19] The new bonding is indicated as dashed lines in Fig. XI.

H CH$_3$ —120°C→ H CH$_3$ (27)

98.5%

$[\sigma 2_a + \pi 2_s]$ or $[\sigma 2_s + \pi 2_a]$

The endo isomer required a higher temperature and was not as cleanly in accord with the selection rules, but it gave mostly allowed products, including some migration of the CH_2 bridge:

CH$_3$ H —150°C→ CH$_3$ H + H CH$_3$ + CH$_3$ H (28)

56.5% 25% 18%

A concerted [1,5] shift of carbon is allowed suprafacial with retention of configuration. A [1,7] shift of hydrogen is thermally allowed if it can

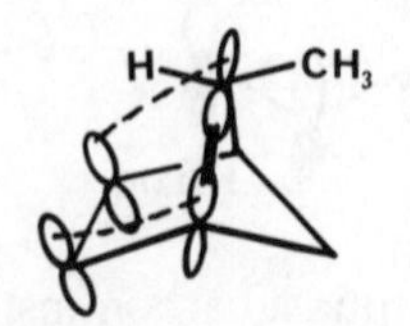

Figure XI. A sigmatropic shift with inversion on carbon.

reach the opposite face of the π system (antarafacial):

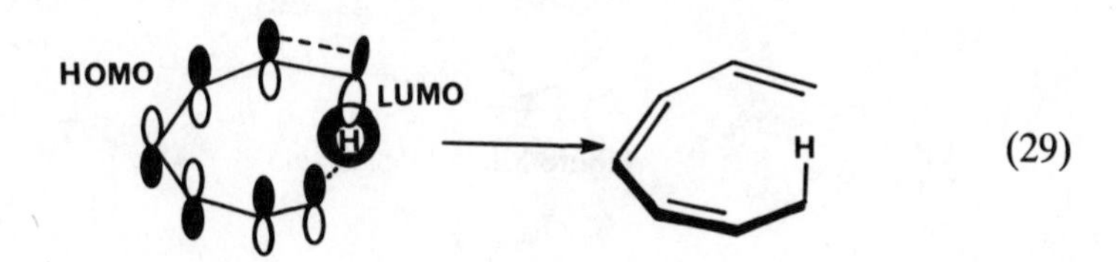

(29)

The [3,3] shift of Eq. 22 is actually a combination of three parts: a σ bond and two separate π bonds. First a HOMO may be drawn that is a combination of the σ bond and one π bond as in Eq. 30 and the closure to a cyclic transition state is the addition of the second π bond.

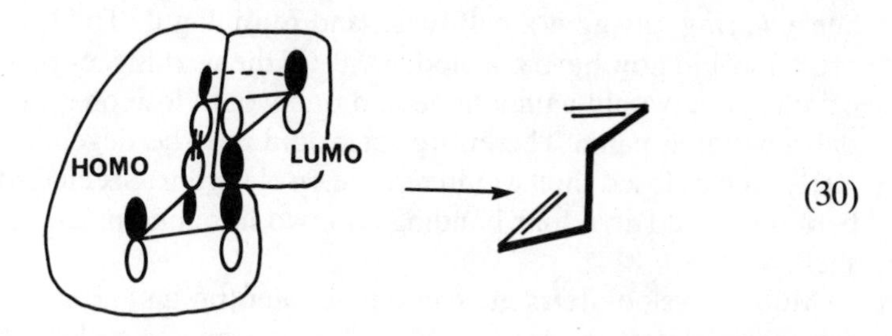

(30)

The HOMO involving four carbons has one node as in butadiene even though it is part σ and part π.[20] This gives in-phase overlap with the ethylene LUMO, and the reaction is thermally allowed, suprafacial on both halves.

The selection rules for sigmatropic reactions of neutral molecules may be summarized as follows. If the number of electrons involved in reorganization (the sum of the numbers in brackets) is $4n + 2$, the reaction is thermally allowed suprafacially without inversion. The common cases are [1,5] and [3,3]. If the number of electrons is $4n$, the reaction is thermally allowed with an inversion or an antarafacial movement. The common cases are [1,3] and [1,7]. Remember that hydrogen cannot invert. The photochemical cases are the opposite of the thermal. Some structural features may make allowed reactions geometrically impossible.

There are more rules covering rearrangements of carbocations and carbanions beyond the scope of this book.[7]

8.3.4 Aromatic Stabilization of Transition States

In Section 8.2 aromatic character was connected with closed-shell electron occupation. In the usual cyclic p-orbital overlap there is a lowest-energy

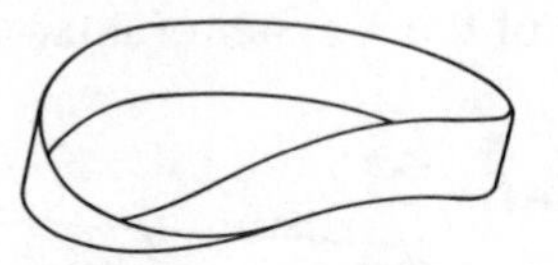

Figure XII. A Möbius strip.

molecular orbital and above this, degenerate pairs of orbitals. Filling the lowest plus one or more degenerate pairs requires $4n + 2$ electrons. These are called *Hückel molecular orbitals*.

A Möbius strip is a band with a 180° twist in it such that the outside and inside surface are continuous and one (Fig. XII). Imagine opening a benzene ring, giving it a half twist, and rejoining it. The lowest molecular orbital would now have one node in it. Of the next higher pair of molecular orbitals, one would gain a node and one would lose one, and likewise for the remaining pairs. The entire set would now be degenerate pairs (Fig. XIII), and a closed shell would require $4n$ electrons. Six electrons in Möbius benzene would give four bonding and two nonbonding and an incomplete shell.

Möbius cyclootatetraene can now be neutral and aromatic (Fig. XIV), with eight electrons. Actually a twisted benzene or cyclooctatetraene ring would have very poor overlap from both lobes of each p orbital. Anything smaller than a 22-membered ring would be poor. However, special stability appears yet even when part of the ring involves only one lobe overlap.

Instead of ground-state molecules, let us now examine the cyclic transition states of several kinds of concerted reactions. Reconsider the disrotatory electrocyclic closure of a hexatriene (Eq. 1). The transition state of this reaction (and the reverse) is shown in Fig. XVa, where a complete loop of overlap is followed with a line from atom to atom through all six. The lower lobes could also be connected around part of the ring, but we will concentrate on one complete loop. This as drawn with a minimum of

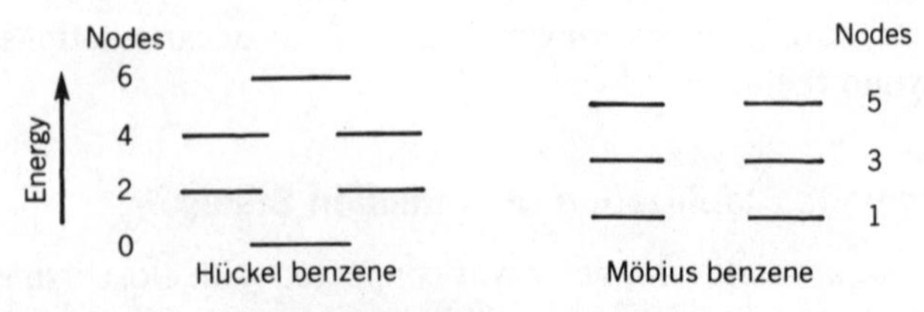

Figure XIII. Molecular orbitals.

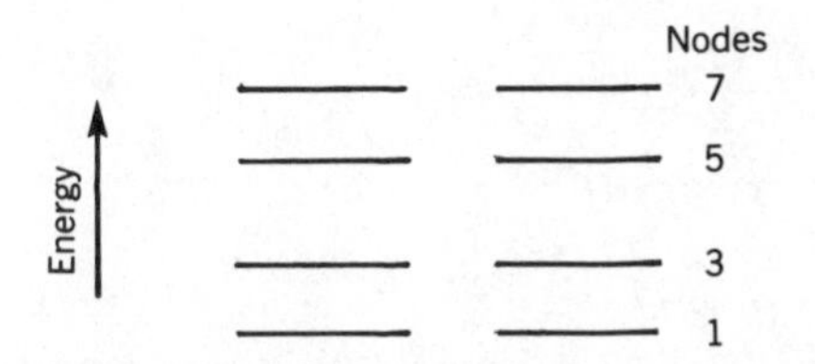

Figure XIV. Möbius cyclooctatetraene orbitals.

nodes resembles the lowest molecular orbital of aromatic benzene since there are six electrons and no nodes. This transition state is favored by aromatic stabilization[21,22] and is reached only by disrotation. Conrotation would have given a transition state with one phase discontinuity (node) (Fig. XV*b*) that would be part of a Möbius orbital set, but six electrons do not give a closed shell here and the transition state does not have aromatic stabilization and is not allowed.

The transition state for the closure (or the reverse) of a butadiene is shown in Fig. XVI. The disrotatory closure allows a loop with no nodes (a Hückel orbital), but four electrons cannot give a closed-shell occupation (as with cyclobutadiene; Section 8.2) and the transition state is not stabilized (is forbidden). However, the conrotatory closure can be drawn with one node and thus belongs to the Möbius group, where the four electrons will give a closed shell with aromatic stabilization; That is, this transition state is stabilized and allowed.

In Fig. XVII we see the transition states for a [4 + 2] and a [4 + 4] cycloaddition. The loop can be drawn with no nodes for each of them, meaning that they are Hückel-type orbitals where six electrons give closed shells [4 + 2] and a thermally allowed reaction, but eight [4 + 4] do not, and that reaction does not have an aromatic stabilized transition state and is not allowed.

In Fig. XVIII we see transition states for four possible sigmatropic shifts. The advantage of this transition state analysis is that all types of con-

Figure XV. Transition states for closure of a hexatriene: (*a*) disrotatory; (*b*) conrotatory.

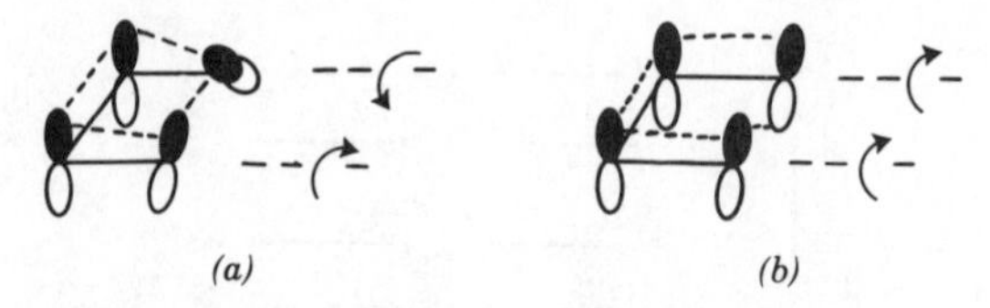

Figure XVI. Transition states for closure of a butadiene: (*a*) disrotatory; (*b*) conrotatory.

certed reactions are covered by basically one selection rule. If a continuous loop of overlap through all the atoms involved in the transition state can be drawn with no nodes, the reaction will be allowed thermally if it involves $4n + 2$ electrons. If the continuous loop requires one node, the reaction will be thermally allowed if it involves $4n$ electrons. Photochemical reactions are the opposite.

More strictly speaking, we are looking at the overlap of a basis set of atomic orbitals contributing to the molecular orbitals of the transition state. These may be drawn in any arbitrary phase orientation. In this way a Hückel set is recognized when strung together by the presence of an even number of nodes and a Möbius set is recognized by the presence of an odd number of nodes.

Finally it must be kept in mind that although these theories have very broad predictive powers, there are many limitations. Other factors such as structural constraints can prevent a reaction that is allowed on an orbital basis. Which direction an allowed reaction will go is not predicted by this theory. Often more than one allowed reaction is possible from given starting materials, and again, these theories usually do not make a distinction. Furthermore, nonconcerted mechanisms may operate to give products that are forbidden by these theories.

Elaborations on the frontier orbital theory taking into account the coefficients at each atom of interacting systems have extended the predictive power into regioselectivity.[23]

Figure XVII. Transition states for [4 + 2] and [4 + 4] cycloadditions.

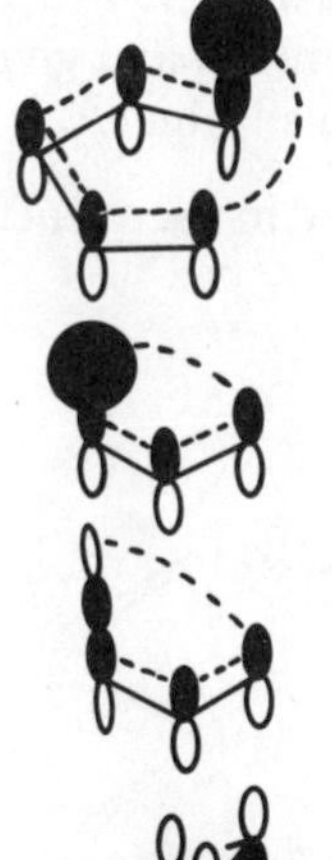

[1,5] suprafacial shift of hydrogen, no nodes necessary therefore Hückel; six electrons, therefore, aromatic, allowed

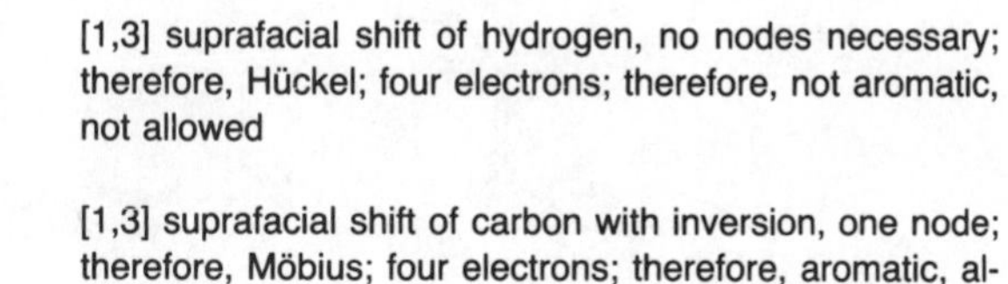

[1,3] suprafacial shift of hydrogen, no nodes necessary; therefore, Hückel; four electrons; therefore, not aromatic, not allowed

[1,3] suprafacial shift of carbon with inversion, one node; therefore, Möbius; four electrons; therefore, aromatic, allowed

[3,3] suprafacial, suprafacial shift, no nodes; therefore, Hückel; six electrons; therefore, aromatic, allowed

Figure XVIII. Transition states for sigmatropic shifts.

PROBLEMS

1. Three of the following structures were products from heating 2-chlorotropone with cyclopentadiene at 105°C. Which are the likely ones? Explain.[24]

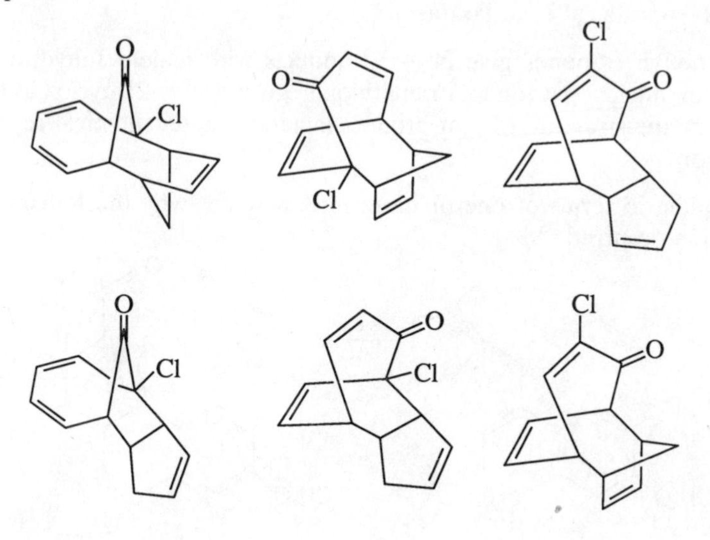

2. *cis*-3,4-Dichlorocyclobutene is isolable from a synthesis at 150°C, while the trans isomer does not survive. Explain the difference and give the product expected from the attempt to make the trans isomer.[25]

3. Which isomer of 9,10-dihydronaphthalene is produced in each reaction? Explain.[26]

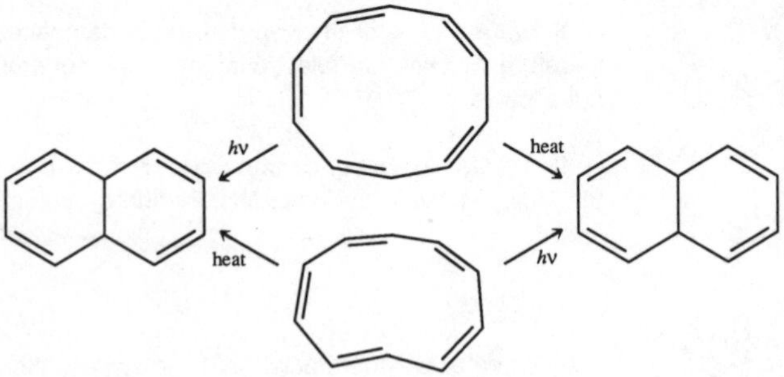

hν heat heat hν

4. Heating the following acid ester gave two acyclic products. Draw the likely structures.[27]

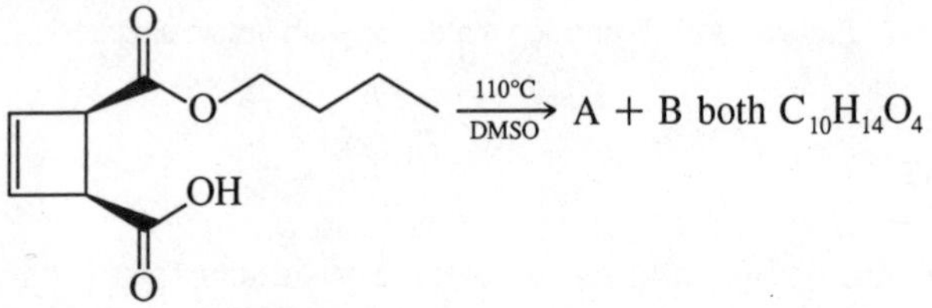

O O OH O

$\xrightarrow[\text{DMSO}]{110^\circ\text{C}}$ A + B both $C_{10}H_{14}O_4$

5. The shift that follows Eq. 9 gives 3,6-diphenyl-1,3,5-octatriene. Is that shift suprafacial or antarafacial?

6. Although tropones give [4 + 2] adducts with maleic anhydride, the sulfur analog, cycloheptatrienethione, gives [8 + 2] cycloaddition.[28] Draw the structure of that product including stereochemical representation.

7. Explain in terms of one or more intermediates why the following reaction occurred:[29]

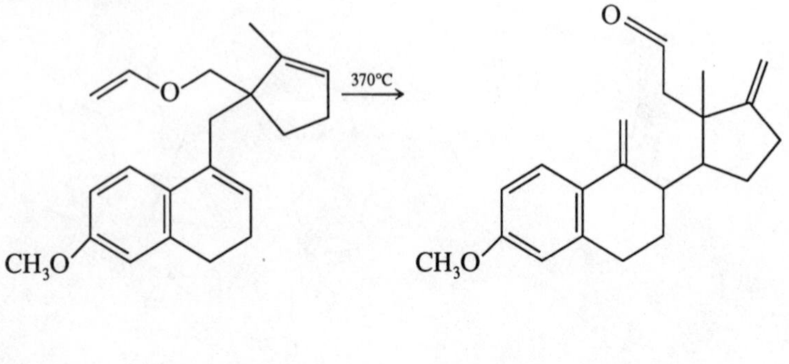

O CH_3O $\xrightarrow{370^\circ\text{C}}$ O CH_3O

8. Draw the intermediate(s), and predict the complete stereochemistry of the product of the following reaction:[30]

REFERENCES

1. Zimmerman, H. E. *Quantum Mechanics for Organic Chemists,* Academic, New York, 1975.
2. Fukui, *Angew. Chem. Internatl. Ed.* **1982,** *21,* 801.
3. Sondheimer, F. *Acc. Chem. Res.* **1972,** *5,* 81.
4. Goldberg, S. Z.; Raymond, K. N.; Harmon, C. A.; Templeton, D. H. *J. Am. Chem. Soc.* **1974,** *96,* 1348.
5. Vogel, E. Altenbach, H.-J.; Cremer, D. *Angew. Chem. Internatl. Ed.* **1972,** *11,* 935.
6. Swinborne-Sheldrake, R.; Herndon, W. C.; Gutman, I. *Tetrahedron Lett.* **1975,** 755.
7. Woodward, R. B.; Hoffmann, R. *Angew. Chem. Internatl. Ed.* **1969,** *8,* 781.
8. Fukui, K. *Acc. Chem. Res.* **1971,** *4,* 57.
9. Fleming, I. *Frontier Orbitals and Organic Chemical Reactions,* Wiley-Interscience, London, 1976.
10. Marvell, E. N. *Thermal Electrocyclic Reactions,* Academic, New York, 1980.
11. Marvell, E. N.; Caple, G.; Schatz, B.; Pippin, W. *Tetrahedron* **1973,** *29,* 3781.
12. Winter, R. E. K. *Tetrahedron Lett.* **1965,** 1207.
13. Doorakian, G. A.; Freedman, H. H. *J. Am. Chem. Soc.* **1968,** *90,* 5310, 6896.
14. Courot, P.; Rumin, R. *Tetrahedron Lett.* **1970,** 1849.
15. Saito, K.; Ida, S.; Mukai, T. *Bull. Chem. Soc. Jpn.* **1984,** *57,* 3483.
16. Farrant, G. C.; Feldman, R. *Tetrahedron Lett.* **1970,** 4979.
17. Chapman, O. L.; Sieja, J. B.; Welstead, W. J., Jr. *J. Am. Chem. Soc.* **1966,** *88,* 161.
18. Bellus, D.; Kearns, D. R.; Schaffner, K. *Helv. Chim. Acta* **1969,** *52,* 971.
19. Roth, W.; Friedrich, A. *Tetrahedron Lett.* **1969,** 2607.
20. Fleming, I., ref. 9, p. 102.
21. Dewar, M. J. S. *Tetrahedron* (Suppl.) **1966,** *8,* 75.
22. Zimmerman, H. E. In *Pericyclic Reactions,* Vol. 1, Marchand, A. P.; Lehr, R. E., Eds., Academic, New York, 1977, Chapter 2.

23. Houk, K. N. In *Pericyclic Reactions,* Vol. II, Marchand, A. P.; Lehr, R. E., Eds., Academic, 1977, Chapter 4.
24. Ito, S.; Sakan, K.; Fujise, Y. *Tetrahedron Lett.* **1969,** 775.
25. Avram, M.; Dinulescu, I.; Elian, M.; Farcasiu, M.; Marcia, E.; Mateescu, G.; Nenitzescu, C. D. *Chem. Ber.* **1964,** *97,* 372.
26. Masamune, S.; Hojo, K.; Hojo, K.; Bigam, G.; Rabenstein, D. L. *J. Am. Chem. Soc.* **1971,** *93,* 4966.
27. Trost, B. M.; McDougal, P. G. *J. Org. Chem.* **1984,** *49,* 458.
28. Machiguchi, T.; Hoshino, M.; Ebine, S.; Kitahara, Y. *J. Chem. Soc. Chem. Commun.* **1973,** 196.
29. Ziegler, F. E.; Lim, H. *J. Org. Chem.* **1982,** *47,* 5229.
30. Kametani, T.; Suzuki, K.; Nemoto, H. *J. Chem. Soc. Chem. Commun.* **1979,** 1127.

9

Physical Influences on Reactions

The experimental section of a journal article is the core of factual observation. It remains of value even when the explanations and theories are revised. This is the part of the article that is consulted in detail when you want to prepare a reported compound or a closely related one. From introductory texts it is easy to get the impression that simply mixing starting materials gives the product. Some cases are as easy as that, but many require particular conditions or techniques to stimulate or control the process.*

The conditions are chosen for efficiency and selectivity. The reactants may be capable of giving many products, but with appropriate handling the desired one may predominate. Proper techniques make the reaction easy to conduct, without temperature extremes or lengthy times.

The larger part of the effort is usually the separation and purification of the products, but these procedures are found in laboratory texts and will not be covered here.

9.1 UNIMOLECULAR REACTIONS

Many eliminations and rearrangements require only heat to proceed, and the simplest procedure is heating with no solvent. Heating dicyclopentadiene at 200–210°C gives the retro Diels–Alder product cyclopentadiene,

* There are many safety considerations in the design of experiments, and this should be done under the supervision of experienced personnel.

which escapes as a gas and is condensed.[1] Heating a chlorooxazolidinone neat gives HCl elimination directly (Eq. 1).[2]

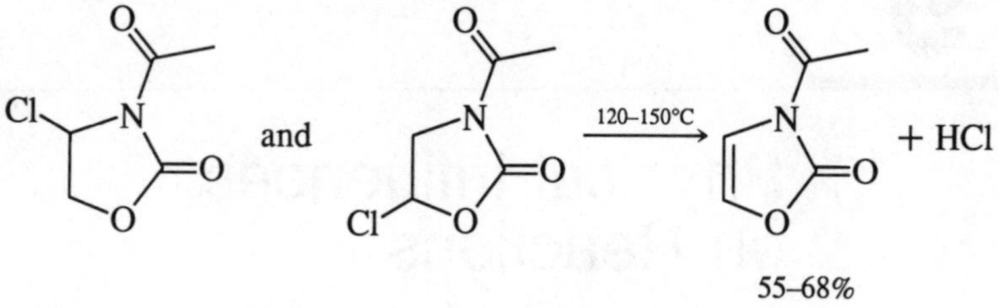

Cl, N, O, O, O and N, O, O, Cl, O 120–150°C N, O, O, O + HCl

55–68%

(1)

More often reactions are carried out in a solvent, which may serve several purposes. A solvent with an appropriate boiling point may be heated at reflux to provide a constant temperature for the reaction. If it is particularly exothermal, the solvent can be a heat sink to dissipate the exotherm rather than allowing the reactant to rise precipitously to a high temperature. If there is a competing bimolecular reaction between molecules of the reactant, the solvent is a diluent that lessens the collisions and favors the unimolecular reaction.

An alternative that avoids the expense of a solvent and its removal is heating in the gas phase. This avoids bimolecular reactions because in the low density of a gas (especially under vacuum) molecular collisions are relatively infrequent. Short reaction times at extremely high temperatures may be used where warranted. For example, *cis*-4b,8a-dihydrophenanthrene was prepared in high yield by passing the benzotricyclic compound (Eq. 2) through a tube at 1 mm pressure and 550°C, even though the product disproportionates (bimolecular reaction) in the liquid state at 150°C.[3]

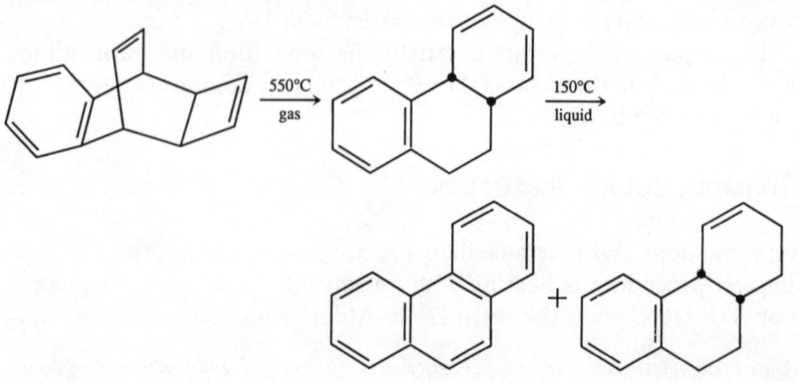

550°C gas 150°C liquid

+ (2)

9.2 HOMOGENEOUS TWO-COMPONENT REACTIONS

Efficient interaction of two or more components occurs if they are combined in a single liquid phase. Cyclooctatetraene and maleic anhydride may be heated together at 165–170°C for 30 min to give a high yield of the Diels–Alder adduct.[4]

If one or more of the reactants remain solid at the desired temperature, the reaction zone is often limited to the surface of the solid and may soon be blocked by a layer of reaction product. A solvent that dissolves both reactants can alleviate this problem and also aid in temperature control. More solvent will give less frequent collisions between reactants and slow the progress. This may be desirable for moderation of very fast reactions or minimized for slower ones. In some cases one reactant may be used in excess as the solvent to maintain the high collision frequency.

9.3 TEMPERATURE OPTIMIZATION

The temperature selected for a reaction is a compromise, arrived at by experimentation using similar cases as guides. A lower temperature will give a longer reaction time, while a higher temperature will bring on competing reactions that will lessen the yield of the desired product and complicate the purification process.

Careful selection of temperatures is necessary in the coupling reactions of alkyl cuprates with primary halides (Eq. 3).

$$RLi + CuI \xrightarrow{THF} R_2CuLi \xrightarrow{R'X} R{-}R' + RCu + LiX \tag{3}$$

The cuprates are thermally unstable and are prepared at low temperatures and used promptly. Starting with *n*-butyllithiuum the sequence can be done at −78°C, but little coupling is found with *sec*-butyl and *tert*-butyllithium. If a temperature of 0°C is used briefly for the first step and then the coupling done at −78°C, good yields are found with all three lithium reagents.[5,6]

9.4 PRESSURE EFFECTS

High pressures are commonly used to increase the solubility of gases in reaction solvents as in hydrogenation in order to increase the rate of reactions. High pressures are incidental to reactions at temperatures above the boiling point of reactants, when they are done in sealed tubes to maintain a liquid phase.

Very high pressures can have a marked influence on reactions in a liquid phase alone.[7] Some reactions have negative activation volumes ($\Delta V^{\ddagger}$); that is, the transition state, including solvation, occupies a smaller volume than the starting molecules. Such reactions will be considerably accelerated by very high pressures. For example, the esterification of acetic acid with ethyl alcohol at 50°C is five times faster at 2000 atm (compared to 1 atm) and 26 times faster at 4000 atm.[8]

Negative $\Delta V^{\ddagger}$ values are common among quaternization of amines and phosphines, hydrolyses and esterifications, Claisen and Cope rearrangements, nucleophilic substitutions, and Diels–Alder reactions.[9]

On a laboratory scale pressures of up to 15,000 atm are readily developed hydraulically in suitable vessels.

Very high pressure is especially valuable where heat alone leads to alternate reaction products. For example, the phosphonium salt in Eq. 4 was prepared in excellent yield at 20°C and 15,000 atm. At 1 atm and 20°C no reaction occurs, and at 80°C only decomposition products are found.[10]

$$\text{(2-(2-bromoethyl)-2-methyl-1,3-dioxolane)} + Ph_3P \xrightarrow[\text{15,000 atm}]{\text{benzene–toluene, 20°C}} \text{(dioxolane)}\text{–}CH_2CH_2\overset{+}{P}Ph_3\ Br^- \quad (92\%) \tag{4}$$

Diels–Alder reactions are significantly improved by high pressure also,[11] as in the case shown in Eq. 5. The cyclic acetal was prepared in 27% yield at 160–180°C and 1 atm, but at 50°C and 20,000 atm the yield was 85%.[12,13]

$$\text{(1-methoxybutadiene, } OCH_3\text{)} + CH_3COCO_2C_2H_5 \longrightarrow \text{(2-methoxy-6-methyl-dihydropyran-6-carboxylate ethyl ester, } OCH_3\text{)} \tag{5}$$

Besides accelerating reactions, high pressure may be used to shift the position of an equilibrium. At 100°C and 1 atm the reaction of naphthalene and excess maleic anhydride reaches equilibrium at only 1% conversion. At 100°C and 10,000 atm the yield is 80%.[14]

9.5 POLAR SOLVENTS

It is frequently necessary to treat an organic compound with an ionic reagent. The ionic reagents are not appreciably soluble in nonpolar organic solvents; therefore polar solvents are used.[15] The ions are solvated by coordination with the oppositely charged end of the solvent dipole or by specific hydrogen bonding. The hydrogen bonding (protic) solvents give higher rates in S_N1 reactions compared to aprotic solvents because they aid the departure of anionic leaving groups by hydrogen bonding with them. The more frequently used S_N2 reactions are rarely aided by these protic solvents because they cluster around the nucleophile anion rendering it less reactive. The smaller the anion, the more concentrated the charge and the more tightly it will be solvated by protic solvents. Thus in methanol the order of nucleophilicity of halides toward iodomethane is $I^- > Br^- > Cl^- > F^-$.[16]

The S_N2 reactions are greatly aided by dipolar aprotic solvents such as DMSO, DMF, hexamethylphosphoric triamide, or tetramethylurea. In these solvents the positive end of the dipole is relatively encumbered while the negative end is exposed and available for association with cations. The anions are thus relatively little solvated and exceptionally reactive. For example, NaCN in DMSO reacts with primary and secondary alkyl chlorides to give nitriles in 0.5 to 2 h, while 1 to 4 days are required in aqueous alcohol.[17,18] In these dipolar aprotic solvents the relative nucleophilicity of anions follows charge density and is the reverse of that found in protic solvents. The displacement on *n*-butyl tosylate in DMSO gives the order $F^- > Cl^- > Br^- > I^-$.[19]

An empirical scale of solvent polarity has been developed on the basis of shifts of a UV/visible absorption maximum of a pyridinophenylate indicator,[20] which should be useful for predicting solvent effects on rates.

9.6 REACTIONS WITH TWO LIQUID PHASES

The water solubility, low volatility, and reactivity of the dipolar aprotic solvents can be disadvantageous; therefore, another approach is of value. Many inorganic reagents are soluble in water and are used with an organic solution, with vigorous stirring to promote a reaction at the surface between the water and organic phases. However, the frequency of successful collision on a surface is far less than in the bulk of a homogeneous solution. Fortunately, in a great many cases this difficulty is readily overcome by adding a small amount of a tetraalkylammonium or phosphonium salt. The quaternary cations with sufficiently large alkyl groups have an affinity for organic solvents and will carry reactive anions with them into solution in the organic layer. These anions are particularly reactive because they carry only a small hydration shell. Some stirring is still necessary because the quaternary salts are used in catalytic amounts and must repeatedly exchange product anions for reactant anions at the phase boundry. This is called *phase-transfer catalysis*.[21-23]

Another mode of operation of these catalysts occurs in the formation of carbanions using concentrated aqueous NaOH. A proton is removed from a precursor by the OH^- ion at the interface, and the carbanion moves into the bulk of the organic phase accompanied by the lipophilic cation. This allows formation of substantial concentrations of carbanions that are ordinarily more basic than aqueous OH^- and yet avoids the difficulties of using stronger bases such as sodamide or LDA under anhydrous conditions.

Similar effects are found with the more expensive macrocyclic polyethers (crown ethers) that complex alkali metal cations, giving them sufficient lipophilicity to dissolve in the organic phase, bringing along the reactive anion.

Alkyl halides and sulfonates will undergo nucleophilic substitution by these reactive inorganic anions or carbanions in the organic layer. The product halide ions are nucleophilic and may compete with reactant anions that have leaving group ability and reach an equilibrium condition that is dependent largely on the relative amounts of the two anions in the organic phase. Chloride or bromide ions cannot displace iodide or tosylate ions efficiently because the iodide and tosylate anions are relatively lipophilic and remain in the organic phase. One can convert organochloride to bromide or iodide and convert organibromide to iodide in good yield. Mesylates are good leaving groups with low lipophilicity and are displaceable by all halides:[24]

CH_3SO_2O ... Cl + KCl $\xrightarrow[\text{benzene–water, reflux}]{n\text{-}C_{16}H_{33}\text{–}n\text{-}Bu_3P^+Br^-}$ Cl ... Cl (6)

70%

Other nucleophilic ions that are not themselves good leaving groups, such as cyanide, phenoxides, carboxylates, carbanions, alkoxides (Chapter 4, Eq. 34), or sulfenates (Eq. 7)[25] can give high yields of substitution products.

SO_2Na + Br $\xrightarrow[\text{benzene–water–acetone, heat}]{n\text{-}Bu_4N^+Br^-}$ SO_2 (7)

85%

Oxidation reactions using permanganate (p. 64), dichromate, or hypochlorite are very effective with phase-transfer catalysis.[26] Borohydride reductions are facilitated also. Dihalocarbenes are often best prepared via phase transfer catalysis as exemplified in Eq. 8.[27]

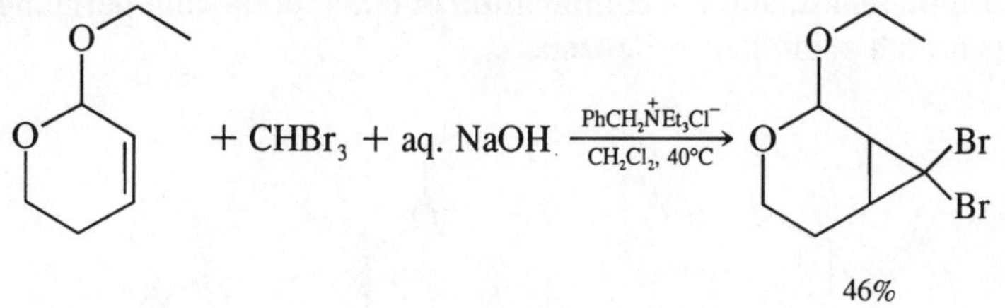

$$+ \ CHBr_3 + \text{aq. NaOH} \xrightarrow[CH_2Cl_2,\ 40°C]{PhCH_2\overset{+}{N}Et_3Cl^-} \qquad (8)$$

46%

9.7 REACTION OF A SOLID WITH A LIQUID

If an ionic inorganic reagent is combined with a low polarity organic solution without a water layer, insolubility is still the problem. An obvious resort is to use a more organic soluble salt of the same reactive anion. For example, although sodium borohydride has very low solubility in dichloromethane, ether, or THF, tetrabutylammonium borohydride has high solubility in methylene chloride. This salt is useful for the reduction of aldehydes and ketones that are not soluble in hydroxylic solvents or that react with those solvents to give hydrates, acetals, or ethers.[28]

Potassium permanganate is insoluble in organic solvents, and when used in water, it decomposes, requiring a large excess. Tetrabutylammonium permanganate is easily prepared and may be used in stoichiometric amount in pyridine solution at room temperature to give oxidation products quickly and in high yield.[29]

The quaternary salts can be used in catalytic amount in truly heterogeneous conditions. For example, solid potassium phthalimide was used in toluene in a Gabriel synthesis of amines as follows:[30]

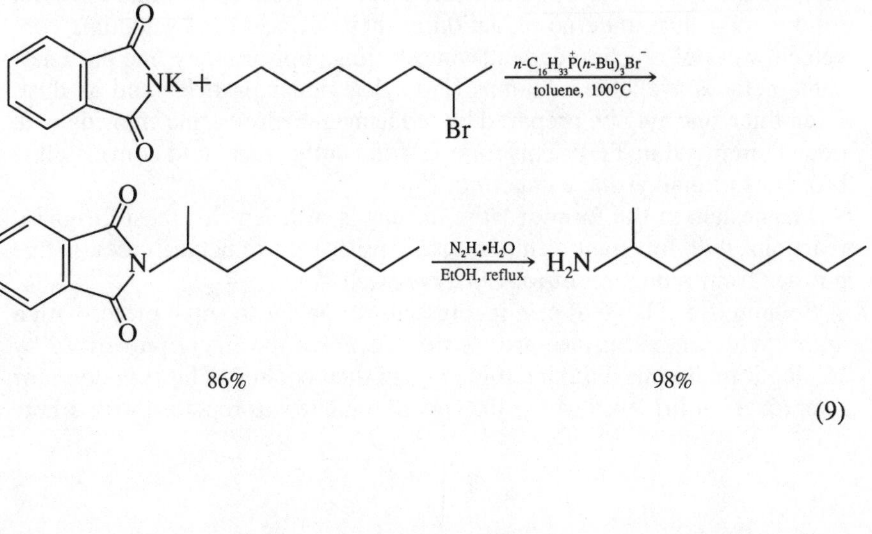

$$\text{NK} + \underset{Br}{} \xrightarrow[\text{toluene, 100°C}]{n\text{-}C_{16}H_{33}\overset{+}{P}(n\text{-Bu})_3Br^-}$$

$$\text{N} \xrightarrow[\text{EtOH, reflux}]{N_2H_4 \cdot H_2O} H_2N$$

86% 98%

(9)

Alkali metal salts become soluble in low-polarity organic solvents when those cations are coordinated by close-fitting cyclic ethers, giving them an organic compatible exterior. Dicyclohexyl-18-crown-6 is particularly suitable for potassium, and the combination of it and potassium permanganate (**1**) has a high solubility in benzene.

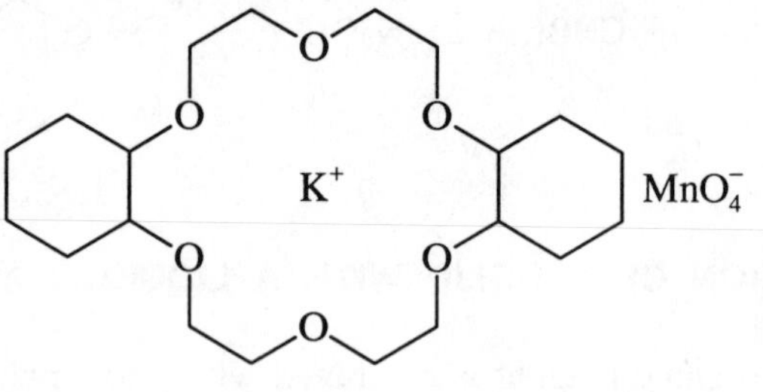

1

On a practical scale the crown ether can be used in catalytic amount in a two-phase process, but the precipitating MnO_2 coats the solid $KMnO_4$ and requires continuous pulverizing in a ball mill.[31] Alkenes are converted to acids and/or ketones but not to diols.

Substitution reactions may be carried out in high yield by using 18-crown-6 with solid potassium cyanide and an acetonitrile or benzene solution of primary or secondary chlorides with vigorous stirring.[32] Dry solid potassium acetate was used in the same way with primary and secondary bromides.[33] The nucleophilic reactivity of this "naked" acetate is very high compared with acetate in hydroxylic solvents.

The term "phase-transfer catalysis" is applied to these solid–liquid reactions, as well as to the liquid–liquid cases in Section 9.6.

In the above cases the intent is to bring at least small amounts of the reagent anions into the organic solution for reaction. In contrast to these, solid metals must undergo oxidation at their surfaces, and obtaining convenient rates of reaction depends on selecting appropriately fine particles. Zinc metal is available as lumps (mossy) granular particles and as dust. Even finer zinc may be prepared by reducing anhydrous zinc chloride with potassium metal in THF. This form is sufficiently reactive to convert alkyl bromides to dialkylzinc compounds.[34]

Magnesium in the form of lathe turning is sufficient for most Grignard reactions, but for some temperature-sensitive or reluctant cases a fine powder from reduction of $MgCl_2$ serves well.[34]

Sodium is available as cast ingots that can be cut to small pieces with a knife. When more surface area is needed, it can be finely pulverized by stirring it molten in refluxing toluene and then cooling. This was done for the process in Eq. 10,[35] where the cooled toluene was replaced with ether.

$$\mathrm{ClCH_2CH_2CO_2Et} + \mathrm{Na} + \mathrm{ClSiMe_3} \xrightarrow{\text{ether}} \text{1-ethoxy-1-(trimethylsiloxy)cyclopropane} \quad (61\%) \tag{10}$$

All of these metals and more can be finely pulverized or made to react dramatically faster using ultrasound (sonication). Placing a flask containing a piece of potassium and dry toluene (but not THF) in an ultrasonic cleaning bath at 10°C gave a silvery blue colloidal suspension of the metal in a few minutes.[36] Sodium was dispersed similarly in xylene. The colloidal potassium is useful for making ketone enolates or for Dieckmann condensations. An ultrasonic probe can be inserted directly in a reaction mixture also. The oscillator is typically 60 to 250 W at 20–50 kHz.

A mixture of reactive metal, organohalide, and solvent may be ultrasonically irradiated to give an organometallic reagent. The cavitation effect of the ultrasound causes rapid erosion of the metal surface. Electrophiles may often be included in the original reaction mixture to react with the organometallic reagent as it is formed. The Barbier reaction proceeds well with lithium metal and ultrasound[37] as exemplified in Eq. 11.[38]

$$\text{bicyclic enone} + \mathrm{Br(CH_2)_2CH{=}C(CH_3)_2} + \mathrm{Li} \xrightarrow[\text{dry THF, 45 min, 0°C}]{\text{ultrasound}} \xrightarrow{\mathrm{H_2O}} \text{tertiary alcohol (OH)} \quad (89\%) \tag{11}$$

Diorganozinc reagents can be prepared from lithium wire, $ZnBr_2$, and organohalide with sonication. These reagents were used for conjugate additions catalyzed by nickel acetonylacetonate.[39] The Reformatksy reaction, in the presence of iodine, proceeds in high yield in minutes at 25–30°C with ultrasonic erosion of ordinary zinc dust in dioxane (but not in ether or benzene).[40]

Conjugate additions of organocuprates can be done in one pot starting with lithium sand, the organohalide, copper salt, and α,β-unsaturated ketone in ether–THF at 0°C with ultrasound.[41]

Triorganoboranes can be prepared directly from a mixture of magnesium turnings, organohalides, and $BF_3 \cdot OEt_2$ in ether in 10–30 min by ultrasound.[42]

9.8 REACTIONS ON INORGANIC SOLID SUPPORTS

Numerous reactions that are inefficient in solution will proceed at lower temperatures and with higher selectivity in the adsorbed state on porous inorganic solids.[43,44] Just as polar solvents have substantial effects on reactions in homogeneous solution, the very polar adsorbants can affect reactants. They modify the dipolar character of the molecules, or they may hold molecules in a reactive orientation in pores in the solid. In some cases a solvent is present to deliver and remove molecules to and from the solid surface, and in other cases no solvent is present at all.

Ozone, because of its very low solubility in ordinary organic solvents, is suitable only for very fast reactions as with alkenes. Benzene rings react at 10^{-5} times the rate of alkenes and are not conveniently ozonolyzed in solution. Silica gel will adsorb up to 4.7% by weight of ozone at −78°C. If an organic substrate is first adsorbed on the silica gel and then ozone added at −78°C, followed by warming to room temperature, oxidation of benzene rings, and also tertiary hydrogen sites occurs conveniently and in good yield:[45,46]

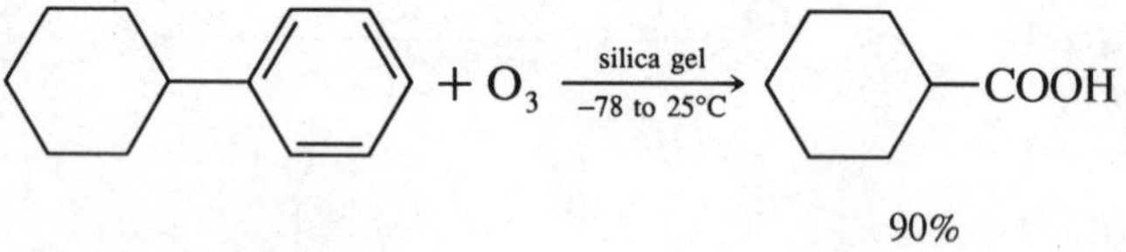

$$\text{C}_6\text{H}_{11}\text{-C}_6\text{H}_5 + O_3 \xrightarrow[\text{-78 to 25°C}]{\text{silica gel}} \text{C}_6\text{H}_{11}\text{-COOH} \quad (12)$$

90%

$$\text{dimethylcyclohexane} + O_3 \xrightarrow[\text{-78 to 25°C}]{\text{silica gel}} \text{dimethylcyclohexanol (OH)} \quad (13)$$

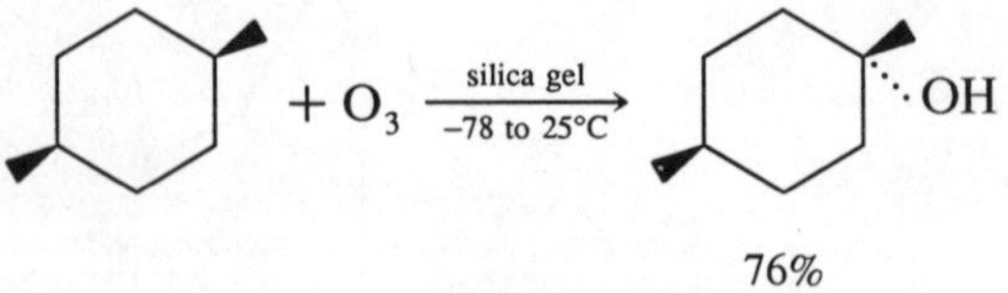

76%

Amines are cleanly converted to nitro compounds similarly.[47]

Nitroalkanes are converted to ketones and aldehydes by a variety of methods in solution involving strong acids or redox reactions. A very simple alternative is to adsorb the material on silica gel rendered basic with sodium methoxide. Elution with ether then gives the carbonyl compound in high yield and purity.[48]

The Diels–Alder reaction shown in Eq. 14 requires heating at 96°C for several hours, but on silica gel it occurs at room temperature.[49]

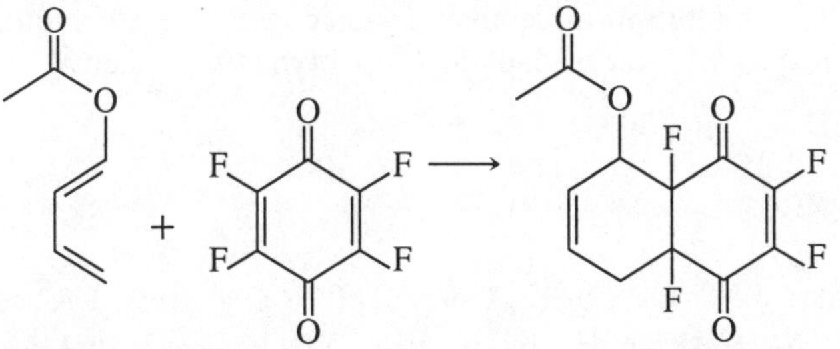

(14)

The chlorination of alkenes with *tert*-butyl hypochlorite on silica gel is clean, fast, and selective,[50] as exemplified in Eq. 15.[51] In the absence of silica gel, no reaction occurs under these conditions.

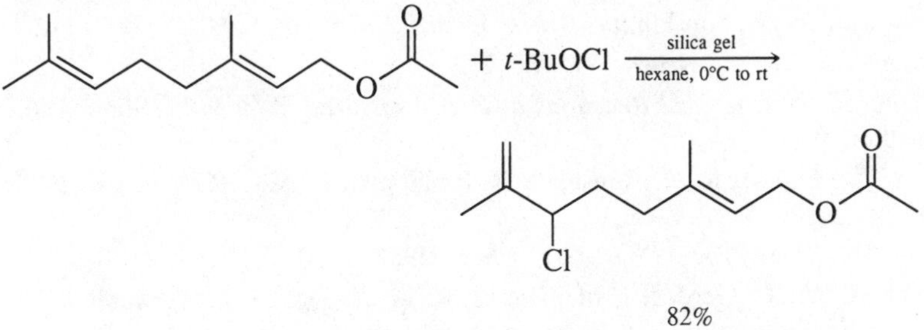

+ *t*-BuOCl $\xrightarrow[\text{hexane, 0°C to rt}]{\text{silica gel}}$

Cl

82%

(15)

A great many other reactions are promoted by solid adsorbants, including silica gel, alumina, clay, carbon, and other materials.[43,44]

PROBLEMS

1. The Grignard reagents from 2-(2-bromoethyl)-1,3-dioxolane and 2-(3-chloropropyl)-1,3-dioxolane are thermally unstable and decompose during their preparations under ordinary conditions.[52,53] What can be done to overcome this problem?[54]

2. Ozone was passed into a solution of 26 g of *cis*-decalin in CCl_4 at 0°C for 147 h. This gave 7 grams of decahydronaphth-9-ol.[55] What could be done to improve the yield and shorten the reaction time?

3. Heating 1-phenyl-1-benzoylcyclopropane with excess ethyl bromoacetate and 20-mesh zinc at reflux for 17 h gave a 47% yield of the Reformatzky product along with 55% recovered ketone.[56] What could be done to improve this reaction?

4. The oxidation of alkynes to α-diketones with potassium permanganate is a well known reaction, but the early examples are carboxylate salts

which are soluble in aqueous permanganate. What conditions would you choose to oxidize 1-phenyl-1-pentyne to 1-phenyl-1,2-pantanedione?[57]

REFERENCES

1. Partridge, J.; Chadha, N. K.; Uskovic, M. R. *Org. Synth.* **1985,** *63,* 44.
2. Scholz, K.-H.; Heine, H.-G.; Hartmann, W. *Org. Synth.* **1984,** *62,* 149.
3. Paquette, L. A.; Kukla, M. J.; Stowell, J. C. *J. Am. Chem. Soc.* **1972,** *94,* 4920.
4. Reppe, W.; Schlichting, O.; Klager, K.; Toepel, T. *Annalen Chem.* **1948,** *560,* 66.
5. Schwartz, R. H.; San Filippo, J., Jr. *J. Org. Chem.* **1979,** *44,* 2705.
6. Posner, G. H. *Org. React.* **1975,** *22,* 253.
7. Weale, K. E. *Chemical Reactions at High Pressures,* E. & F. N. Spon, Ltd., London, 1967.
8. P'eng, S.; Sapiro, R. H.; Linstead, R. P.; Newitt, D. M. *J. Chem. Soc.* **1938,** 784.
9. LeNoble, W. J. *Progr. Phys. Org. Chem.* **1967,** *5,* 207.
10. Dauben, W. G.; Gerdes, J. M.; Bunce, R. A. *J. Org. Chem.* **1984,** *49,* 4293.
11. Dauben, W. G.; Krabbenhoft, H. O. *J. Org. Chem.* **1977,** *42,* 282.
12. Jurczak, J.; Chmielewski, M.; Filipek, S. *Synthesis* **1979,** 41.
13. Jurczak, J.; Tkacz, M. *Synthesis* **1979,** 42.
14. Jones, W. H.; Mangold, D.; Plieninger, H. *Tetrahedron* **1962,** *18,* 267.
15. Reichardt, C. *Solvent Effects in Organic Chemistry,* Verlag-Chemie, Weinheim, 1979, pp. 144–155.
16. Pearson, R. G.; Sobel, H.; Songstad, J. *J. Am. Chem. Soc.* **1968,** *90,* 319.
17. Smiley, R. A.; Arnold, C. *J. Org. Chem.* **1960,** *25,* 257.
18. Friedman, L.; Shechter, H. *J. Org. Chem.* **1960,** *25,* 877.
19. Fuchs, R.; Mahendran, K. *J. Org. Chem.* **1971,** *36,* 730.
20. Reichardt, C. *Angew. Chem. Internatl. Ed.* **1979,** *18,* 96.
21. Makosza, M. "Two Phase Reactions in Organic Chemistry," in *Survey of Progress in Chemistry,* Vol. 9, Scott, A. F., Ed., Academic, New York, 1980.
22. Starks, C.; Liotta, C. *Phase Transfer Catalysis, Principles and Techniques,* Academic, New York, 1978.
23. Weber, W. P.; Gokel, G. W. *Phase Transfer Catalysis in Organic Synthesis,* Springer Verlag, Berlin, 1977.
24. Orsini, F.; Pelizzoni, F. *J. Org. Chem.* **1980,** *45,* 4726.
25. Crandall, J. K.; Pradat, C. *J. Org. Chem.* **1985,** *50,* 1327.
26. Lee, D. G. *Oxidation in Organic Chemistry,* Vol. 5-D, Trahanovsky, W. S., Ed., Academic, New York, 1982, p. 147–206.

27. Porter, N. A.; Ziegler, C. B., Jr.; Khouri, F. F.; Roberts, D. H. *J. Org. Chem.* **1985,** *50,* 2252.
28. Raber, D. J.; Guida, W. C. *J. Org. Chem.* **1976,** *41,* 690.
29. Sala, T.; Sargent, M. V. *J. Chem. Soc. Chem. Commun.* **1978,** 253.
30. Landini, D.; Rolla, F. *Synthesis* **1976,** 389.
31. Sam, D. J.; Simmons, H. E. *J. Am. Chem. Soc.* **1972,** *94,* 4024.
32. Cook, F. L.; Bowers, C. W.; Liotta, C. L. *J. Org. Chem.* **1974,** *39,* 3416.
33. Liotta, C. L.; Harris, H. P.; McDermott, M.; Gonzales, T.; Smith, K. *Tetrahedron Lett.* **1974,** 2417.
34. Rieke, R. D. *Acc. Chem. Res.* **1977,** *10,* 301.
35. Salaun, J.; Marguerite, J. *Org. Synth.* **1985,** *63,* 147.
36. Luche, J.-L.; Petrier, C.; Dupuy, C. *Tetrahedron Lett.* **1984,** *25,* 753.
37. Luche, J.-L.; Damiana, J. C. *J. Am. Chem. Soc.* **1980,** *102,* 7926.
38. Uyehara, T.; Yamada, J.; Ogata, K.; Kato, T. *Bull. Chem. Soc. Jpn.* **1985,** *58,* 211.
39. Petrier, C.; deSouza Barbarosa, J. C.; Dupuy, C.; Luche, J.-L. *J. Org. Chem.* **1985,** *50,* 5761.
40. Han, B-H.; Boudjouk, P. *J. Org. Chem.* **1982,** *47,* 5030.
41. Luche, J.-L.; Petrier, C.; Gemal, A. L.; Zikra, N. *J. Org. Chem.* **1982,** *47,* 3805.
42. Brown, H. C.; Racherla, U. S. *J. Org. Chem.* **1986,** *51,* 427.
43. McKillop, A.; Young, D. W. *Synthesis* **1979,** 401, 481.
44. Posner, G. H. *Angew. Chem. Internatl. Ed.* **1978,** *17,* 487.
45. Klein, H.; Steinmetz, A. *Tetrahedron Lett.* **1975,** 4249.
46. Cohen, Z.; Keinan, E.; Mazur, Y.; Varkony, T. H. *J. Org. Chem.* **1975,** *40,* 2141.
47. Mazur, Y.; Keinan, E. *J. Org. Chem.* **1977,** *42,* 844.
48. Keinan, E.; Mazur, Y. *J. Org. Chem.* **1977,** *99,* 3861.
49. Hudlicky, M. *J. Org. Chem.* **1974,** *39,* 3460.
50. Sato, W.; Ikeda, N.; Yamamoto, H. *Chem. Lett.* **1982,** 141.
51. Novak, L.; Poppe, L.; Szantay, C. *Synthesis* **1985,** 939.
52. Forbes, C. P.; Wenteler, G. L.; Wiechers, A. *J. Chem. Soc. Perkin Trans. I* **1977,** 2353.
53. Eaton, P. E.; Mueller, R. H.; Carlson, G. R.; Cullison, D. A.; Cooper, G. F.; Chou, T. C.; Krebs, E.-P. *J. Am. Chem. Soc.* **1977,** *99,* 2751.
54. Bal, S. A.; Marfat, A.; Helquist, P. *J. Org. Chem.* **1982,** *47,* 5045.
55. Durland, J. R.; Adkins, H. *J. Am. Chem. Soc.* **1939,** *61,* 429.
56. Bennett, J. G.; Bunce, S. C. *J. Org. Chem.* **1960,** *25,* 73.
57. Lee, D. G.; Chang, V. S. *Synthesis* **1978,** 462.

10

Interpretation of NMR Spectra

Most organic compounds are colorless solids or liquids of rather similar appearance, and unknown samples present a puzzle for identification. Physical measurements such as melting point, boiling point, and refractive index are useful for matching against lists of values for limited numbers of known compounds. Actual structural information is readily obtained by means of various spectroscopic methods (or ultimately by X-ray crystallography). Ultraviolet–visible spectroscopy gives information on the extent, shape, and substituents of π-conjugation in molecules.[1–3] It is a measure of the energy gaps between the electronic ground and excited states. Infrared spectroscopy is particularly useful for determining the presence and identity of functional groups.[1–3] This is a measure of the frequency of bending and stretching of bonds where the bond dipole changes with the movement. The stretching vibrations of double and triple bonds in alkenes and alkynes involve small bond dipole changes and give weak or no infrared absorptions. For these cases laser Raman spectroscopy gives strong, informative signals.[1,4]

Mass spectra do not involve electromagnetic radiation as the others do. A molecule is ionized by an electron beam and the resulting ionic fragments are sorted by mass, and their abundance and masses are measured. This is useful for determining what elements are present, and much structural information.[1–3,5] It has the special value of requiring as little as 1 μg of sample.

All of these are important, but here we select to elaborate on the currently most heavily used technique, nuclear magnetic resonance (NMR) spectroscopy. Elemental nuclei that have an odd mass number and/or an odd atomic number have a magnetic moment, and many of these can be observed in an NMR spectrometer. Those that show the most practical value thus far are ^{1}H, ^{13}C, ^{19}F, and ^{31}P. The spectra are readily obtained on modern equipment, and they also interact; for example, a fluorine in a molecule will split proton signals and vice versa.

10.1 INTERPRETATION OF PROTON NMR SPECTRA

In your introductory course you learned to use four basic kinds of information from ^{1}H NMR spectra: number of signals, chemical shift values, integrated signal area, and splitting patterns. We will now delve further into some of these, particularly the splitting patterns.[1-3]

10.1.1 Magnitude of Coupling Constants

The size of the magnetic influence a proton receives from a neighboring proton depends on the distance between them, and the intervening bonds and their angular relationships. The coupling constant J is a measure of this effect. It is specified in hertz, which is directly proportional to the field strength by Eq. 1, where ν is in hertz and H_0 in gauss and the proportionality constant is for protons only.

$$\nu = 4257 H_0 \quad (1)$$

Typical values of coupling constants for various neighboring relationships are given in Table I.

Consideration of nonequivalent neighbors allows a simple explanation of multiplets as, for example, in Fig. I, in the 60-MHz ^{1}H NMR spectrum of 2-(2-iodoethyl)-1,3-dioxolane. The triplet at 4.9 ppm is from the one hydrogen on the ring at position 2. It is split by the neighboring CH_2 group with $J = 4.0$ Hz. The triplet at 3.2 ppm is from the CH_2 group attached to iodine. It is split by the neighboring CH_2 group with $J = 7.8$ Hz. The signal at 2.2 ppm is from the CH_2 group neighbor to both of the above. It is called a *doublet of triplets* (dt) or *triplet of doublets*. The 4.0- and 7.8-Hz coupling constants may both be measured in it. Coupled neighbors must split each other by the same J value because the distance and intervening bonds are identical in either direction. The multiplet at 3.9 ppm is

Table I. Coupling Constants for Neighboring Hydrogens (Hz)

Structure	*J*	Structure	*J*	Structure	*J*
X—C(H)—C(H)—	6–8	H(X)C=CH (trans H, H)	18	cyclohexane, H axial / H axial	9
—C(H)=C—C(H)—	2	HC=CH (cis H, H)	11	cyclohexane, H equatorial / H axial	3
—C(H)—C(=O)H	3	X-benzene, H / H ortho	8	cyclohexane, H equatorial / H equatorial	3
—C(H)—C—C(H)—	0	X-benzene, H / H meta	2–3	>C(H)H	0–20
X(C=C)H, H (geminal)	1	X-benzene, H / H para	0		

the overlapping signals for the remaining ring hydrogens, cis and trans to the iodoethyl group. Their difference is too small to give a simple pattern.

The assignment of signals to hydrogens on aromatic rings is often facilitated by considering the magnitude of the coupling constants as, for example, in 2-chloro-5-methoxyphenylhydrazine (Fig. II). The hydrogen giving the doublet at 6.6 (J = 3 Hz) has a meta neighbor but not ortho and must, therefore, be on C-6. Its coupling partner gives the signal at 6.2 ppm (dd, J = 3 and 9 Hz), and it must be on C-4. The larger coupling constant indicates a neighbor ortho and matches the splitting in the signal at 7.1 ppm, which must, in turn, be from a hydrogen on C-3.

10.1.2 Spin Decoupling

In some spectra it may not be possible to distinguish coupling constants, or several *J* values will be essentially the same and neighboring relationships

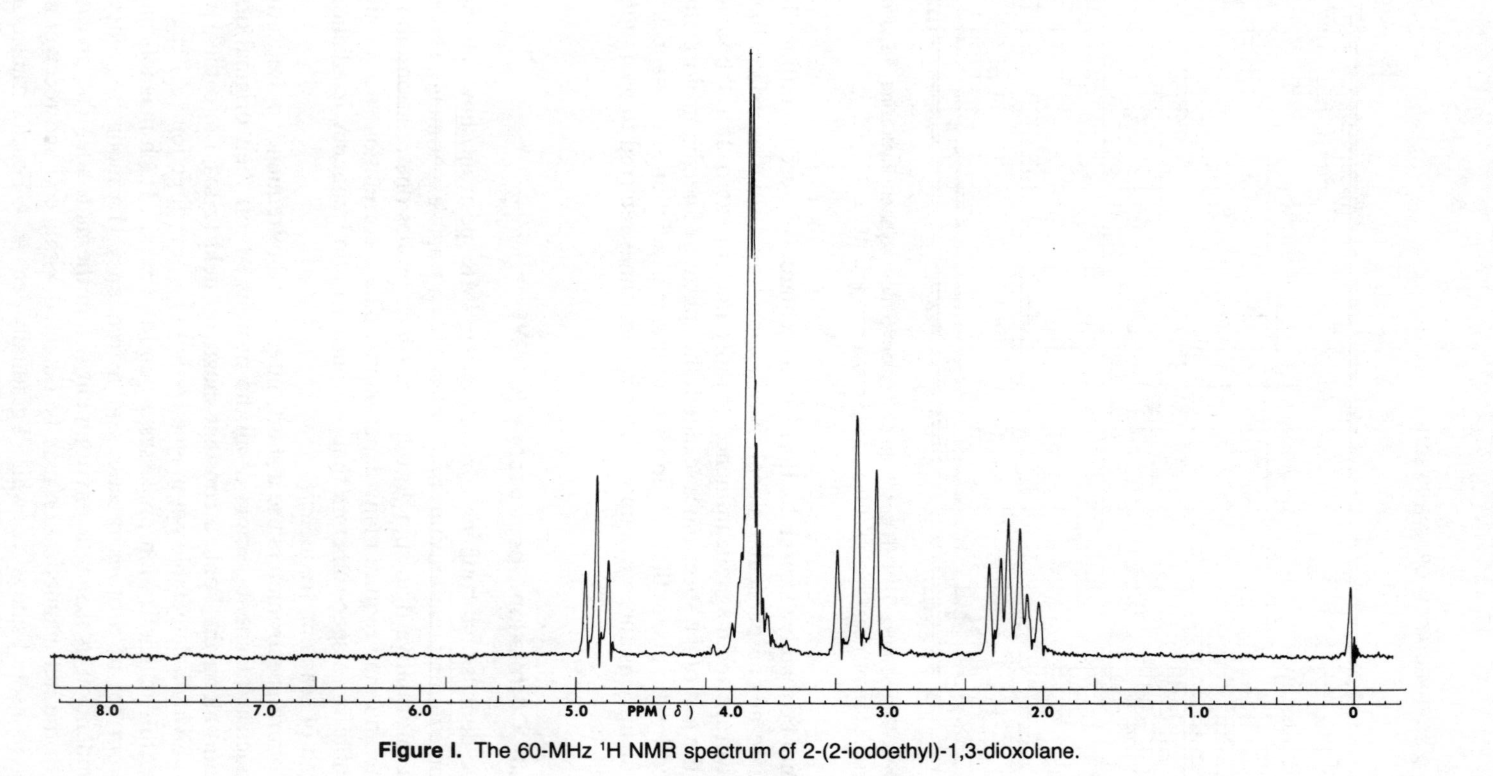

Figure I. The 60-MHz ^1H NMR spectrum of 2-(2-iodoethyl)-1,3-dioxolane.

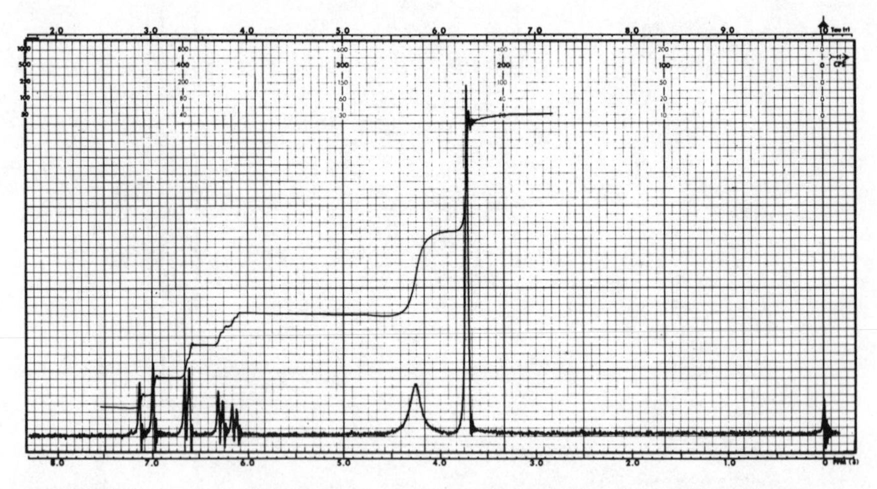

Figure II. The 60-MHz ^1H NMR spectrum of 2-chloro-5-methoxyphenylhydrazine. ©Sadtler Research Laboratories, Division of Bio-Rad Laboratories, Inc., 1972.

may not be readily discernible. If a second oscillator is placed to continually irradiate at one particular signal, it will cause rapid spin inversions that will average the magnetic influence of those nuclei to zero. If at the same time the rest of the spectrum is scanned, the splitting of neighbors normally caused by the nuclei that are under continuous irradiation will disappear, thus identifying those neighbors. This is sometimes referred to as *double resonance*.

10.1.3 Correlation Spectroscopy (COSY)

A two-dimensional coupling correlation ^1H NMR spectrum shows all the proton coupling relationships in a molecule in a single experiment. This is more informative than spin decoupling, which probes one relationship at a time and may require many experiments. These correlation spectra are available from spectrometers that use pulsed radio-frequency irradiation instead of scanning frequencies.

Modern spectrometers use a short, intense radio-frequency pulse to tip the nuclear magnetic vectors of all the protons 90° off their original orientation along the applied constant magnetic field (z axis) into the x,y-plane giving a detectable net magnetization in a new direction.[6] After the pulse, this magnetization precesses around the z axis. Each magnetically different set of protons precesses at a frequency (Larmour frequency) dependent on its magnetic environment within the molecule. The precession of the net magnetization may be detected with a coil and receiver as a sum of a set of sine waves while it gradually decays. A Fourier transform computation converts the sum into individual frequencies to give a plot like that from scanning frequency. The whole spectrum is recorded simultaneously in a fraction of a second to a few seconds. Weak spectra are enhanced by repeating this process many times at very short intervals and summing the results. The random noise partially cancels while the signals grow. The signal-to-noise ratio (S/N) increases with the square root of the number of spectra added.

Two-dimensional correlations are brought out with multipulse irradiations. A second radio-frequency pulse slightly delayed from the first will transfer magnetization from nuclei to others with which they are spin-coupled in the molecule. The transferred magnetization is detected and recorded as off-diagonal intensity in a two-dimensional plot[7] (Fig. III).

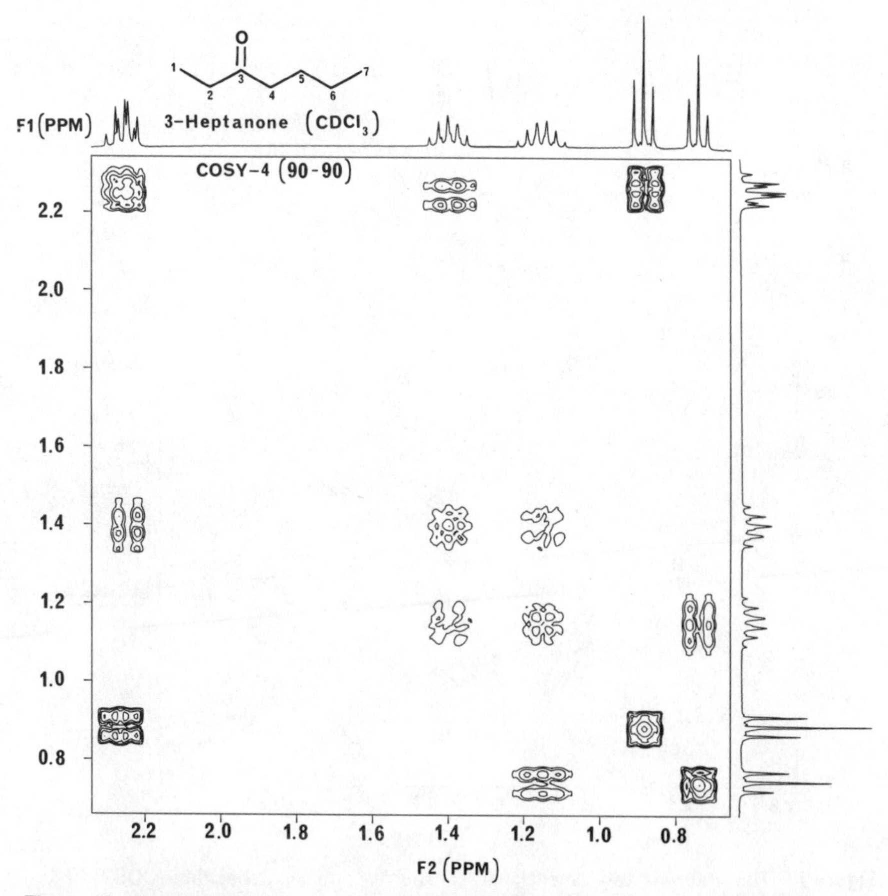

Figure III. The 300-MHz two-dimensional ^1H–^1H shift correlation spectrum (COSY) of 3-heptanone. Spectrum courtesy of Varian Associates.

How such plots are generated will not be covered here, but their ready availability from automated spectrometers and the simplicity of interpretation makes them very useful. Strong off-diagonal signals indicated by several intensity contour lines indicate the large *J* value couplings of adjacent neighbors, while weak spots (Fig. IV) indicate long range, for example, allylic coupled neighbors.

In the COSY spectrum of 3-heptanone (Fig. III) all the neighboring relationships may be traced by locating the two chemical-shift values correlated by the off-diagonal signals. For instance, the contours at $F_1 = 2.3$ and $F_2 = 0.9$ show that the triplet of 0.9 ppm is the methyl group neighboring one of the downfield CH_2 groups adjacent to the carbonyl group; that is to say, the signal at 0.9 ppm is for hydrogens on carbon-1 of

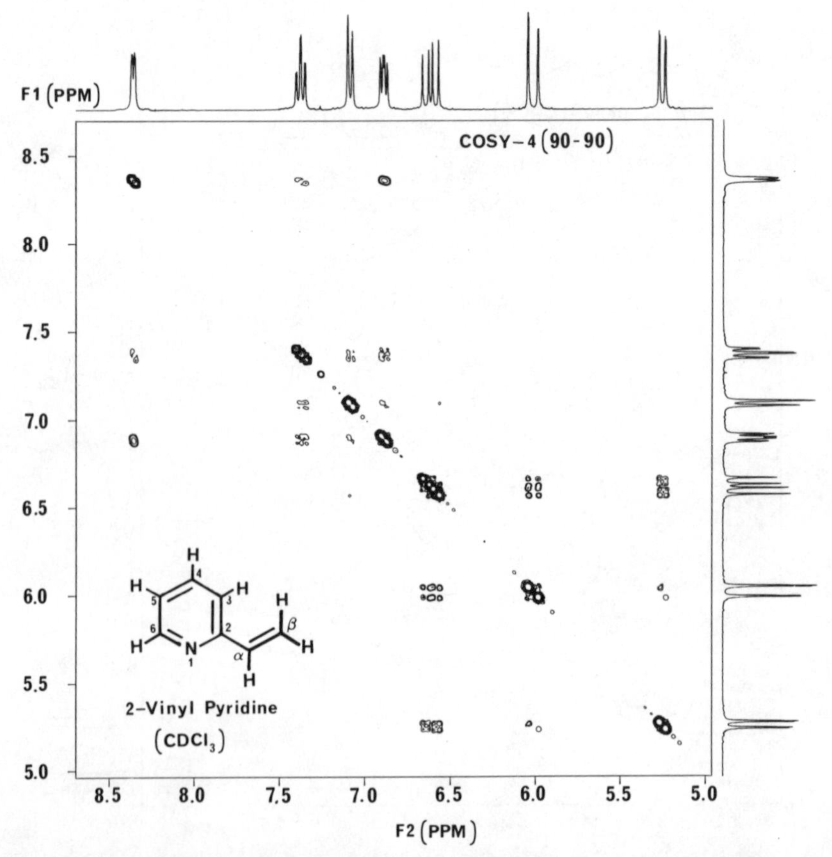

Figure IV. The 300-MHz two-dimensional ^1H–^1H shift correlation spectrum (COSY) of 2-vinylpyridine. Spectrum courtesy of Varian Associates.

3-heptanone. Of course, the assignments in 3-heptanone could have been made with only a one-dimensional spectrum, but this simple example was chosen for illustration. COSY spectra are more useful for complex molecules as illustrated in many recent journal articles.[8]

10.1.4 Non-First-Order Signals

When we see a simple triplet, we conclude that the hydrogen(s) giving that signal have two equivalent neighboring hydrogens. When we see a dd, we conclude that the hydrogen(s) giving that signal have two nonequivalent neighbors. This is called *first-order analysis* and is possible when the chem-

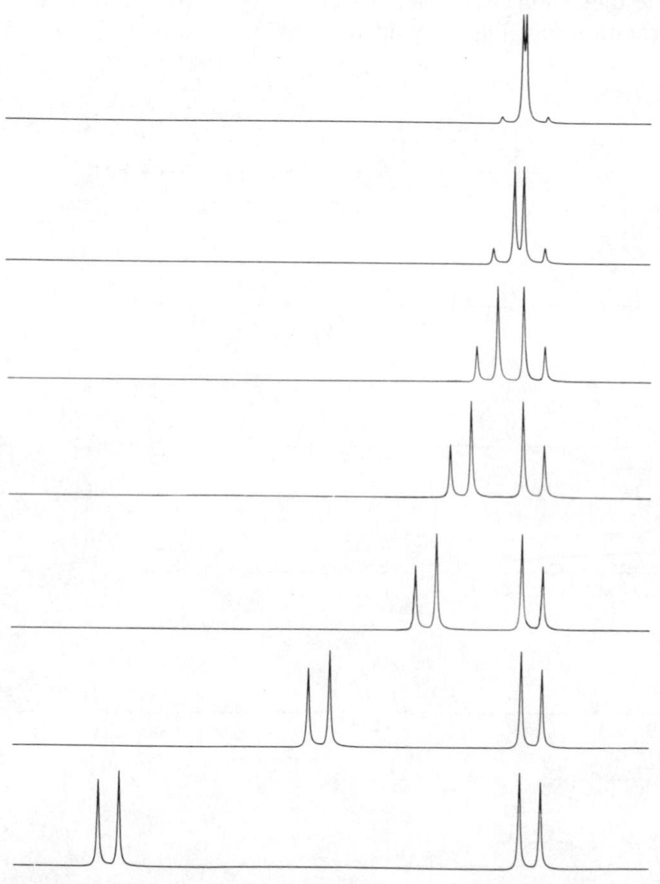

Figure V. Simulated AB splitting pattern with $J/\Delta\nu$ = 1.5, 1, 0.5, 0.3, 0.2, 0.1, and 0.05. Courtesy of R. F. Evilia, University of New Orleans.

ical-shift difference between neighboring hydrogens is much larger than the *J* value. Often this is not the case. If the chemical-shift difference is less than 10 times the *J* value, the signals are distorted from simple first-order expectations. This is illustrated with computer simulations in Figs. V–VII.

In Fig. V we have a single hydrogen with its single neighbor (AB). In the bottom spectrum we recognize a pair of doublets. As the chemical-shift difference becomes smaller, proceeding to the top of Fig. V, the inner half of each doublet becomes taller while the outer diminishes. The extreme would be identical neighboring hydrogens, which would, of course, be one singlet, which these spectra approach. No attempt is made here to give the theoretical reasoning or predictive methods,[9,10] but you should be able to *recognize* these non-first-order patterns as you analyze spectra and know what structural meaning they have.

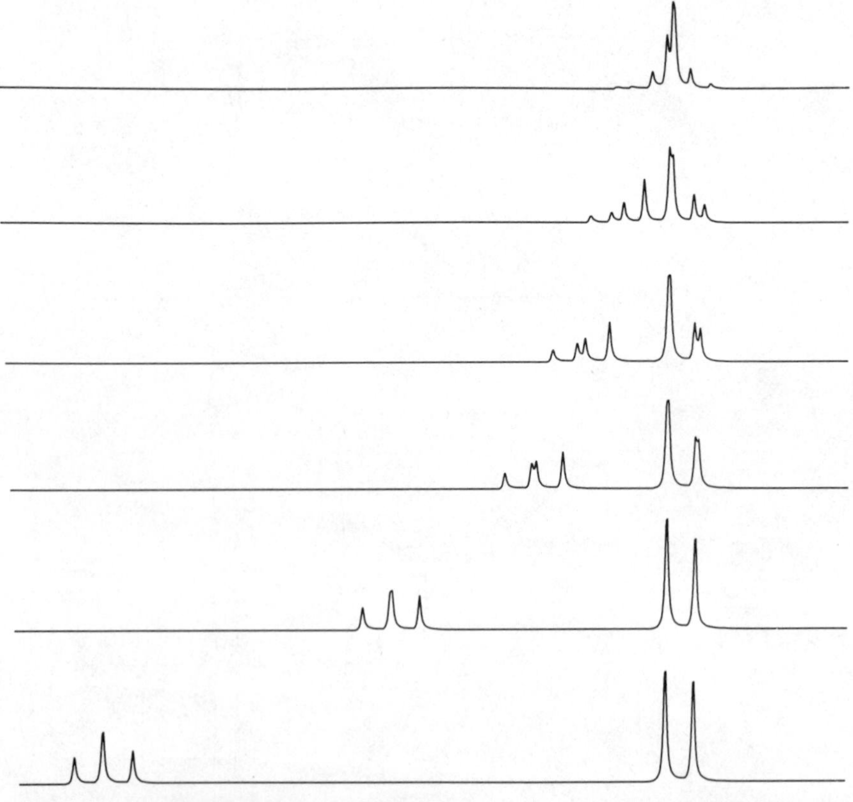

Figure VI. Simulated A_2B splitting pattern with $J/\Delta\nu$ = 1, 0.5, 0.3, 0.2, 0.1, and 0.05. Courtesy of R. F. Evilia, University of New Orleans.

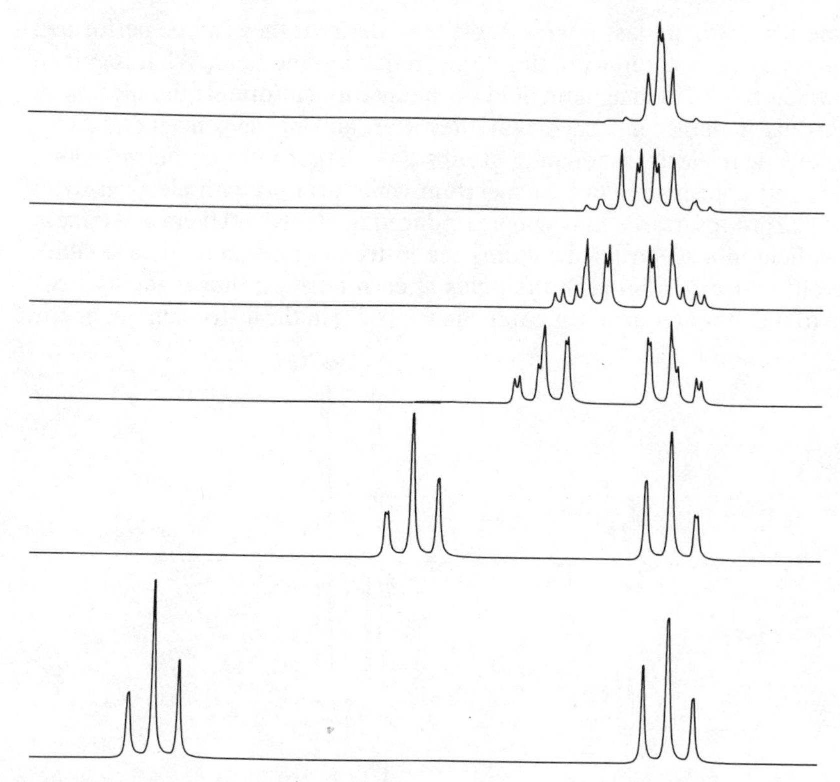

Figure VII. Simulated A_2B_2 splitting pattern with $J/\Delta\nu$ = 1, 0.5, 0.3, 0.2, 0.1, and 0.05. Courtesy of R. F. Evilia, University of New Orleans.

In Fig. VI at the bottom we see a first-order spectrum for a CH_2 group and a single neighbor (A_2B). In the progression toward smaller chemical-shift differences we see distortions of size and increasing numbers of peaks, again trending at the top toward a singlet. In Fig. VII we see a similar series for two neighboring CH_2 groups (A_2B_2). Notice that the left and right halves of the signals are mirror images, which makes this pattern quickly recognizable in spectra. Similar but more complicated patterns involving larger numbers of hydrogens have been analyzed and computer-simulated.

10.1.5 Spectra at Higher Magnetic Fields

Equation 1 shows the direct proportionality between the magnetic field strength and the oscillator frequency for proton resonance. Routine ^1H NMR spectra are usually run at 60 MHz in a magnetic field of 14,092 G.

If these spectra are less interpretable than desired, they can be performed in instruments with up to 10 times this frequency and field. What is gained by doing this? The magnetic field of the instrument orients the motion of electrons in molecules such that they give an opposing magnetic field. Protons near electron donating groups have larger induced fields around them and appear upfield in a spectrum while protons with electron withdrawing groups nearby have smaller induced fields around them and appear downfield in a spectrum. Doubling the instrument magnetic field doubles the difference ΔH between the fields at each nucleus; thus scanning from one to the other requires a greater change (ΔH) in the instrument magnetic

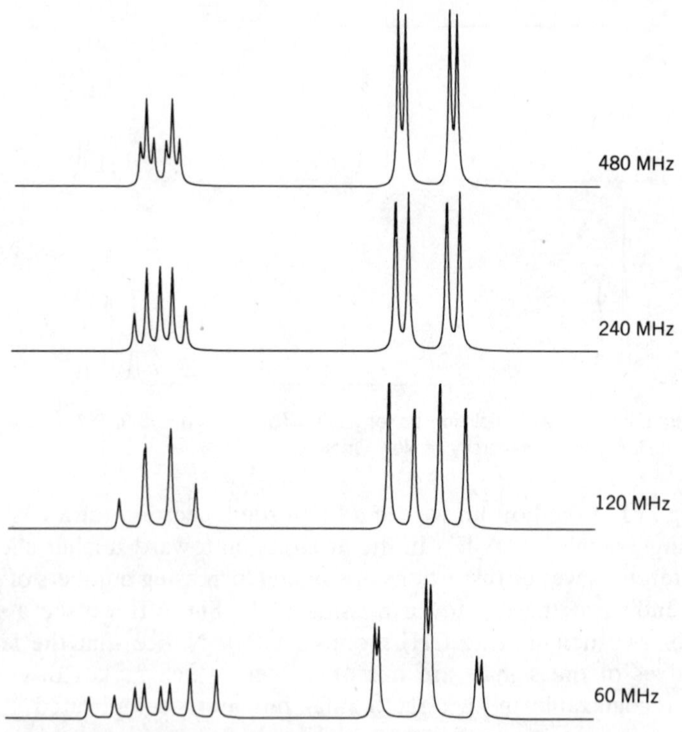

Figure VIII. Simulated spectra of two A_2B patterns with doubling field strengths. This is an arbitrary selection of two A_2B patterns that overlap with coincidences at lower field strengths. The lowest trace is the appearance at 60 MHz where peaks are coinciding in the A_2 portion. The distortion from first order is equivalent to that in the $J/\Delta\nu = 0.2$ trace in Fig. VI. The B portion is clearly two intermingled patterns, each the same as that in Fig. VI. The 120-MHz trace is first order interpretable, but the two triplets in the B portion are now doubly coinciding. The 240-MHz trace lessens the overlaps, and finally the 480-MHz spectrum (top) is clear with no overlaps. All four spectra are plotted to the same ppm scale. These simulations were provided by R. F. Evilia, University of New Orleans.

field, and the spectrum is more spread out. If we simply stretch a 60-MHz spectrum out horizontally, we will not see more peaks or eliminate overlaps; however, in the high-field spectra we will lessen the overlaps. The width of a multiplet is determined by the magnetic field effects of nearby nuclei, the size of which is fixed and independent of the instrument magnetic field. Therefore, as the spectrum is spread with higher magnetic fields, the individual multiplets do not widen but come away from each other, diminishing their overlaps.

The chemical-shift axis of the spectra is not generally in units of field strength but in parts per million. That is (ΔH between signals) $\times$ 10^6 $\div$ (total magnetic field strength). Thus when we use a magnet of twice the field strength, the ΔH between signals doubles but the dividend remains the same. This is done so that chemical-shift differences have the same ppm values regardless of the applied magnetic field. On the other hand, the J values remain the same in hertz or gauss as we go from one instrument to another. Therefore, on the ppm scale the separation of peaks within a multiplet shrinks to half as we go to an instrument with twice the magnetic field strength. Thus it appears that with increasing magnetic field, the multiplets are narrowing but staying centered on the same ppm values (Fig. VIII). Spectacular detail has been resolved in spectra of complex molecules at high fields.

There is another major change in spectra on going to higher-field instruments. The ratio of chemical shift to J increases for all coupled pairs; therefore, the signals become more nearly first order and interpretation is simplified.

10.1.6 Stereochemical Effects

Diastereomers are chemically different materials with different physical properties, and they give different spectra. The differences may be small but are often sufficient to allow the use of integration to measure the percentage of each isomer.

Diastereotopic groups reside in diastereomeric environments in molecules (Sec. 3.9) and thus give separate signals, and each can split the signal of the other if they are close together. In Fig. IX notice that the CH_2 groups do not give a simple quartet as they do in acetone diethylacetal. The complex pattern, although not all discernible at 60 MHz, consists of a total of 16 peaks: two doublets of quartets. The geminal hydrogens are diastereotopic, and each gives a separate but overlapping signal about 0.16 ppm apart. Each of these splits the other to distorted doublets as in Fig. V, which are all, in turn, split to quartets by the CH_3 group. A similar pattern for acetaldehyde diethylacetal is shown clearly at 270 MHz in Ault's

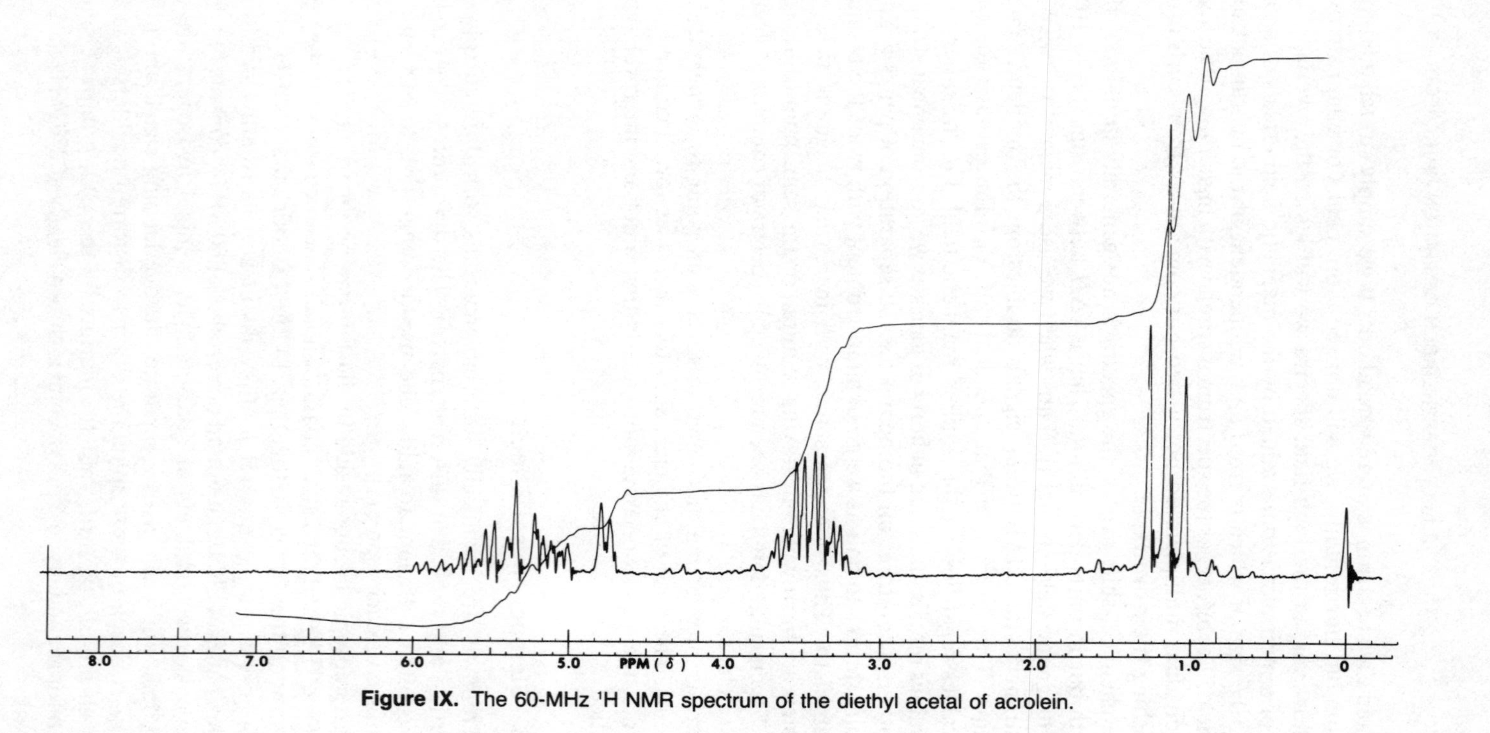

Figure IX. The 60-MHz ^1H NMR spectrum of the diethyl acetal of acrolein.

collection.[11] When a signal appears more complex than it ought to on simple considerations and is also nearly symmetrical, consider diastereotopic groups. If they are geminal hydrogens, they should split each other with considerable distortion, and you should look for the small outer peaks that complete the pattern.

Enantiomeric compounds give identical spectra in ordinary solvents, but diastereomeric complexes may form in the presence of a single enantiomer of a chiral complexing agent. These may give separate measurable signals as discussed in Section 3.5.

10.2 INTERPRETATION OF CARBON NMR SPECTRA

The abundant ^{12}C isotope has no nuclear magnetic moment, but the 1.1% natural abundance ^{13}C isotope does.[2,3,12,13] The signal from these carbons, however, is only $\frac{1}{64}$th the intensity from the same number of hydrogens. These two factors cause carbon spectra to be about $\frac{1}{6000}$th of the intensity of a hydrogen spectrum of the same sample. This necessitates the use of pulsed irradiation and Fourier transform methods and summation of many repeated spectra (Section 10.1.3). Typically the spectra are obtained at 2–10-s intervals and, for a routine concentrated sample, less than 30 min is required to accumulate a good summation.

10.2.1 General Characteristics

Chemical Shift Range. The signals for carbon occur 2.98×10^6 ppm upfield of those from protons. At 21,138 G the frequency for carbon resonance is 22.628 MHz, and for hydrogen it is 90 MHz. The chemical shift-values are generally recorded as ppm downfield from the carbons of tetramethylsilane and extend over a range of about 230 ppm. Since this is more than 10 times the range found for hydrogens, the problem of obscuring overlaps is much less here. The sp^3 hybrid carbons bonded only to hydrogens and other carbons generally occur in the 6–50-ppm range. Those carrying electronegative atoms give signals 10–60 ppm lower field. The sp^2 hybrid carbons of alkenes and aromatic rings occur in the 100–165-ppm range, and those in carbonyl groups are at 150–220 ppm. In Section 10.2.2 we will see how to make good numerical predictions for specific structures.

Number of Signals. Each structurally different carbon in a molecule gives a separate signal. Identical carbons will, of course, coincide. A *tert*-butyl group gives two signals, one for the three equivalent CH_3 carbons and one for the central carbon. *p*-Nitroanisole gives five signals, two of which represent two carbons each.

Splitting of Signals. The low natural abundance of ^{13}C makes it highly improbable that two would be side by side in a molecule; therefore, ^{13}C–^{13}C splitting is not observed in routine spectra. Protons split the signals of the carbon to which they are bonded by 100–240 Hz. Protons on an adjacent or the next farther carbon give only 4–6-Hz splitting. The upper spectrum in Fig. X (isobutyl alcohol) illustrates proton splitting. The quartet centered at 18.9 ppm indicates three bonded hydrogens, that is, a methyl group. The broadening of the peaks is caused by the small coupling with hydrogen on α and β carbons. A doublet similarly broadened occurs at 30.8 ppm and overlaps part of the quartet. This indicates a CH group. Finally, the triplet at 69.4 ppm indicates a CH_2 group. Without resolving the very small splitting, one may simply generalize that quartets are CH_3 groups, triplets are CH_2 groups, doublets are CH groups, and singlets are carbons bearing no hydrogens. This is very simple compared to proton spectra, but the large coupling constants make it difficult to know which peaks belong together as multiplets in complex spectra. Routinely, broad irradiation (spin decoupling) of all the protons is used to reduce the ^{13}C spectrum to all singlets, as in the lower trace in Fig. X.

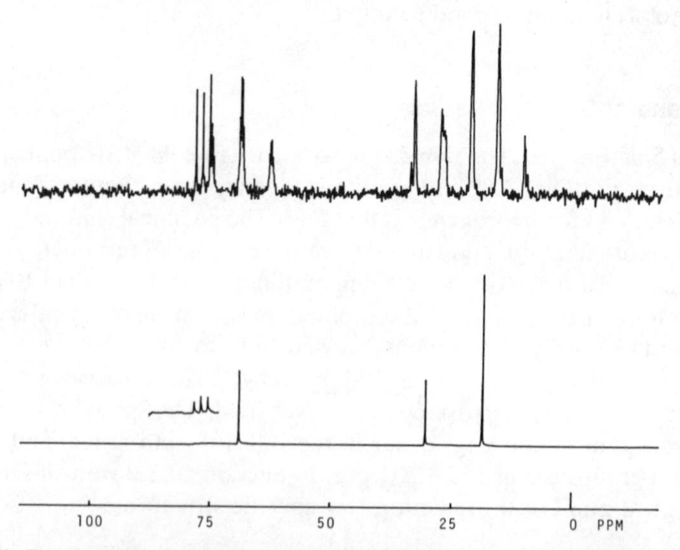

Figure X. The 22.5-MHz ^{13}C NMR spectrum of isobutyl alcohol. The upper trace includes proton coupling, while the lower trace is proton decoupled. The signal for the deuterochloroform solvent is weak in the lower trace; therefore, it was retraced at 18 times the amplitude. Spectrum courtesy of R. F. Evilia, University of New Orleans.

Area of Signals. The broad irradiation of the proton resonances in the decoupled spectra is particularly valuable because it hastens relaxation of the excited ^{13}C nuclei to which the protons are bonded so that the signal is stronger in quickly repeated pulses. This signal enhancement is called the *nuclear Overhauser effect.* This gives sharper and stronger spectra in a shorter time, but the effect varies from carbon to carbon and the ^{13}C signal areas under these conditions are not quite proportional to the numbers of carbons represented. Thus under routine ^{1}H decoupled conditions the integrated ^{13}C signal areas are not so useful as they were in ^{1}H spectra. Qualitatively, the carbons bearing no hydrogens are noticeable as unusually weak signals, and the signals for two equivalent carbons are usually somewhat larger than those for single carbons in a molecule. If necessary, long pulse intervals may be used to obtain signal areas nearly proportional to the ratio of carbons represented.

In Fig. XI we see a routine ^{1}H decoupled ^{13}C NMR spectrum of an acetal. Those signals representing one carbon each vary in size, but the one representing a pair of equivalent carbons is notably larger than the rest.

The instruments maintain a constant calibration by using a deuterium signal to lock the field–frequency ratio. Therefore, deuterated solvents are necessary. Deuterochloroform is most frequently used. The carbon in this solvent gives a weak signal that is split by the deuterium (which has three spin states) into three equal peaks at 75.6, 77.0, and 78.4 ppm downfield from tetramethylsilane (Fig. X). These are seen in many spectra, especially where the solution is dilute.

The individual signals in a proton spin decoupled ^{13}C spectrum may be identified as arising from a CH, CH_2, or CH_3 carbon with a distortionless enhancement of polarization transfer (DEPT) experiment. A DEPT experiment is a five-pulse sequence that generates individual ^{13}C CH, CH_2, and CH_3 subspectra.[14]

10.2.2 Calculation of ^{13}C Chemical-Shift Values

The chemical-shift value for a carbon moves about 8 ppm downfield for each carbon bonded to it or to an adjacent carbon. An electronegative atom substantially changes the chemical shift of the carbon to which it is attached but has a relatively small effect on nearby carbons. Predictions of the chemical-shift values can be made by using tabulated additive corrections and are useful for assigning signals to particular carbons in a structure. They can help to determine which of several structures is in accord with the spectrum. The corrections are empirically derived from measured values of many known compounds. The chemical-shift values

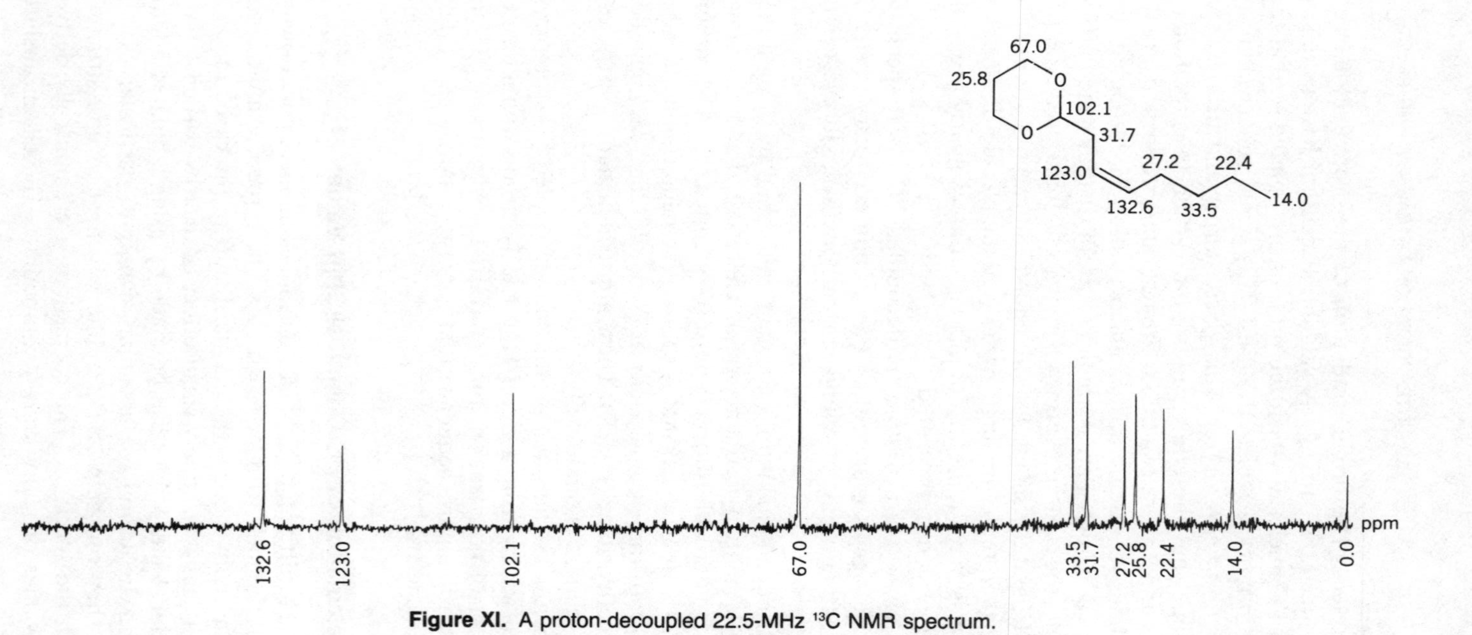

Figure XI. A proton-decoupled 22.5-MHz ^{13}C NMR spectrum.

Table II. Structural Parameters for Calculating ^{13}C NMR Shifts of Alkanes

$$\copyright\!-\overset{\alpha}{C}-\overset{\beta}{C}-\overset{\gamma}{C}-\overset{\delta}{C}$$

Carbon to Be Predicted	Add	For Each Carbon of the Indicated Type, Add the Listed Increment				
CH_3	6.8	α CH_3	0	β 0	γ −3.0	δ 0.5
		α CH_2	9.6			
		α CH	17.8			
		α C	25.5			
CH_2	15.3	α CH_3	0	β 0	γ −2.7	δ 0.3
		α CH_2	9.8			
		α CH	16.7			
		α C	21.4			
CH	23.5	α CH_3	0	β 0	γ −2.1	δ 0
		α CH_2	6.6			
		α CH	11.1			
		α C	14.7			
C	27.8	α CH_3	0	β 0	γ −0.9	δ 0
		α CH_2	2.3			
		α CH	4.0			
		α C	7.4			

for each carbon in acyclic alkanes and alkyl parts of functional molecules may be calculated by use of the increments in Table II.[15] The error is usually less than 1 ppm.

To illustrate the use of Table II, the values expected from 3-methylpentane are calculated as follows. The C-1 is a CH_3 group, and there is one α CH_2, one β carbon, two γ carbons, and one δ carbon.

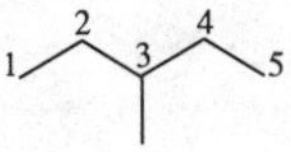

	Increments	Calculated	Measured
C-1:	6.8 + 9.6 + 2(−3.0) + 0.5	= 10.9	11.4
C-2:	15.3 + 16.7 − 2.7	= 29.3	29.3
C-3:	23.5 + 2(6.6)	= 36.7	36.4
3-CH_3:	6.8 + 17.8 + 2(−3.0)	= 18.6	18.8

2,3-Dimethylbutane is calculated similarly:

C-1:	6.8 + 17.8 + 2(−3.0)	= 18.6	19.5
C-2:	23.5 + 11.1	= 34.6	34.0

The vinyl carbons of alkenes are calculated according to Table III. The error is usually less than 3 ppm. The allylic carbons of an alkene are calculated according to Table II for the saturated analog; then, if the alkene is cis, add −3, or trans, add +3. If both (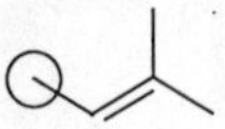), add 0. The remainder of the carbons are calculated by Table II for the saturated analog. *cis*-2-Heptene is predicted as follows:

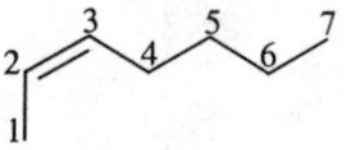

	Increments	Calculated	Measured
C-1:	6.8 + 9.6 − 3.0 + 0.5 − 3	= 10.9	12.6
C-2:	123.3 + 10.6 − 7.9 − 1.8 + 1.5 − 1.1	= 124.6	123.5
C-3:	123.3 + 10.6 + 7.2 − 1.5 − 7.9 − 1.1	= 130.6	130.8
C-4:	15.3 + 2(9.8) − 2(2.7) − 3	= 29.5	26.7
C-5:	15.3 + 2(9.8) − 2.7	= 32.2	32.0
C-6:	15.3 + 9.8 − 2.7 + 0.3	= 22.7	22.4
C-7:	6.8 + 9.6 − 3.0 + 0.5	= 13.9	13.9

Table III. Parameters for Calculating ^{13}C NMR Shifts of Vinyl Carbons

$\overset{\gamma}{C}-\overset{\beta}{C}-\overset{\alpha}{C}-Ⓒ=C-\overset{\alpha'}{C}-\overset{\beta'}{C}-\overset{\gamma'}{C}$

Carbon to Be Predicted	Add	For Each Carbon of the Indicated Type, add the Listed Increment	
>C=	123.3	α 10.6 β 7.2 γ −1.5 α′ −7.9 β′ −1.8 γ′ +1.5	
		Steric corrections	
		α,α′ trans	 0
		α,α′ cis	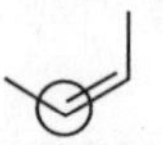 −1.1
		α,α	 −4.8
		α′,α′	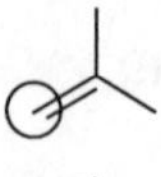 +2.5
		β,β	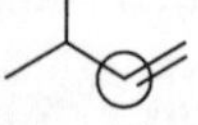 −2.3

2,4,4-Trimethyl-1-pentene is also predicted as follows:

		Calculated	Measured
C-1:	123.3 + 2(−7.9) − 1.8 + 3(1.5) + 2.5	= 112.7	114.4
C-2:	123.3 + 2(10.6) +7.2 + 3(−1.5) − 4.8	= 142.4	143.7
C-3:	15.3 + 16.7 + 21.4	= 53.4	52.2
C-4:	27.8 + 2.3 + 2(−0.9)	= 28.3	31.6
C-5:	6.8 + 25.5 − 3.0 + 2(0.5)	= 30.3	30.4
2-CH_3:	6.8 + 17.8 − 3.0 + 3(0.5)	= 23.1	25.4

Replacement of a hydrogen with a heteroatom or a carbonyl group causes large shifts in the carbon to which it is attached and lesser changes

Table IV. Chemical-Shift Changes on Replacement of a Hydrogen with Group X

X—C(C)—C—C (α, β, γ)

X	On Primary or Secondary Carbon	α	β	γ
—OH	1°	48	10	−5
—OH	2°	41	8	−5
—OR	1°	58	8	−4
—OR	2°	51	5	−4
—NH_2	1°	29	11	−5
—NH_2	2°	24	10	−5
—F	1°	68	9	−4
—F	2°	63	6	−4
—Cl	1°	31	11	−4
—Cl	2°	32	10	−4
—Br	1°	20	11	−3
—Br	2°	25	10	−3
—I	1°	−6	11	−1
—I	2°	4	12	−1
—COOH	1°	21	3	−2
—COOH	2°	16	2	−2
—COOR	1°	20	3	−2
—COOR	2°	17	2	−2
—COR	1°	30	1	−2
—COR	2°	24	1	−2
—Ph	1°	23	9	−2
—Ph	2°	17	7	−2
—NO_2	1°	63	4	—
—NO_2	2°	57	4	—

to farther carbons. These effects are listed in Table IV.[13] To use this table, first calculate the analogous hydrocarbon using Table I, and then make the changes to the α, β, and γ carbons. For example, 3-methyl-1-butanol is assigned by first calculating the shifts of 2-methylbutane:

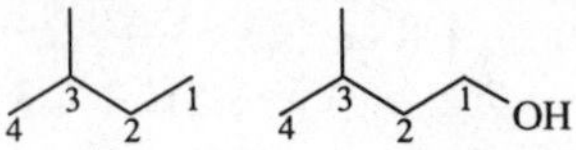

		Calculated	Measured
C-1:	6.8 + 9.6 + 2(−3.0) + 48	= 58.4	60.7
C-2:	15.3 + 16.7 + 10	= 42.0	41.7
C-3:	23.5 + 6.6 − 5	= 25.1	24.8
C-4:	6.8 + 17.8 − 3.0	= 21.6	22.6

The carbons of benzene rings may be assigned by starting with the value for benzene (128.5 ppm) and adding the parameters in Table V.[13] As you might expect, electron-donating, ortho, para-directing groups resonance-shield the ortho and para carbons, and the effects of any substituent are small at the meta position. Where more than one substituent is on the ring, the parameters are additive, as shown for 4-nitroanisole:

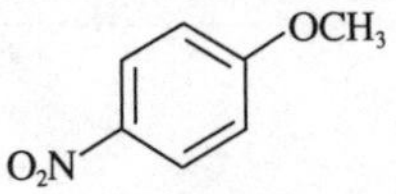

		Calculated	Measured
C-1:	128.5 + 31.4 + 5.8	= 165.7	164.7
C-2:	128.5 − 14.4 + 0.9	= 115.0	114.0
C-3:	128.5 + 1.0 − 4.8	= 124.7	125.7
C-4:	128.5 − 7.7 + 20.0	= 140.8	141.5
OCH_3:			55.9

In the proton decoupled ^{13}C spectrum the 1 and 4 carbons are apparent because they are quite small owing to the lack of nuclear Overhauser effect.

The partial plus charge on carbonyl carbons gives them very low field signals as shown in Table VI. In conjugated carbonyl compounds the partial plus charge is delocalized and the signals are 8–13 ppm higher field than for nonconjugated carbonyl compounds.

More extensive tables are available, as well as other methods of calculating values.[1,13,16] A good collection of 500 ^{13}C NMR spectra is also available.[17]

Alicyclic ring carbons are not well predicted by the methods covered

Table V. Chemical-Shift Changes at Each Ring Carbon on Replacement of a Hydrogen with Group X.

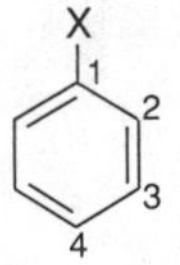

X	1	2	3	4
$—CH_3$	9.3	0.8	0	−2.9
$—CH_2CH_3$	15.6	−0.4	0	−2.6
$—COOCH_3$	2.1	1.1	0.1	4.5
$—COCH_3$	9.1	0.1	0	4.2
$—OCH_3$	31.4	−14.4	1.0	−7.7
$—NH_2$	18.0	−13.3	0.9	−9.8
$—NO_2$	20.0	−4.8	0.9	5.8
—Cl	6.2	0.4	1.3	−1.9
—Br	−5.5	3.4	1.7	−1.6

above, but they may be estimated by empirical correction methods based on analogous pairs of related molecules.[18]

10.3 CORRELATION OF ^1H AND ^{13}C NMR SPECTRA

A sequence of pulses at the proton and carbon frequencies can give magnetic polarization transfer from protons to be detected in the carbons. This is plotted as a two-dimensional spectrum (heteronuclear coupling correlation, HETCOR) with the proton spectrum along one axis and the ^{13}C spectrum along the other. Contoured peaks correlate the signal for a proton with the signal for the carbon to which it is attached. The HETCOR spectrum for 3-heptanone is shown in Fig. XII. Once again, the signals in

Table VI. Chemical-Shift Values (in ppm) of Carbonyl Carbons and Nitriles

Nitriles	117–121
Carbonates	155–156
Amides	165–175
Esters	165–175
Anhydrides	165–175
Acids	175–185
Aldehydes	200–210
Ketones	205–220

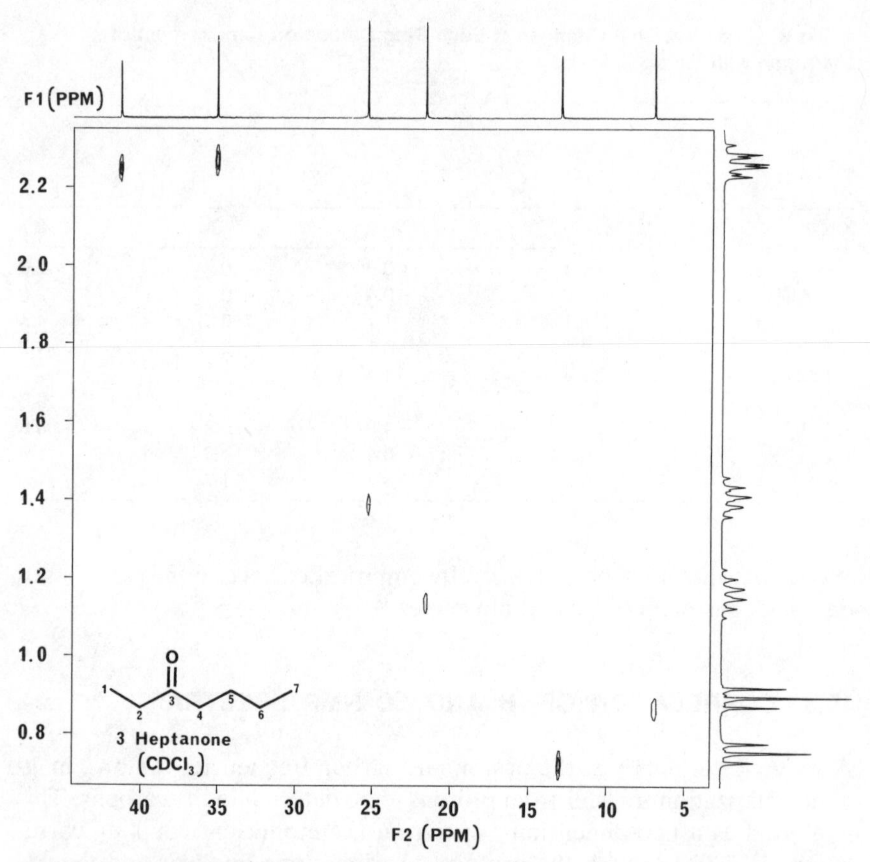

Figure XII. Two-dimensional ^1H–^{13}C heteronuclear shift correlation spectrum (HETCOR) of 3-heptanone. Spectrum courtesy of Varian Associates.

3-heptanone are assignable without HETCOR, but this simple example clearly shows the effect.

PROBLEMS

1. Both ^{13}C spectra in Fig. X were obtained from the same sample. Why is the signal for the solvent $CDCl_3$ considerably smaller in the ^1H-decoupled spectrum?
2. The ^{13}C NMR of 4-methyl-2-pentanol shows six signals at 22.4, 23.1, 23.9, 24.8, 48.7, and 65.8 ppm. Why should we not expect five signals as in 2-methylpentane?

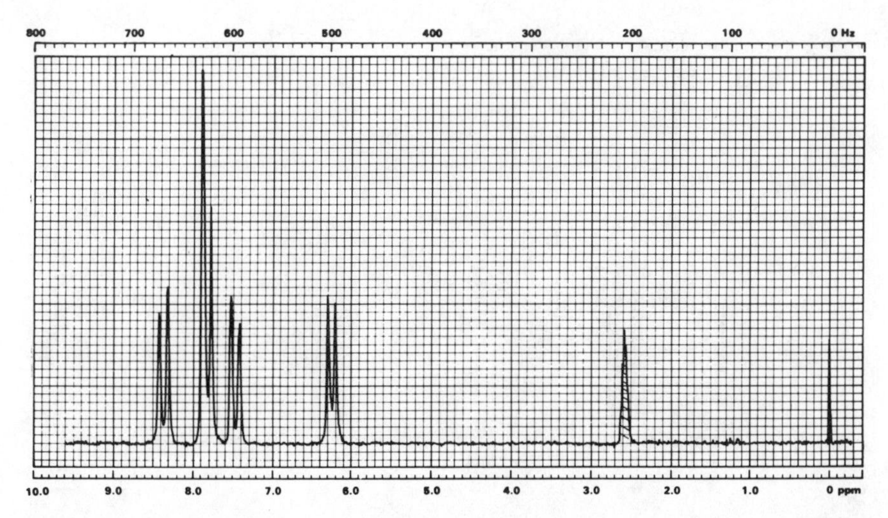

Figure XIII. The ^{1}H NMR spectrum for problem 3.

3. Figure XIII shows the ^{1}H NMR spectrum of **1** with signals centered at 6.16 ppm (1H, J = 7 Hz); 7.47(1H, J = 9 Hz); 7.78–7.90(2H), 8.36(1H, J = 9 Hz). The signal for the OH is absent. This spectrum was run at 80 MHz. Consider the structure, and then calculate and graph the part of the spectrum from 7.6 to 8.0 ppm as it should appear at 270 MHz.

OH

CF_3 N

1

4. Figure XIV shows the ^{1}H NMR spectrum of **2**. Assign all the signals to protons in the structure, and explain the splitting pattern in the multiplet at 3.0 ppm.

O

OH

HN O

HO

2

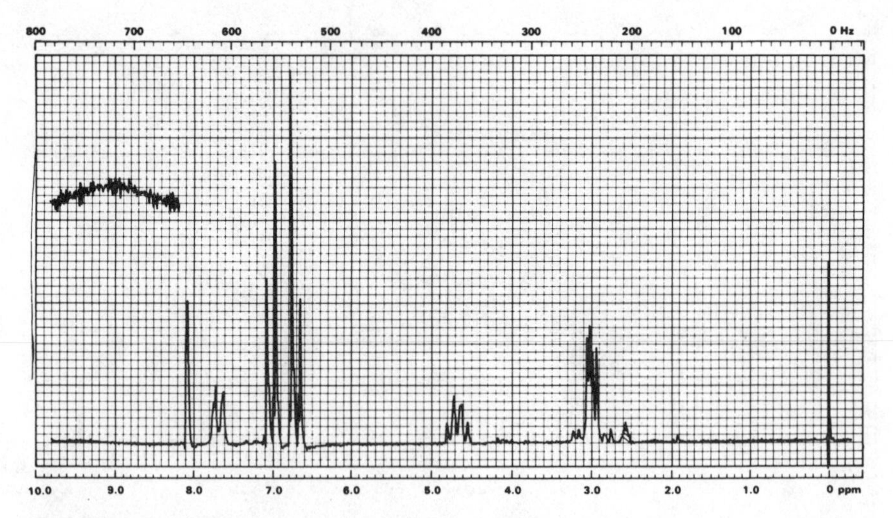

Figure XIV. The ^{1}H NMR spectrum for problem 4. ©Sadtler Research Laboratories, Division of Bio-Rad Laboratories, Inc., 1977.

5. Draw the structure of the compound $C_{10}H_{12}O$ that gives the 60-MHz ^{1}H NMR spectrum in Fig. XV and comment on the splitting pattern.

6. Figure XVI shows the 60-MHz ^{1}H NMR spectrum of **3**. Assign all the signals to particular hydrogens in the structure, and explain the splitting patterns. The offset peak is at 12.2 ppm.

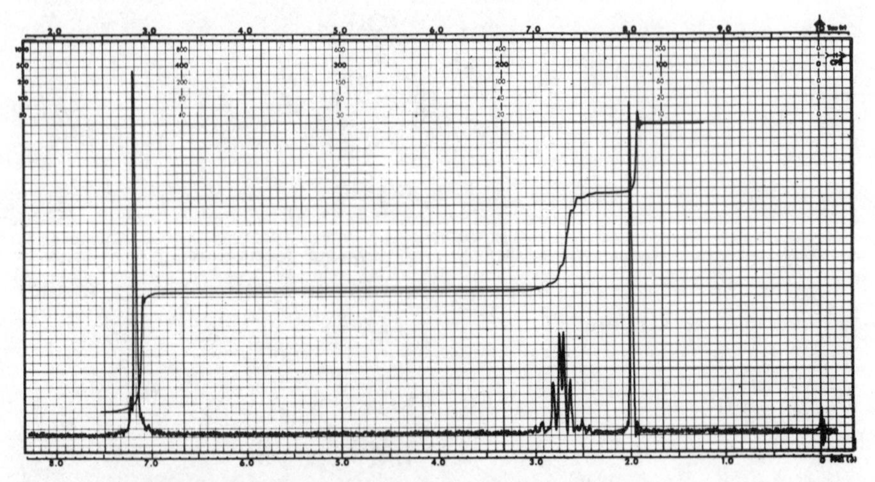

Figure XV. The ^{1}H NMR spectrum for problem 5. ©Sadtler Research Laboratories, Division of Bio-Rad Laboratories, Inc., 1971.

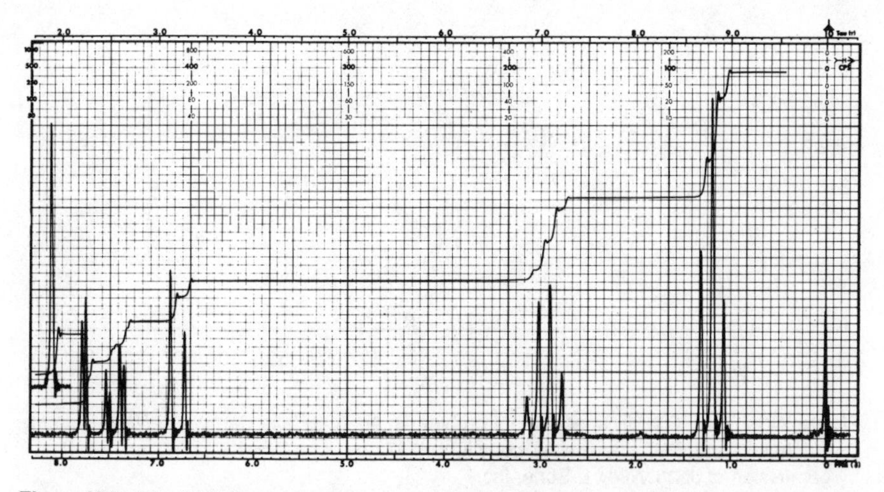

Figure XVI. The ^{1}H NMR spectrum for problem 6. ©Sadtler Research Laboratories, Division of Bio-Rad Laboratories, Inc., 1970.

Br O OH

3

7. The following reaction gave many products, including three compounds with the boiling points and ^{13}C NMR spectra listed below.[19]

HO, OH + pyridinium (N^+–H) Cl^- $\longrightarrow$ 1, 2, 3, etc.

Give the structure of each and assign as many signals as possible to particular carbon atoms in the structures.

1: bp 162.5°C 22.3, 27.3, 37.7, 109.6, 145.8 ppm
2: bp 164°C 17.7, 22.4, 25.7, 27.8, 28.0, 37.6, 109.7, 124.5, 131.1, 145.6 ppm
3: bp 165.5°C 17.7, 25.7, 28.6, 124.6, 131.1 ppm

8. Give the structure of the compound that gives the ^{13}C NMR spectrum in Figure XVII.

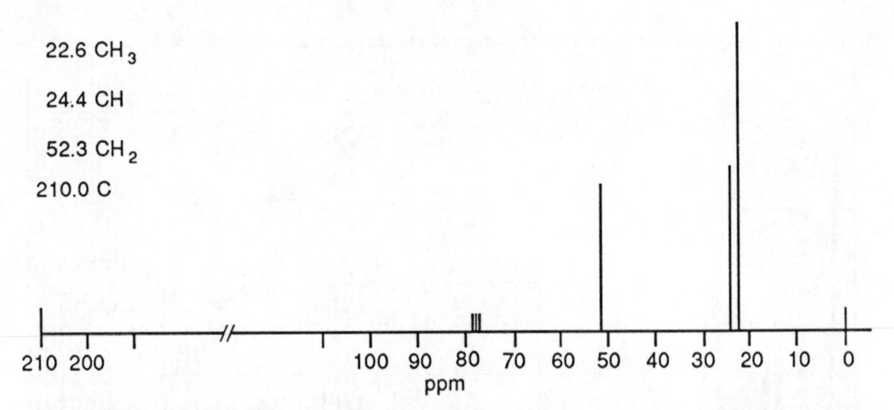

Figure XVII. The ^{13}C NMR spectrum for problem 8. Adapted from *Carbon-13 NMR Spectra* L. F. Johnson and W. C. Jankowski, copyright ©1972 by John Wiley & Sons, Inc. Reprinted by permission of John Wiley & Sons, Inc.

9. A certain compound of molecular formula $C_9H_{18}O_3$ dissolves in dilute sulfuric acid on prolonged boiling and gives off CO_2 gas. The proton-decoupled ^{13}C NMR spectrum of the compound is shown in Fig. XVIII. What is the structure?

10. Figure XIX shows the proton-decoupled ^{13}C NMR spectrum of a compound of molecular formula $C_8H_9ClO_2$. Give the structure of the compound, and assign as many signals as possible to particular carbons in the structure.

11. Examine the ^1H splitting pattern and the COSY spectrum for 2-vinylpyridine (Fig. IV), and assign all the ^1H signals to particular protons

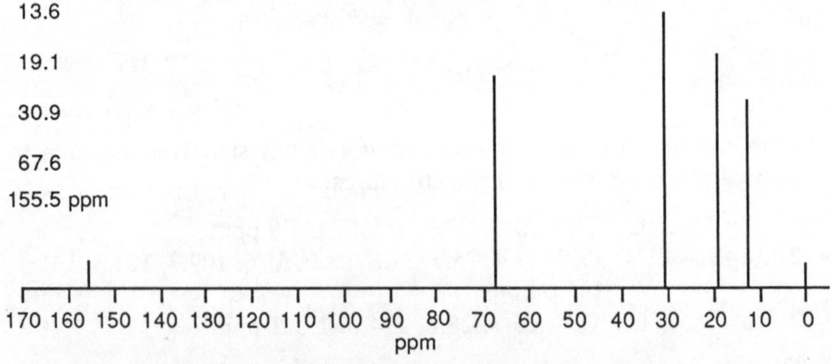

Figure XVIII. The ^{13}C NMR spectrum for problem 9. Adapted from *Carbon-13 NMR Spectra,* L. F. Johnson and W. C. Jankowski, copyright© 1972 by John Wiley & Sons, Inc. Reprinted by permission of John Wiley & Sons, Inc.

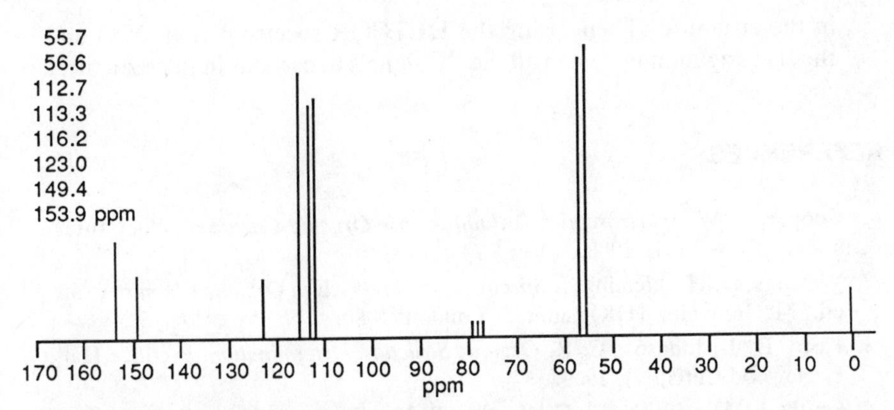

Figure XIX. The ^{13}C NMR spectrum for problem 10. Adapted from *Carbon-13 NMR Spectra*, L. F. Johnson and W. C. Jankowski, copyright© 1972 by John Wiley & Sons, Inc. Reprinted by permission of John Wiley & Sons, Inc.

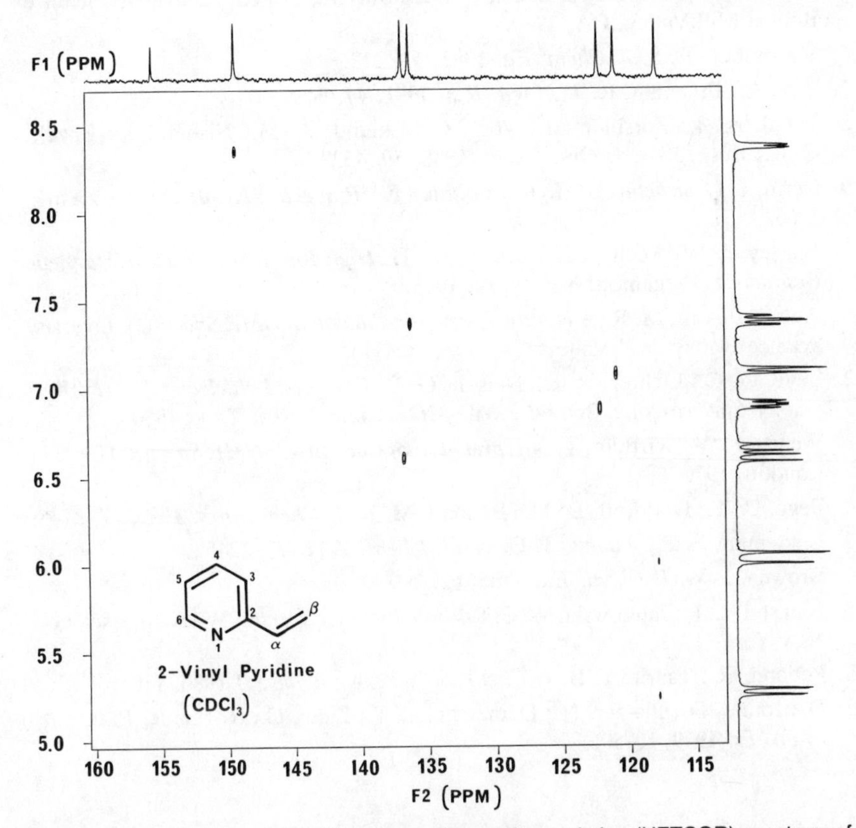

Figure XX Two-dimensional ^{1}H–^{13}C heteronuclear shift correlation (HETCOR) spectrum of 2-vinylpyridine for problem 11. Spectrum courtesy of Varian Associates.

in the structure. Then, using the HETCOR spectrum (Fig. XX) and the ^{1}H assignments, assign all the ^{13}C signals to carbons in the structure.

REFERENCES

1. Cooper, J. W. *Spectroscopic Techniques for Organic Chemists,* Wiley-Interscience, New York, 1980.
2. Williams, D. H.; Fleming, I. *Spectroscopic Methods in Organic Chemistry,* 3rd ed., McGraw-Hill, (UK) Limited, London, 1980.
3. Pasto, D. J.; Johnson, C. R. *Organic Structure Determination,* Prentice Hall, Englewood Cliffs, NJ, 1969.
4. Smith, L. M.; Smith, R. G.; Loehr, T. M.; Daves, G. D., Jr.; Daterman, G. E.; Wohleb, R. H. *J. Org. Chem.* **1978,** *43,* 2361 (an excellent example of the value of laser Raman spectroscopy).
5. McLafferty, F. W. *Interpretation of Mass Spectra,* 3rd ed., University Science Books, Mill Valley, CA, 1980.
6. Macomber, R. S. *J. Chem. Ed.* **1985,** *62,* 213.
7. Bax, A.; Freeman, R. *J. Magn. Res.* **1981,** *44,* 542.
8. Wahlberg, I.; Forsblom, I.; Vogt, C.; Eklund, A.-M.; Nishida, T.; Enzell, C. R.; Berg, J.-E. *J. Org. Chem.* **1985,** *50,* 4527.
9. Corio, P. L. *Structure of High-Resolution NMR Spectra,* Academic, New York, 1966.
10. Emsley, J. W.; Feeney, J.; Sutcliffe, L. H. *High Resolution Nuclear Magnetic Resonance,* Pergamon, New York, 1965.
11. Ault, A.; Ault, M. R. *A Handy Systematic Catalog of NMR Spectra,* University Science Books, Mill Valley, CA, 1980, p. 340.
12. Levy, G. C.; Lichter, R. L.; Nelson, G. L. *Carbon-13 Nuclear Magnetic Resonance Spectroscopy,* 2nd ed., Wiley-Interscience, New York, 1980.
13. Wehrli, F. W.; Wirthlin, T. *Interpretation of Carbon-13–NMR Spectra,* Heyden, London, 1976.
14. Pegg, D. T.; Doddrell, D. M.; Bendall, M. R. *J. Chem. Phys.* **1982,** *77,* 2745.
15. Lindeman, L. P.; Adams, J. Q. *Anal. Chem.* **1971,** *43,* 1245.
16. Brown, D. W. *J. Chem. Ed.* **1985,** *62,* 209.
17. Johnson, L. F.; Jankowski, W. C. *Carbon-13 NMR Spectra,* Wiley-Interscience, New York, 1972.
18. Petiaud, R.; Taarit, Y. B. *J. Chem. Soc. Perkin Trans. II* **1980,** 1385.
19. Dauzonne, D.; Platzer, N.; Demerseman, P.; Lang, C.; Royer, R. *Bull. Soc. Chim. Fr.* **1979,** II–506.

Index

KRAUSKOPF LIBRARY
547 St79i
STACKS
3 1896 00029 0985